T0155457

Productionizing AI

How to Deliver AI B2B Solutions with Cloud and Python

Barry Walsh

Apress®

Productionizing AI: How to Deliver AI B2B Solutions with Cloud and Python

Barry Walsh
Ely, UK

ISBN-13 (pbk): 978-1-4842-8816-0
ISBN-13 (electronic): 978-1-4842-8817-7
https://doi.org/10.1007/978-1-4842-8817-7

Managing Director, Apress Media LLC: Welmoed Spahr
Acquisitions Editor: Celestin John Suresh
Development Editor: Laura Berendson
Coordinating Editor: Mark Powers

Cover designed by eStudioCalamar

Cover image by Vackground on Unsplash (www.unsplash.com)

Distributed to the book trade worldwide by Springer Science+Business Media New York, 1 New York Plaza, Suite 4600, New York, NY 10004-1562, USA. Phone 1-800-SPRINGER, fax (201) 348-4505, e-mail orders-ny@springer-sbm.com, or visit www.springeronline.com. Apress Media, LLC is a California LLC and the sole member (owner) is Springer Science + Business Media Finance Inc (SSBM Finance Inc). SSBM Finance Inc is a **Delaware** corporation.

For information on translations, please e-mail booktranslations@springernature.com; for reprint, paperback, or audio rights, please e-mail bookpermissions@springernature.com.

Apress titles may be purchased in bulk for academic, corporate, or promotional use. eBook versions and licenses are also available for most titles. For more information, reference our Print and eBook Bulk Sales web page at http://www.apress.com/bulk-sales.

Any source code or other supplementary material referenced by the author in this book is available to readers on GitHub (github.com/apress). For more detailed information, please visit http://www.apress.com/source-code.

Printed on acid-free paper

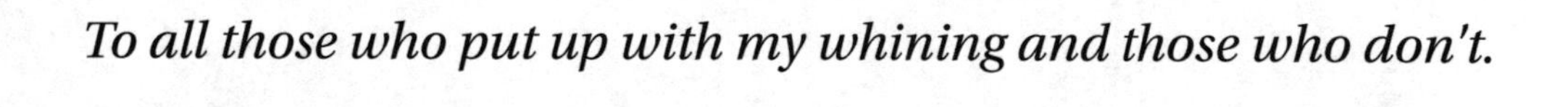

To all those who put up with my whining and those who don't.

Table of Contents

About the Author

Barry Walsh is a software delivery consultant and AI trainer at Pairview with a background exploiting complex business data to optimize and de-risk energy assets at ABB/Ventyx, Infosys, E.ON, Centrica, and his own start-up ce.tech. Certified as an Azure AI Engineer and Data Scientist as well as an AWS Cloud Practitioner, he has a proven track record of providing consultancy services in Data Science, BI, and Business Analysis to businesses in Energy, IT, FinTech, Telco, Retail, and Healthcare, Barry has been at the apex of analytics and AI solutions delivery for over 20 years. Besides being passionate about Enterprise AI, Barry spends his spare time with his wife and 8-year-old son, playing the piano, riding long bike rides (and a marathon on a broken toe this year), eating out whenever its possible or getting his daily coffee fix.

About the Technical Reviewer

Pramod Singh works at Bain & Company as a Senior Manager, Data Science, in their Advanced Analytics group. He has over 13 years of industry experience in Machine Learning (ML) and AI at scale, data analytics, and application development. He has authored four other books including a book on Machine Learning operations. He is also a regular speaker at major international conferences such as Databricks AI, O'Reilly's Strata, and other similar conferences. He is an MBA holder from Symbiosis International University and a data analytics certified professional from IIM Calcutta. He lives in Gurgaon with his wife and five-year-old son. In his spare time, he enjoys playing the guitar, coding, reading, and watching football.

Preface

When Barry approached me to write a preface for this book, it filled me with delight, not just because I was writing a preface for his new book, although it is always nice to be asked, but primarily because I was being asked by someone who I have watched flourish under my leadership at Pairview for the last few years.

The work we do at Pairview has helped shape thousands of fledgeling careers across the UK and around the world for more than thirteen years at the time of writing. Fresh graduates struggling to find high value jobs, returning to work mums who want flexible work with high pay, and those who have reached the end of a career cycle and have no real prospect for further growth have all turned to Pairview to kickstart their careers. The company was created to help close the talent gap that was growing in the data space. This gap has since spread to most other aspects of technology as new technology innovation comes around and companies continue to grapple with the ultimatum to adopt technology-induced change or sink.

Barry's passion for Artificial Intelligence combined with his vast knowledge of the subject matter and the work he has done for our numerous clients over the years makes him an authority on the subject. Barry has acquired first-hand exposure to the opportunities, challenges and the risks associated with AI ecosystems and development processes. He knows what it takes to get buy-ins and drive organisation wide-adoption of AI capabilities and the business impact. Barry has over the years become an evangelist of the value of AI to business while being fully aware of the enormous investment of time, resources and scrutiny required to get AI right first time and drive it through the digital transformation value chain.

Although Artificial intelligence has been with us for many decades, in recent times new capabilities have emerged in the constellation; enabling more complexity in sensing, comprehending, acting and learning with human-like levels of intelligence. With mathematical technologies like machine learning and natural language processing the landscape of AI continues to expand and increasingly co-exist with humans, enabling businesses to dare to digitalise with never-before levels of accuracy, consistency and availability. A combination of data analytics, machine learning and deep learning embedded with ever-more powerful computing power such as quantum and edge, AI is being used to deliver next generation capabilities across all aspects of human reach.

While we have seen many organisations invest quite heavily in machine learning and analytics capabilities, bringing the insight and models from machine learning developments into production continues to pose a significant challenge for many business leaders, particularly CDOs, CEOs and Data and AI leads responsible for the embedding and delivery or AI enabled products to markets at the speed and scale required for success. This is why this book is a must read for leaders of these organisations, programmes or products. It provides a strong framework for planning, developing and deploying enterprise AI to production at scale.

Frank Abu
Director
Pairview Limited

Prologue

Out of a wasteland of failed Data Science projects and mounting technical debt,[1] organizations today are attempting to redress the AI landscape and enforce a broader, more considered Design/System Thinking approach which cuts through the hype. The imperative is to ensure that Data/AI solutions are built with multiuser engagement at the outset (both technical and nontechnical) and have system-wide ecosystem, enterprise data center, infrastructure/integration, and end-to-end process in mind.

Applications of Artificial Intelligence leveraging the latest technological advances are firmly at the top of this hype cycle.[2] With the productivity benefits of AI difficult to ignore, Covid and galloping digitalization have given rise to a vicious culture of disruption, the accelerated uptake forcing fragile companies to build or buy solutions cheaper, smaller, and more open sourced. This convergence of forces means that demand for rapid prototyping and accelerated AI solution delivery is high.

At the same time, not every company is cognizant of what AI can do, or what it means. Often these are corporates burdened by legacy tools and poor innovation practices. Some are fearful of the impact on jobs; others have ethical concerns. But what has become clear for most C-Levels is that AI implementation must fit an "Enterprise AI" vision – a $341b market[3] – and the prevalent trend is toward a platform of disparate but highly integrated "best-of-breed" AI solutions.

For many of us where data and digital are inextricably tied up in our professional life and associated opportunities in the job market, these meaningful, high-value AI solutions are what employers are targeting. And the constant focus on delivering an Return On Investment (ROI) – understanding and, more importantly, delivering tangible outcomes using AI, typically machine and deep learning – is what keeps us employed.

---

[1] The rush to digital means 60% of businesses (in Europe and globally) will find themselves with more technical debt post-pandemic, not less (Source: Forrester).

[2] `https://www.gartner.com/en/articles/what-s-new-in-artificial-intelligence-from-the-2022-gartner-hype-cycle`. See also `https://tinyurl.com/3c7pcpfm`

[3] `www.idc.com/getdoc.jsp?containerId=prUS48127321`

This can be exhausting and part of the challenge for business leaders is that highly technical skillsets have never led to particularly visual, communicable (and understandable) results for the rest of the organization – most Data Scientists are not good at BI or recruited for their soft skills. Employers in the job market today are increasingly looking for more rounded/broader "end-to-end" skills which translate to better visualization, front-end features, and integration. The opportunity for Data Professionals is to give themselves an "edge" by addressing technical debt and possessing the ability to deliver full-stack data solutions.

A great deal of this "opportunity" depends on cloud computing; AI or Data Science is not just a Python notebook and a flip chart – it demands identifying and ingesting suitable datasets and leveraging services on cloud to scale from sandbox to proof-of-concept to prototype to minimum viable product. Ultimately today it is Enterprise AI that most companies and organizations are targeting. But for many individuals employed in roles and for many noncorporate companies, leveraging cloud is a minefield of sometimes unclear, poorly documented artifacts and hidden costs. Enterprise AI is far from affordable.

Successful (and affordable) Enterprise AI project delivery requires a healthy amount of Emotional Intelligence,[4] an agile mindset, robust data-fed pipelines and a whole lot of workarounds to design, scope, and achieve integration across people, processes, and tools – with all of us dependent on the three main hyperscalers/cloud service providers (CSPs): Amazon Web Services, Microsoft Azure, and Google Cloud Platform,[5] bound to their data centers, scalable storage, and compute instances.

Agile is important, but so are hybrid/agnostic solutions, multiskill, T-shaped capabilities, and results-based delivery. Above all, senior stakeholders/managers should not be taking credit for being spiderman or spiderwoman implementing "agile" project methodologies if they are just orchestrating chaos or spinning webs that attribute project failures/blame elsewhere. Agile only works if it follows project design through to a standard of delivery which meets the overall project vision, not if under the surface it's just a pile of shit.[6]

---

[4] Amit Ray: "as more and more artificial intelligence is entering into the world, more and more emotional intelligence must enter into leadership."

[5] All firmly in the world's top 10 companies by market capitalization: `https://companiesmarketcap.com/`

[6] Forrester: "while investments in automation will continue to rise rapidly across Europe, many companies have historically lacked a coherent data/AI strategy, with a patchwork of siloed digital services that have left their IT function in a mess and customers frustrated."

With many organizations struggling to operationalize AI, this book is intended to go a step beyond a simple Python scripting and address the "so what" of current Data Science delivery.

This doesn't mean python is passed over – far from it. Python skills are more than ever in demand in the workplace, with popularity up in September 2022 and still number 1 on the TIOBE Programming Community index.[7] Developing in Python is central to the aims of this book but here we take a top-down, "goal-focused" approach, preferring Python-based and low-code frameworks for accelerating the full-stack development process, rather than stream of consciousness Python back-ends. The narrative of the book is also aligned with DataOps best practice and an awareness of meeting KPIs around shifting requirements, slipped schedules, flexibility improvements, and disappointed users.

Hands-on labs are populated throughout the book and are intended to present "how-to" examples, not just utilizing Python code, but as much with critical cloud services for AI and no-/low-code interfaces for key nontechnical stakeholders, often the main decision makers in whether to commission, implement, or scale an AI project.

Enterprise AI is about embedding AI into an organization's data strategy, and an entire workforce needs to be aligned and trained on a company's tangible (and intangible/hidden) data and AI assets. Everyone therefore should get to use and understand AI, and by developing the solutions in this book it is hoped the reader is better equipped to deliver a company's "bigger picture" goals.

[7] www.tiobe.com/tiobe-index/

CHAPTER 1

Introduction to AI and the AI Ecosystem

In the age of the Fourth Industrial Revolution that we live in today, AI is a central focus of innovation, along with IoT and genetic engineering among others. But it comes with a lot of hype, in particular in relation to job loss and job creation as much of the world's businesses and organizations hyperspeed transforming their operational model in the Digital Age. Much of the job creation is centered around the booming career of the Data Scientist, a role which has really only been mainstream since about 2012. But the role itself and the scope of skills it entails has not always evolved in that relatively short period of time to accommodate growing needs across a company's key assets: people, processes, and tools. In particular, much money and time has been spent on poorly planned Data Science projects where nothing other than an R or Python script has been developed, often with highly sophisticated coding, but of which very few people in the organization are cognizant. In effect, while these solutions may be robust and relatable to a group of Data Scientists, they are not "fit for purpose" in a large organizational/enterprise-wide structure with many different people, processes, and tools interfacing and competing for internal budgets.

This chapter introduces our take on the current mismatch in capabilities and requirements through the lens of those topics which we consider most relevant to the successful delivery of an AI project today.

The goal of this chapter, from an initial whistle-stop historical tour, is to define key AI concepts at a fairly high level and introduce the reader to current and emerging trends in AI, including the hype and the pitfalls. Ultimately the chapter leads to how many businesses and organizations are struggling to get machine and deep learning operationalized in today's workplaces. The topics discussed in this chapter are then elaborated in greater detail in the subsequent chapters.

B. Walsh, *Productionizing AI*, https://doi.org/10.1007/978-1-4842-8817-7_1

In this chapter, we will scratch the surface first, before embarking on the more involved hands-on practice and applications of later chapters. The intention is to provide readers with the tools in this chapter to go forward, providing concise context and definitions around the AI Ecosystem, the main applications of AI, data ingestion and data pipelines, and machine and deep learning with neural networks before wrapping up on productionizing AI.

The AI Ecosystem

Our first section sets the scene for AI today – starting with a look at the hype cycle before a retrospective on how AI has evolved to this stage. We also introduce some definitions, cloud computing as the enabler for scalable AI, the ecosystem for "full-stack" AI, and discuss growing ethical concerns.

The Hype Cycle

While there is considerable hype around AI, it is generally perceived as now delivering on its potential. The benefits for businesses and organizations are tangible, accelerated by Covid-digitalization and evidenced by the proliferation of chatbot support during the pandemic, deep-learning supported healthcare diagnostics, usage of computer vision for social-distancing measures, and machine learning modeling of the effects from reopening economies.

In the workplace, while AI is maturing as a service, application has largely involved only a handful of IT experts. Democratization of AI is a big focus in 2022 and shifting from expert/niche knowledge to achieving buy-in across the wider ecosystem of key stakeholders (all employees, customers, and business partners). Industrialization of AI is also a dominant trend today with employers pushing for "smarter" implementation of AI projects; focusing on reusability, scalability, and safety of AI at a Design Thinking stage, rather than as an afterthought.

We start this first section with a look at how Artificial Intelligence has evolved and matured into this current status quo best encapsulated by a need to scale up from "standalone AI" to "Enterprise AI."

Historical Context

Depending on your perspective, Artificial Intelligence has its roots in the age of computing in the 1950s or in the "automata" of ancient philosophy.

Modern AI probably originated in classical philosophy in references to human thought as a mechanical process so before we start, it's useful to establish some of this context for AI today.[1] Table 1-1 provides an overview of Artificial Intelligence's evolution.

Table 1-1. *The evolution of Artificial Intelligence*

Date	Event
5th century BC	First records of mechanical robots: Chinese Daoist Philosopher Lao Tzu accounts of a life-sized, human-shaped mechanical automaton
c. 428–347 BC	Greek scientists create "automata" – specifically Archytas creates a mechanical bird
9th century	First recorded programmable complex mechanical machine
1833	Charles Babbage conceives an Analytical Engine – a programmable calculating machine
1872	Samuel Butler's novel Erewhon includes the notion of machines with human-like intelligence
1st half of 20th century	Science Fiction awareness of AI (Tin Man in Wizard of Oz, Maria robot double in Metropolis)
1950	Alan Turing publishes Computing Machinery and Intelligence – asking "Can machines think?" – or "can machines successfully imitate thought?"
1956	MIT cognitive scientist Marvin Minsky coins the term "Artificial Intelligence"
1974–1993	Long AI "Winter" – lack of tangible commercial success and poor performance of neural networks lead to reduced funding from governments
1997	IBM's Deep Blue defeats Garry Kasparov at Chess
2011	IBM Watson wins the quiz show "Jeopardy!"
2012	ImageNet competition – AlexNet Deep Neural Networks result in significant reduction in error in visual object recognition

[1] www.forbes.com/sites/gilpress/2016/12/30/a-very-short-history-of-artificial-intelligence-ai/?sh=1cfaac1f6fba

AI – Some Definitions

Whether we consider a philosophical automaton or a thinking machine as real "AI" is up for debate but what we can make use of, with the benefit of hindsight, are specific terms to understand and articulate Artificial Intelligence

Terminology	Description	Example
Simple machine	A device that does work (transfer of energy from one object to another)	Wheels, levers, pulleys, inclined planes, wedges, and screws
Complex machines	Combinations of simple machines	Wheelbarrow, bicycle, mechanical robot (Lao Tzu, mechanical bird)
Programmable machines	Receive input, store and manipulate data, and provide output in a useful format	Punched cards, encoded music rolls
Calculating machines	Mechanical device used to perform automatically the basic operations of arithmetic	Abacus, slide rule, difference engine, calculator
Digital machines	Systems that generate and process binary data	Computers

On a timeline, Figure 1-1 shows how this looks.

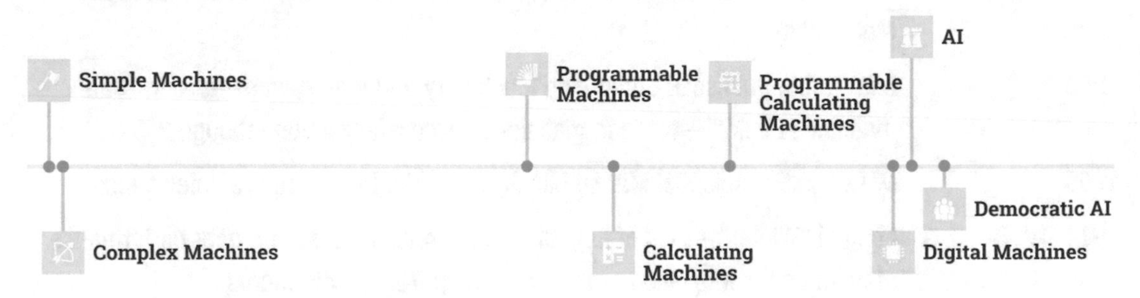

Figure 1-1. *A timeline of AI's evolution*

AI Today

While the question of "what is AI" may be less clear (we will address this in the next section), without Digital Machines or Computers we certainly wouldn't be talking about AI – and as we will see shortly, ultimately it's the growth of cloud computing and the high-performance computing that comes with it that have enabled AI or democratic AI.

When we look at the mechanics and, in particular, the applied use cases of AI today, Machine and Deep Learning are the underpinning techniques of real Artificial Intelligence, rather than any misconceived ideas about a "rise of the machines" partly based on ignorance, part science fiction. AI in the job-world stands for Augmented Intelligence and no-one really wants to see Artificial General Intelligence in the same way as no-one (hopefully) wants a third world war.

Machine Learning

As shown in Figure 1-2, Machine Learning can be thought of as a subset of AI giving computers the ability to learn without being explicitly programmed. Operationally, machine learning is much like how humans learn from experience, that is, if we touch something hot and get burnt, the negative experience is stored in memory and we quickly learn not to touch it again.

We feed a computer data, which represents past experience, and then with the use of different statistical methods, we "learn" from the data and apply that knowledge to future events – these are our model "predictions."

Deep Learning

Deep Learning is often considered a subset of Machine Learning and is distinguished by the fact that deep neural network layers are used to solve predictive problems.

Because of their reliance on Big data and Modelling, all of AI, Machine, and Deep Learning are core techniques for Data Science which combines modeling, statistics, programming, and some domain expertise to extract insights and value from data.

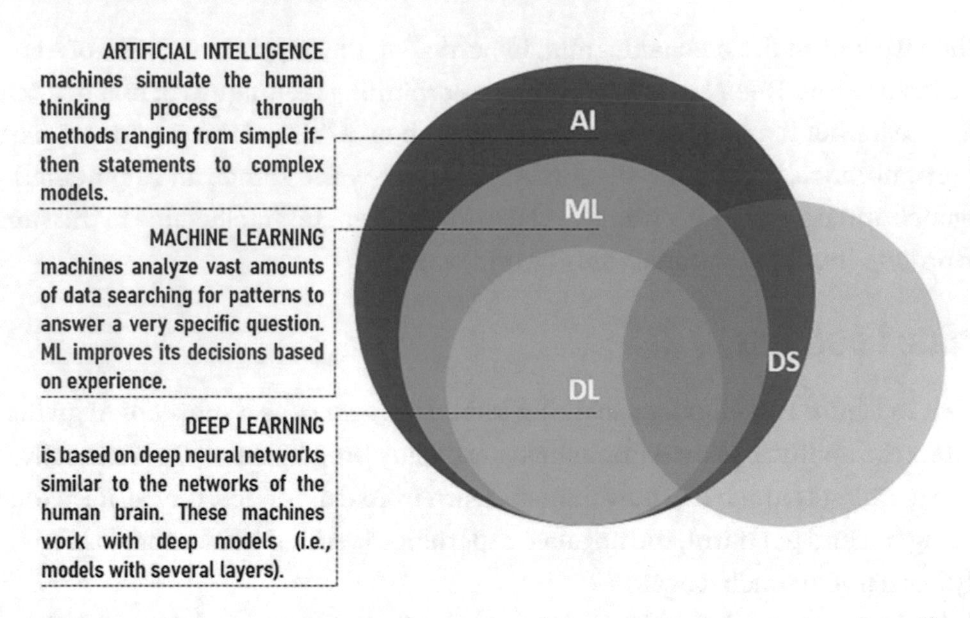

***Figure 1-2.** Deep Learning as a subset of AI (Source: Abris AI in Banking[2])*

What Is Artificial Intelligence

The Artificial Intelligence with which we are concerned with in this book is often confused with the science fiction variety. How do we distinguish the two?

Narrow AI is the current form of AI used in business and organizations, that is, where a machine is designed to perform a single task. The machine gets very good at performing that particular task (such as e.g. Google Translate) but once the machine is trained, it does not generalize to unseen domains.

Artificial General Intelligence (AGI) is the form of AI that can accomplish any human intellectual task, with underlying "conscious" decision making. Although it may constitute an existential threat, it does remain, for now, aspirational due to challenges including scaling of hardware, energy consumption, and catastrophic memory loss which also affects some of the advanced deep learning algorithms today.

Artificial Superintelligence or ASI is the form of AI most close to that used in science fiction movies. ASI in theory has capability more than that of humans.

[2] www.abrisconsult.com/artificial-intelligence-and-data-science/

Cloud Computing

So we are concerned with Narrow AI, at least for now, and by the time AGI or ASI comes around, if we are to believe Elon Musk, we should all hope we are gone.

As mentioned above, the enabler for this form of AI is cloud and successful Narrow AI implementation requires an end-to-end Cloud Infrastructure. It is difficult to underestimate the degree to which cloud computing is a fundamental requirement for any business in 2022. Growth has been explosive with a 33% increase in cloud spend in 2020 driven by intense demand to support remote working and learning, ecommerce, content streaming, online gaming, and collaboration.

Storage and Compute Power are the main Cloud components used for Big Data handling central to AI. While Enterprise Machine Learning projects can run with low overheads on both, Deep Learning projects cannot. Amazon Web Services, Azure, and Google Cloud Platform are the major ("Big Three") cloud service providers (or CSPs), with IBM Cloud and Heroku also used.[3] We will cover hands-on examples of all of these in this book. All CSPs provide a catalog of AI services and tools which greatly simplify the process of building applications.

While cloud is the key enabler of AI, cloud computing only really works for Enterprise, or production-grade AI if the company's Data Strategy is underpinned by rich, BIG data sources and/or training data.

CSPs – What Do They Offer ?

Each of the CSPs has clear differentiators over the others. AWS has the greatest reach, while Azure naturally interfaces well with Windows-based systems. Google Cloud Platform tends to have better support for app-building, as depicted in Figure 1-3.

[3] Ali Baba is the fourth biggest cloud service provider but for now largely confined to the Chinese market

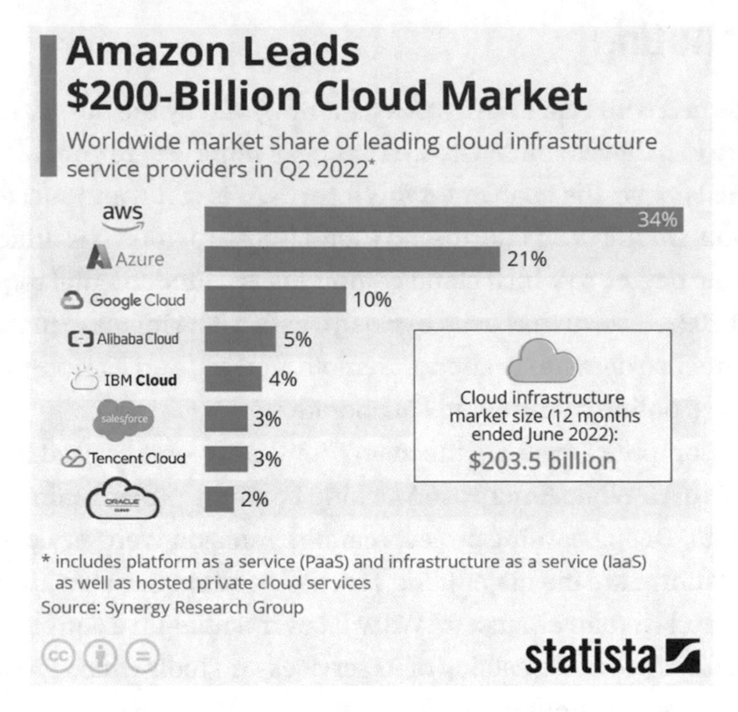

Figure 1-3. *The leading cloud platforms Q222 (Source: Statista)*

Summarized in Figure 1-4 are some of the additional USPs of each of the main cloud platforms.

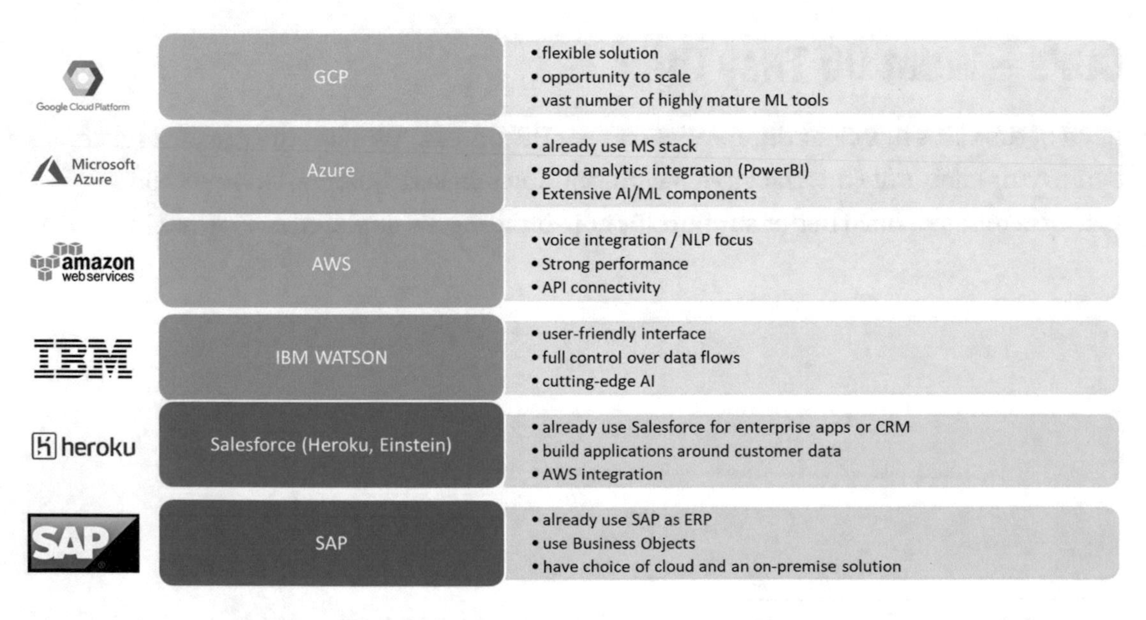

Figure 1-4. *USPs for cloud service providers*

The Wider AI Ecosystem

Despite the dominance of the Big Three CSPs, some companies are starting to move away or influenced by Consultant/SIs away from Big Tech "lock-in" and toward a platform of multivendor, niche AI solutions. These open source, cheaper platforms can also be quicker to implement, are more closely coupled with agile delivery models (more on this in Chapter 2), and are somewhat unburdened by legacy, size, or corporate social responsibility (CSR).

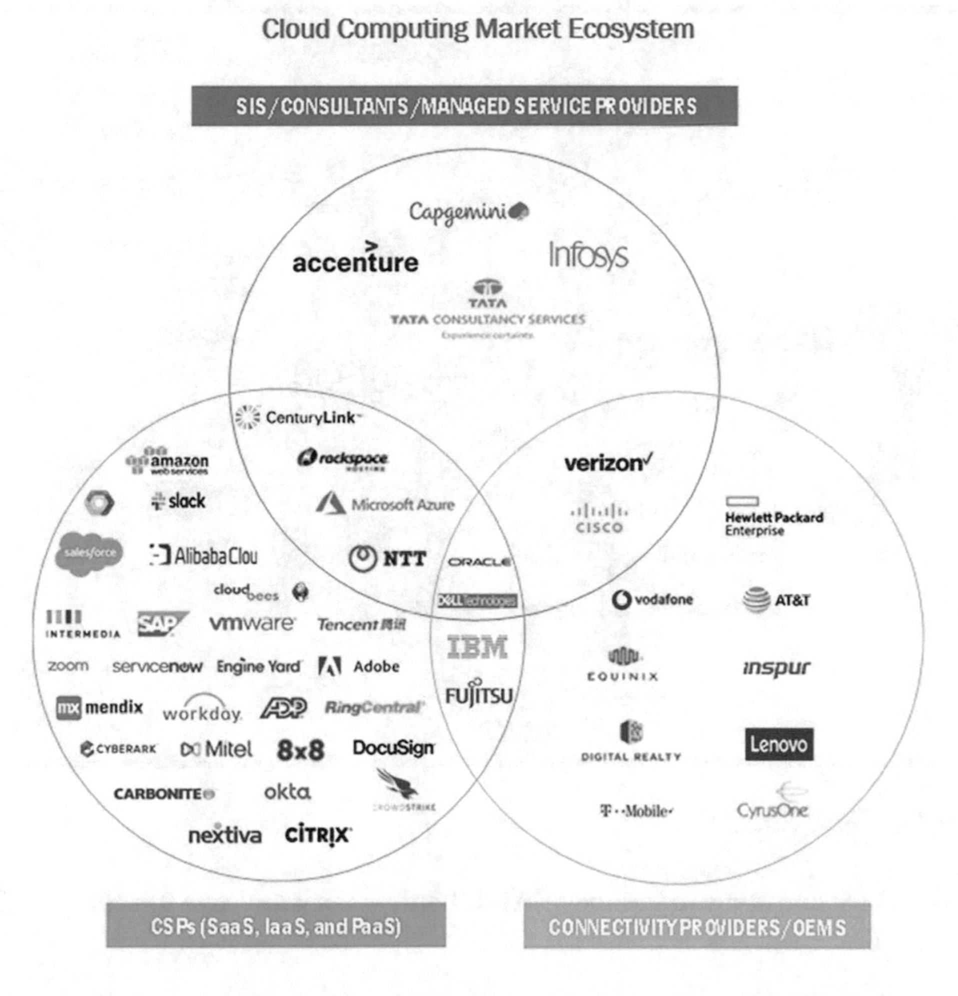

Figure 1-5. *Supporting the wider AI Ecosystem – CSPs, Sis, and OEMs*

Full-Stack AI

It's not all about cloud of course. While cloud provides the platform, a whole host of proprietary and open-sourced tools are used to implement AI from data engineering tooling such as Apache Kafka or AWS Kinesis to NoSQL databases such as mongoDB and AWS DynamoDB through back-end programming languages like Python and Scala[4] coupled with model engines like Apache Spark and front-end BI layers/dashboards like Dash, PowerBI, and Google Data Studio.

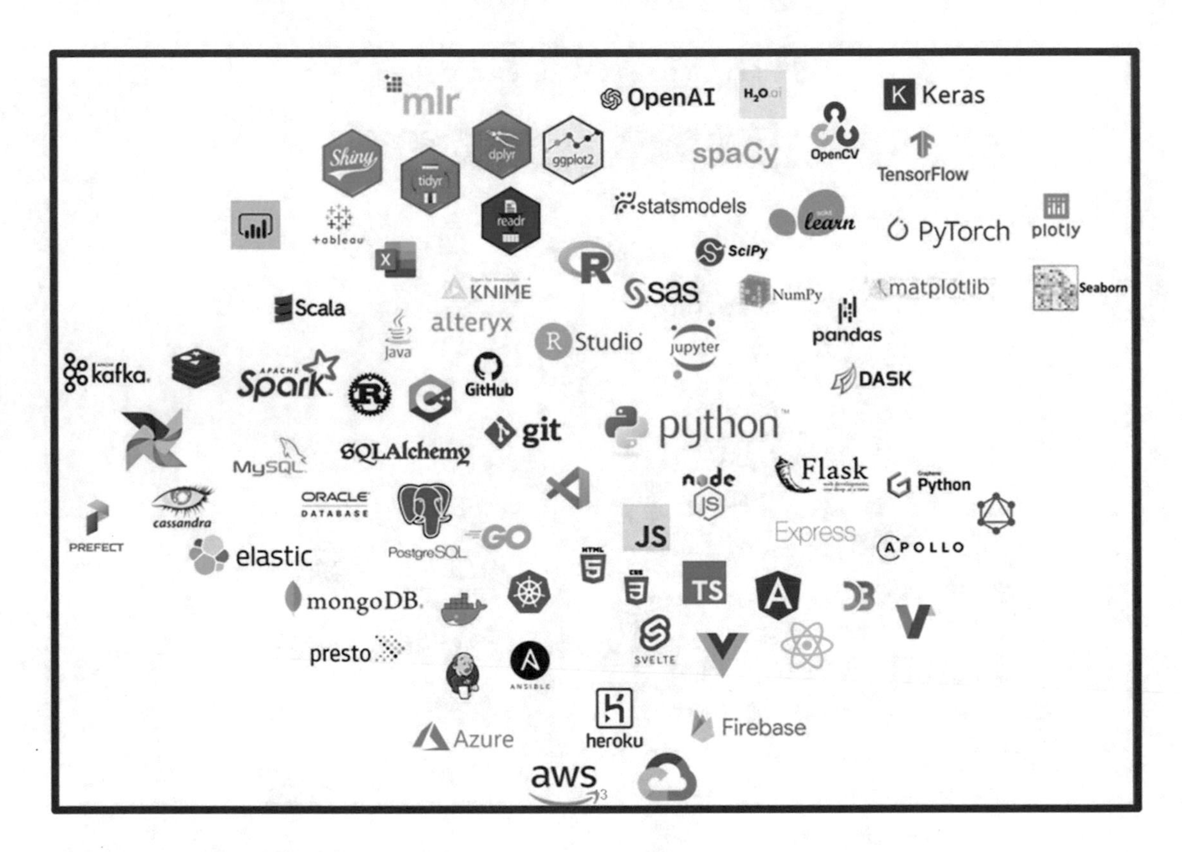

Figure 1-6. *Full-Stack AI*

We will be using many of the above in our hands-on examples in this book.

---

[4] other languages such as R and Go / Golang are of course also used for Data Science, but are some way off the popularity and spread of Python at the moment. We will however touch on Scala further in Chapters 3 and 9.

AI Ethics and Risk: Issues and Concerns

A final word on this first section in this chapter should be reserved for what is rapidly becoming a growing fear in many larger organizations today - the ethical use of data and AI to drive business value.

Many problems today stem from overly hyped early years of commercial AI development where issues around data bias, model risk, "black-box" auditability, and even user consent were largely ignored in order to generate ROI from the outcomes.

Following some high-profile cases in the media, there is a negative perception that the inherent bias in an AI model coupled with the overreliance on automation can discriminate or set poverty traps for the poor. There are in effect two main algorithmic "webs" outlined below:

Credit-reporting algorithms - these affect access to private goods and services like cars, homes, and employment. Of concern are potential ethical breaches such as the Apple Card case where Goldman Sachs (as the card operator) was investigated by regulators[5] for using an AI algorithm that allegedly discriminated against women by granting larger credit limits to men.

Government/public sector agencies - here AI algorithms impact access to public benefits like healthcare, unemployment, and child support services. Another recent case in the United States saw UnitedHealth investigated by regulators for creating an algorithm with racial basis toward white patients over sicker black patients.

The AI ecosystem: Hands-on Practise

AI Ethics and Governance brings to an end our first section in this introductory chapter but in order to get to grips with some of the outcomes from an AI project, its worthwhile taking a look at an important tool from the wider AI ecosystem.

CREATING A COVID DASHBOARD WITH POWERBI

All good AI solutions require a visualisation / business intelligence (BI) layer to ensures solutions are delivered with a compelling dashboard or interface to a dashboard. There are many such tools used for this purpose including AWS QuickSight, Google Data Studio, Cognos,

[5] See, for example, www.standard.co.uk/tech/apple-card-sexist-algorithms-goldman-sachs-credit-limit-a4283746.html

Tableau and Looker[6]. Here we take a look at Microsoft PowerBI – one of the leading BI tools at the moment.

1. Accept cookies and sign up and download PowerBI from the link below:

   `https://powerbi.microsoft.com`

2. Google "John Hopkins Covid data GitHub" to find the latest GitHub data from John Hopkins University. For reference the live csv files for confirmed, recovered and mortality cases are at the link below:

   `https://github.com/CSSEGISandData/COVID-19/tree/master/csse_covid_19_data/csse_covid_19_time_series`

3. Open PowerBI and Go to Get Data > Web. Enter the url for each of the three files separately, selecting "Load Data" to import the data to the PowerBI data model

   NB as the link is to the "live" files, the data shown in the visuals below will automatically update each day

4. Exercise: In the Explorer view of PowerBI recreate the visuals shown in the example dashboard at the link below:

   `https://app.powerbi.com/view?r=eyJrIjoiN2FkNzZlMWQtMmE2OCOoNzRiLWIoZGItNDMzNzZhYTIwYTViIiwidCI6IjhlYTkwMTE5LWUxYzQtNDgyNCO5Njk2LTYONzBjYmZiMjRlNiJ9`

5. Make sure to create the visuals one by one by:

   a. Selecting the correct visual (e.g. card, table, bar chart, area chart, tree chart)

   b. Dragging and dropping the correct dimensions (entities to report on e.g. country) and measures (values to report on e.g. confirmed cases)

   c. Filter if necessary, e.g. on top 10 cases

   d. Finally add slicers on date and country to allow a user to quickly drill-down into Covid cases by date (window) and country

6. Exercise (Stretch) – push the finished PowerBI report to PowerBI Service[7], then proceed to host your dashboard on a public url

---

[6] And DS.js and Apache Superset. We will look into more Python-specific visualisation tools in Chapter 7 AI Application Development

Applications of AI

Taking the context from the previous section and in keeping with the light introductory nature of this first chapter, we will now address in the next section the main AI applications:

- Machine Learning
- Deep Learning including Computer Vision and Portfolio, Risk Management and Forecasting
- NLP including Chatbots
- Cognitive Robotic Process Automation (CRPA)

Machine Learning

Machine Learning is a technique enabling computers to make inferences from complex data and remains the biggest area of AI research today. There are three main types: Supervised, Unsupervised, and Reinforcement Learning,[8] the development and deployment of each we will look at in subsequent chapters. For now, we have provided some basic definitions focused on the inherent difference between these machine learning approaches:

Supervised – training on datapoints where the desired "target" output is known

Unsupervised – no outputs available but machine learning is used to identify patterns in data

Reinforcement learning – training a machine learning model by maximizing a reward/score

A key application of machine learning today is in Fraud Detection, which is often run as both an unsupervised ML and supervised ML problem. The goal is to try and predict patterns in transactional (and customer) data that indicate fraud is taking place.

[7] PowerBI Service requires an organisational email. For assistance on see `https://dash-bi.medium.com/how-to-use-power-bi-service-for-free-without-a-professional-email-in-4-steps-f97dbaf4c51e` or `https://learn.microsoft.com/en-us/power-bi/fundamentals/service-self-service-signup-for-power-bi`

[8] Four if a semi-supervised approach is used – see Chapter 4

It is assumed that most readers are familiar in some way with basic Machine Learning techniques and would direct readers to other books to augment their understanding in case some of the applications discussed in this book go beyond assumed knowledge.

Deep Learning

In many ways Deep Learning is a subset of Machine Learning; essentially extending Machine Learning to hard, "Big Data" problems typically solved using neural networks. Neural networks themselves are inspired by neuron connections in the human brain - when put together we get something like the image on the right (a). This is in fact a multilayer neural network taking four inputs and providing one output after various data transformations carried out within the two "hidden" layers of four nodes (or neurons) each.

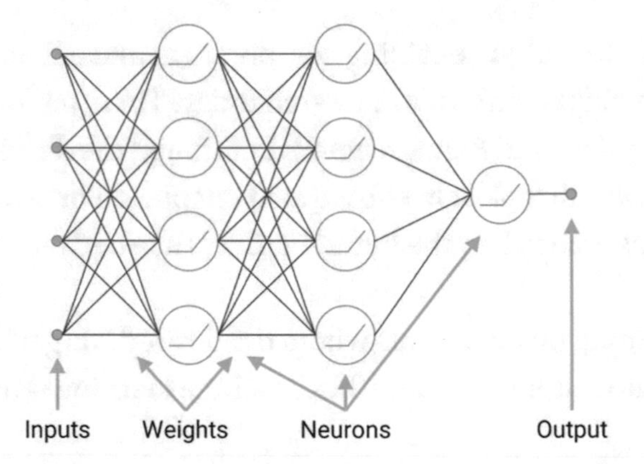

Figure 1-7. *Multilayer PerceptronComputer Vision*

Deep Learning is commonly used in Image Classification or Computer Vision (such as predicting whether an image contains people, buildings, or vehicles) where the inputs are essentially the (training) images converted to pixels, then a machine-readable format (tensors) and the hidden layers act as function mappings to extract "patterns" which tend to prevail in the images.

As illustrated in Figure 1-8, Computer Vision is used for Object Classification, Identification, Verification, Detection, Segmentation, and Recognition and, using the picture as an example, gives rise to responses to questions such as:

- Which type of object is in the image?
- Does the object exist in the image?
- What category of objects are in the image?

The many applications of Computer Vision include facial recognition and surveillance, document search and archiving, healthcare diagnostics, and crop disease prevention.

***Figure 1-8.** Object Classification in Computer Vision (Source: Ilija Mihajlovic, `towardsdatascience.com`)*

Portfolio, Risk Management, and Forecasting

AI, and specifically Deep Learning, is widely used today in risk management and portfolio optimization to increase the efficiency and accuracy of operational workflows. Similarly, the optimization of financial assets, stock price prediction, portfolio, and risk management are all exciting areas of innovation in 2022.

Forecasting is also a big area right now – advances in cloud storage and processing power have exposed weather and demand forecasting, asset optimization, algorithmic trading, and robo-advisors to AI methods. High-accuracy, industrial-scale recurrent neural networks (RNNs), and specialized long short-term memory models (LSTMs) are increasingly able to outperform traditional methods. We will look at some practical hands-on examples of these in Chapter 5 on Deep Learning.

Natural Language Processing (NLP)

The global market for NLP is growing year on year, and is predicted to reach 43 million dollars by 2025 (Statista).

Natural Language Processing is a branch of AI that deals with the interaction between computers and humans using natural language. Essentially NLP applies Machine Learning algorithms to unstructured data and converts it into a form that computers can understand – the objective is to read, decipher, understand, and make sense of language in a manner that is (commercially) valuable.

NLP is the driving force behind language translation apps such as Google Translate. Word Processors such as Microsoft Word and Grammarly also employ NLP to check the grammatical accuracy of texts. The growing adoption of NLP applications in industry means the global market is predicted to reach $43m by 2025.[9]

Chatbots

Probably the most well-known application of NLP is chatbots and personal assistant applications such as Siri and Alexa and Watson Assistant.

Chatbots are essentially software applications to conduct interactive dialogue, with growing text-to-speech and speech-to-text capability.

The technology has vastly improved since the earliest days of Cortana on Windows. Today it has evolved to the extent that today Intelligent Virtual Agents (IVAs or Chatbots 2.0) or Interactive Voice Response (IVR) applications are widely used in call centers to respond to user requests. IVAs have in-built self-learning capability and adapt to context in contrast to earlier dialog rule-based chatbots.

The business value in 2002 is in improving the customer journey and customer experience – with the ability to resolve issues quickly, although not in more complex cases.

[9] Statista

The main techniques in Natural Language Processing are syntactic analysis and semantic analysis. Syntactic analysis focuses on grammar while semantic analysis is concerned with the underlying meaning of text. Both involve a number of underlying subprocesses (such as lemmatization and word disambiguation) important for categorization and more ultimately, insight extraction. We will look at these in more detail in a later chapter.

In its simplest sense, Natural Language Processing in chatbots is invoked to detect, then best-fit user "intents" and "entities" to a preconfigured dialogue "corpus." As user interaction grows, increasingly more data can be used in the training process to improve the matching process, and enhance the dialog.

Cognitive Robotic Process Automation (CRPA)

Cognitive Robotic Process Automation is Robotic Process Automation that leverages AI. While traditional RPA configures computer software to execute a business process, cognitive RPA tools and solutions add predictive capability and enhanced exception handling on top of the underlying business process making use of such techniques as OCR and automated scanning, Text Analytics, Voice to text, and Machine Learning.

While traditional RPA supports automation based on structured data, CRPA more commonly includes unstructured data sources.

Many operational or portfolio management processes are today coupled with Cognitive Robotic Process Automation (CRPA), marrying AI and automation for business problems such as:

- Bulk payment processing
- Transactional risk monitoring
- Auto-completion of forms
- Document Archiving

In risk management, sophisticated AI augmentation of Monte Carlo and VaR approaches are also increasingly common to both benchmark and improve existing risk practices.

Other AI Applications

The above introduction to the main uses of Artificial Intelligence in the workplace is intended to give a flavor of what to expect later on in this book. Specifically, Machine and Deep Learning, Natural Language Processing, and CRPA underpin the industry-specific use cases adopted by the private and public sectors today.

We will describe and walk through many of these later on in our hands-on practice, but for now we have captured in Figure 1-9 a broad segmentation of use cases by sector.

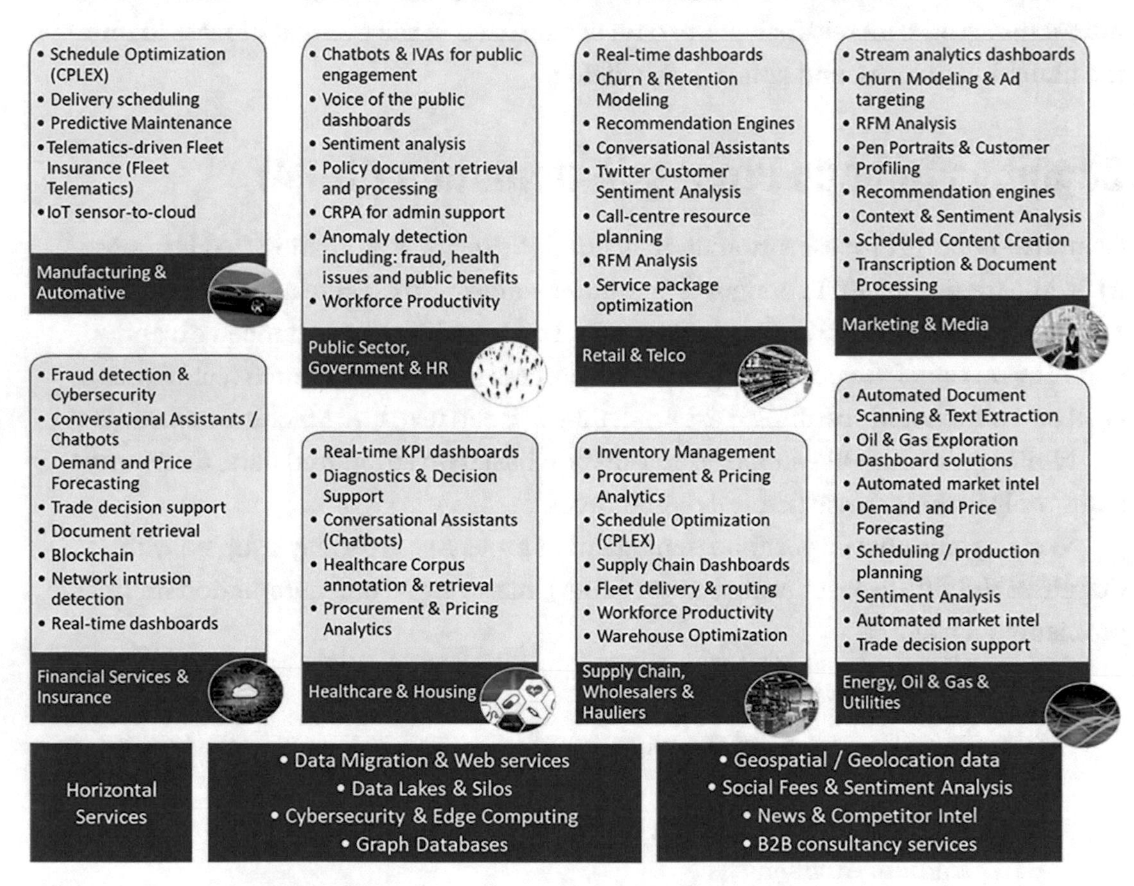

Figure 1-9. *AI applications*

While emerging AI applications may not yet be fully commercialized or indeed in use in most workplaces, we also include below a view from Gartner of what is on the horizon with several of note (e.g., Reinforcement Learning and Edge AI) likely to enter the mainstream in the coming years.

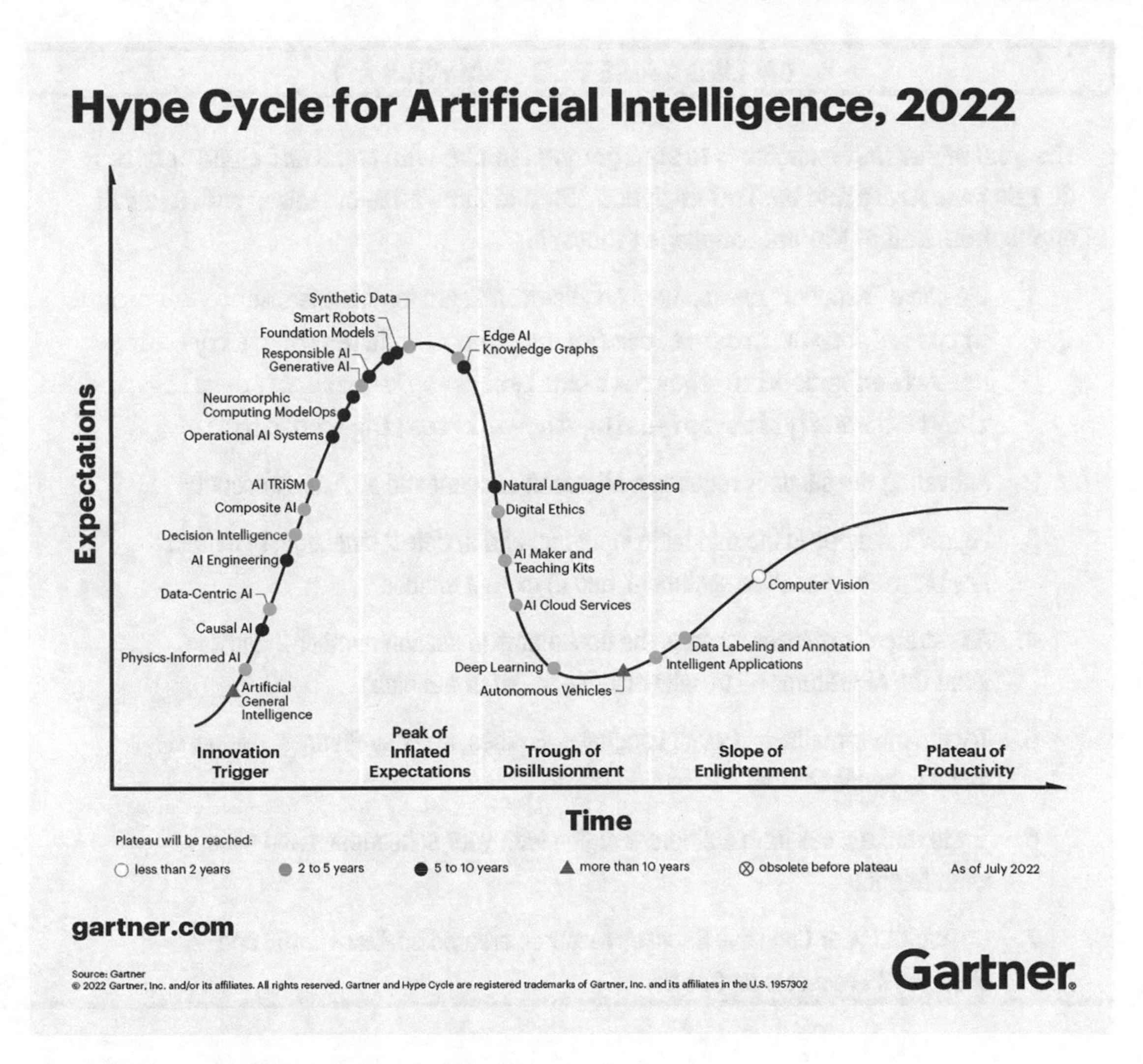

Figure 1-10. *Gartner AI hype cycle*

AI Applications: Hands-on Practice

Now that we have taken a first look at the main applications for Artificial Intelligence in today's companies and organizations, we will now take a look at a hands-on lab example

CALLING AZURE TEXT ANALYTICS API

The goal of our first exercise is to start getting familiar with cloud and cloud services (in this case Azure and the Text Analytics API) and familiarize ourselves with a key AI application, that of Natural Language Processing

1. We use a "sandbox" environment on Microsoft Learn for this exercise: `https://docs.microsoft.com/en-us/learn/modules/classify-user-feedback-with-the-text-analytics-api/3-exercise-call-the-text-analytics-api-using-the-api-testing-console`
2. Activating the sandbox requires a Microsoft account and an Azure account[10]
3. Follow the steps in the tutorial to input text and a) detect language, b) extract key phrases, c) analyze sentiment, and d) extract entities
4. As "stretch" exercises, change the documents in section number 2 and see what the API returns -- try with "Ich bin so sauer auf dich"
5. Try the other methods, Detect Language, Entities, and Key Phrases, using the same subscription key
6. Try to make a call from a different region with your subscription and observe what happens
7. Go back to your Cognitive Services resource created on Azure Portal and observe API requests on free tier

Data Ingestion and AI Pipelines

From key AI application, we move on to our next section on Data Ingestion and automating data pipelines. Chapter 3 will discuss this in far greater detail but for now we introduce AI Engineering as the core skill requirement for delivering tangible,

[10] Free for 30 days, then requires credit card to maintain as pay-as-you-go (as long as resources are deleted immediately after creation, the cost is likely < £2 costs per month or free). You can check costs incurred at the link below `https://portal.azure.com/#blade/Microsoft_Azure_CostManagement/Menu/costanalysis`

production-grade AI solutions[11] along with important definitions for Data Pipelines, pivotal ETL processes, best practice around data wrangling and transformation, and AutoAI.

AI Engineering

Industry research shows that very few AI projects are successful, partly because technically minded, and often junior resources on an AI project tend to forget that AI involves people, processes, and tools. The reality is writing machine or deep learning code is just one small part – and invariable notebooks or scripts fail when not considered complex surrounding infrastructure.

Data pipelines lie behind every successful AI application – from data ingestion through several stages of data classification, transformation, analytics, training machine learning, and deep learning models through inference and retraining/data-drift processes, the goal is to yield increasingly accurate decisions or insights.

Ultimately no AI project can be a success without a Data Strategy in place with a robust, serviced delivery pipeline for ingesting data into downstream modeling and analytics processes. With that in mind, we discuss in this section how data ingestion works and setting up data pipeline necessary for building a successful AI application.

What Is a Data Pipeline?

A Data Pipeline is a flow of data from upstream source(s) to a downstream sink where it can be more easily consumed by end users. More formally, it is an automated set of actions that extract or **ingest** data from various sources, often transform the data, and then place in a data store or repository for downstream analysis.

In the age of cloud computing, there are multiple ways this is done such as shifting data from the web or an app to a data warehouse or from a data lake to a database. What all have in common in 2022 is a growing need to make the tools and transformational processes involved as seamless as possible and ultimately achieve full automation.

[11] Enterprise AI – more on this in Chapter 7

Extract, Transform, and Load (ETL)

Extract, transform, and load, or ETL, is often confused with a data pipeline. While an ETL process refers to the process by which data is extracted from different sources, transformed into a usable format, and then loaded into an end-user system, it is a discrete, limited set of steps. Data pipelines on the other hand are not discrete in time and deal with the continuous flow of data.

Because a data pipeline can in a sense be considered as a continuous loop of ETL steps, it can be helpful to define these three steps.

Extract

Data is extracted from a variety of internal and external sources such as text files, csv, excel, json, html, relational and nonrelational databases, websites, or APIs. More modern formats such as parquet and avro are also increasingly utilized for their efficient compression of datasets.

Transform

Data is transformed in order to make it suitable for, and compatible with, the **schema** of a target end-user system. Transform involves cleansing data to remove duplicates or out-of-date entries; converting data from one format to another; joining and aggregating data; sorting and ordering data among others.

Load

In the final ETL step, data is loaded into a target system such as a data warehouse. Once inside the data warehouse, data can be efficiently queried and used for analytics and business intelligence.

Data Wrangling

The above concepts are very much central to the role of the Data Engineer but there is considerable overlap, particularly in the Transform step in ETL in serving downstream Data Science and Data Analysis processes.

Data Wrangling (or Data Munging) is the main process in Data Science and AI which ensures data is in a fit-for-purpose state to carry out analytics or BI. Many people are familiar with the statistic that cleaning data is 80% of the job of a Data Scientist In truth, Data Wrangling can take up to 80% of a Data Scientist job, and it involves more than just cleaning data: there are many more subprocesses including formatting, filtering, encoding, scaling and normalizing, and shuffling or splitting. And these are not only restricted to structured data, unstructured data (such as text or images) is in scope too for both machine and deep learning

Data Wrangling sits in a process after data acquisition and before modeling/machine or deep learning. It is highly iterative and often coupled with exploratory data analysis (EDA) to understand better the structure of the individual fields (typically columns) within a dataset. While EDA focuses on passively "looking" at the data, Data Wrangling actually actively "changes" the data in some way.

We will discuss Data Wrangling (and ETL processes) in more detail in our chapter on Machine Learning and in particular look to establish best practice techniques for productionizing Machine Learning by looking at key Case Studies such as Fraud Detection.

Performance Benchmarking

While ETL and Data Wrangling deal with preprocessing of data, we also need a means to "score" (ideally continuously) the data flowing into an AI solution.

Building an AI app requires constant training and testing. Knowing how to benchmark performance and which measures to use is a key overhead in both machine and deep learning and needs to be rigorous and adaptive to the evolving (input) data pipeline.

Measures such as accuracy, recall, precision, and a confusion matrix for Supervised Classification problems to understand better the proportion of actual cases we correctly predicted (whether negative or positive). For Supervised Regression problems, root mean squared error and R-squared are used to compare forecasted output with the actual target data. In Deep Learning we may use similar measures to above plus additional specific Deep Learning measures such as loss and cross-entropy.

AI Pipeline Automation – AutoAI

Much of the focus in recent years in AI has been around the potential to automate the entire end-to-end process of performing ETL, Data Wrangling, and Performance benchmarking. Coupled with automation around Data Drift (measuring how much data changes each time we refresh the input pipeline), achieving end-to-end automation with full-scale monitoring and metrics for each step is key to achieving a true Enterprise AI strategy.

While the "vision" for most organizations is to automate everything, in practice, not everything is currently in the scope of AutoAI - instead, we mention in Table 1-2 some of the key variables that can be incorporated into a fully functional and automated Data Pipeline, as evidenced in tools such as IBM Cloud Pak for Data, DataRobot, and Google Vertex AI.

Table 1-2. *Key processes and levers for AI pipeline automation*

Pre-Modeling	Post-Modeling
Raw data import	**Feature engineering**
• Static (batch) files	• Dimensionality reduction
• Multiple SQL queries	• Normalization (scale between 0 and 1)
• Authenticated APIs via Python libraries	• Standardization (scale to mean 0, std 1/fit to a normal distribution)
Data wrangling	**Model tuning**
• Identification + treatment	• Performance benchmarking
• Missing values	• Hyperparameter tuning/grid search
• Encoding	• Algo selection
Data partitioning	**Retraining**
• Identify target variable	• Data drift
• Shuffle and split train/validation	
• Test or k-folds	

Build Your Own AI Pipeline: Hands-on Practice

Having spent some time looking at what constitutes an AI Pipeline, let's take a look at a real lab example – how does this look like in practice?

NO-CODE CLASSIFICATION

The goal of this exercise is to understand the dependency and flow of data on the outcomes from any AI application. The exercise walks through how this is done with a "No-code" binary (supervised) classification model to predict income levels in Microsoft Azure ML Studio

1. If you do not already have a Microsoft account (e.g., a `hotmail.com`, `live.com`, or `outlook.com` account), sign up for one at `https://signup.live.com`
2. Navigate to `https://studio.azureml.net`, click the option to sign up, and choose the Free Workspace option. To test, sign in using your Microsoft account, then sign out again.
3. Follow the steps in the tutorial below to import data, perform basic EDA and Data Wrangling, modeling, and performance benchmarking.
4. `http://gallery.cortanaintelligence.com/Details/3fe213e3ae6244c5ac84a73e1b451dc4`
5. As "stretch" exercises, try to improve the model performance by modifying the input data (features), encoding one of the categorical features, changing the training/test set partition % or changing the algorithm from two class-boosted decision trees.

Neural Networks and Deep Learning

While the Data, or the AI Pipelines we discussed in the previous section are the key means by which AI applications seamlessly integrate with and incorporate evolving data sources, they are wrapped around the engine of AI applications, that is, a machine or deep learning model which lies at the "kernel" **of the application.**

Before embarking on a high-level look at Deep Learning, let's take a quick refresher on Machine Learning. As mentioned in the opening section, it is expected readers have some grounding in Machine Learning already, so we will only address important concepts here that are relevant to implementing an AI solution.

Machine Learning

At a basic level, there are two types of Machine Learning: Supervised and Unsupervised Machine Learning. Reinforcement Learning is sometimes considered a third type, although can equally be considered a type of Unsupervised Learning.

Supervised Machine Learning

What distinguishes Supervised Machine Learning from Unsupervised Machine Learning is the prevalence of "labeled" or "ground truth" data, that is, a specific target field or variable that we wish to train a model to predict.

There are two main types of Supervised Machine Learning:[12] classification and regression. Whereas the label or target variable in classification is discrete (usually binary, but sometimes multiclass), the label in a Supervised Regression problem is continuous. A key industry machine learning application using a classification technique is determining whether a customer is likely to churn (or not), while forecasting customer revenue is an example of a regression technique. In both of these cases, the **features** used to predict or forecast the target variable are typically customer attributes, normally both transactional and demographic but can also include behavioral or attitudinal data such as time spent on a web page or sentiment from social media engagement.

Unsupervised Machine Learning

With unsupervised machine learning, we have no ground truth, so the predictive modeling aspect is instead trying to impose an unseen pattern on the underlying data. Typically some form of clustering is used to group data into clusters, for example, for customer data buried in a CRM platform, an unsupervised machine learning approach might throw up segments which have some degree of commonality (high spend, low to mid income, located in a specific region. etc.).

[12] Or three if Time Series Forecasting is considered as distinct from Regression

Dimensionality reduction is also sometimes considered an unsupervised technique in that a machine learning algorithm is used to simplify the data (often from thousands of underlying features into tens of features if using Principal Component Analysis) into a form which statistically resembles the original data. While this approach can greatly reduce datasets and improve runtime/performance, it is not strictly a Machine Learning modeling technique in the normal sense as the outcome in this case is another dataset, albeit compressed, rather than a trained model.

Reinforcement Learning

Reinforcement learning involves real-time machine (or deep) learning with an agent/environment mechanism which either penalizes or rewards iterations of a model based on real-time feedback from the surrounding environment (how accurate is the model).

While the scope of this book is mainly focused on mainstream business and organizational applications, advances in reinforcement learning are in general where there is considerable hype in the media – essentially this is the underlying technique that drives "industrial-scale" applications such as Google's Search Engine, autonomous vehicles, and robotics.

What Is a Neural Network?

With that brief recap on Machine Learning, let's move onto a high-level overview of Deep Learning.

Artificial Neural Networks (ANNs) are the structures underpinning Deep Learning. Biologically inspired by the brain, ANNs are used for complex problem solving for their ability to extract hierarchical, abstract, or "hidden" features from underlying data.

In general, an Artificial Neural Network consists of an input layer, multiple hidden layers, and an output layer. In a "forward pass" through the network, various data points are taken as inputs (comparable with features in a Machine Learning model), fed through the weighted network to activate various neurons and ultimately produce a numerical output. In the case of a Computer Vison or Image Classification application, iterating over many of these forward passes (an epoch) results in a probabilistic array of values corresponding to the probability the data corresponds to two or more outcomes (in this case images, such as a person or an object).

The Simple Perceptron

The Simple Perceptron is the building block of an Artificial Neural Network – effectively a simplified model of the biological neurons in our brain.

The single neuron (shown in Figure 1-11) has multiple inputs, and based on these inputs the neuron either fires off or it doesn't. If we structure these perceptrons in layers we get a Multilayer Perceptron or Deep Neural Network and something that more closely resembles our brain in function, with multiple neurons firing depending on the input signal.

The "activation" function in a simple perceptron uses the Heaviside step function – a simple on/off switch. Simple perceptrons can also only learn linear functions and as such are only really useful as a performance reference. Multilayer perceptrons (Figure 1-11) on the other hand can learn nonlinear functions by virtue of having more than one neuron and their reliance on more advanced activation functions such as sigmoid, tanh, and ReLu.

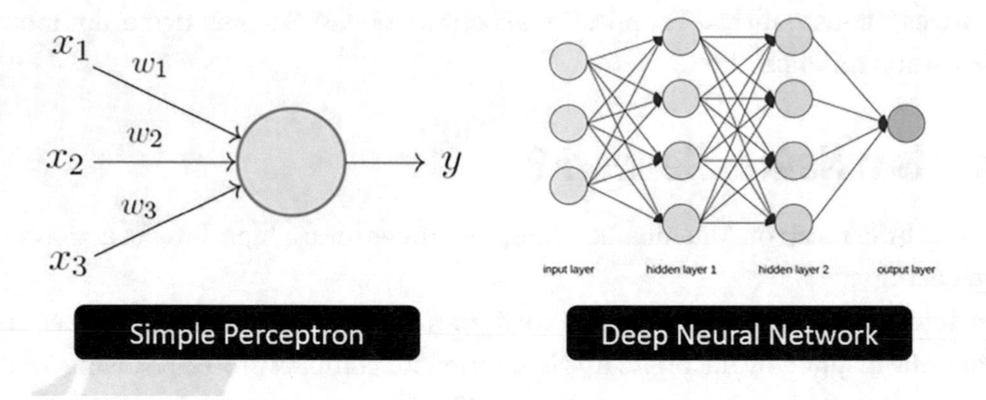

Figure 1-11. *Simple Perceptron vs. Multilayer Perceptron/Deep Neural Network*

Deep Learning

We saw above that Deep Learning refers to applying artificial neural networks with several hidden layers, that is, a deep neural network.

Deep Learning though can use a number of different types of Artificial Neural Network, each of which has a number of key features which contribute to their adoption in solving specific predictive analytics challenges in the workplace. The main types covered in our hands-on labs in this book are listed below.[13]

Convolutional Neural Networks

Convolutional neural networks are a type of neural network that uses convolutions to extract hierarchical patterns from the input data. Convolution is a mathematical way of combining two signals to form a third signal. CNNs are predominantly used in data that has spatial relationships, such as images.

Recurrent Neural Networks

Recurrent neural networks are used to process sequential data or data with underlying order/syntax. As a result, RNNs are typically used in forecasting time series data such as stock market data or IoT/sensor data that has underlying temporal dependencies.

Because grammatical terms follow a certain sequence and order, speech recognition and Natural Language Processing applications such as chatbots and IVAs also typically use a type of Recurrent Neural Network, such as an LSTM (long short-term memory) network to capture nuances, structure and order in the underlying grammar.

Autoencoders and Variational Autoencoders (VAEs)

Variational Autoencoders are an improvised version of autoencoders – unsupervised artificial neural networks that learn how to efficiently compress and encode data. Autoencoders receive a lot of attention in the media because of their ability to generate image data; and are sometimes confused as the technique behind Deep Fakes. Autoencoders and VAEs consist of an encoder (for the input data) and a decoder (for reconstructing the input from the network output). The difference between these two types of neural network techniques is that VAEs produce a distributed (probabilistic) output.

[13] Due to their now rather limited application to e.g. Dimensionality Reduction of datasets, Restricted Boltzmann Machines are not covered in this book.

Generative Adversarial Networks (GANs)

A GAN is a type of CNN that uses a generator continuously generating data while a discriminator learns to discriminate fake from real data. As training progresses, the generator improves at generating fake data that looks real while the discriminator's ability to detect the difference between fake and real also improves. GANs are the real technology behind Deep Fakes.

As a result of this "virtuous cycle," a GAN trained on faces can be used to generate images of faces that do not exist but look very real. Despite the similarity, GANs are different from Autoencoders and VAEs as they work to generate new data which can't be distinguished from real data, rather than reconstructing the (same) input data.

Neural Networks – terminology

Due to its complexity, Deep Learning comes with a bewildering vocabulary of concepts, tools and techniques for model configuration and improvement. We will look at these more closely later and expand on the definitions below, but for now we list below the main concepts fundamental to an understanding of how the model training process works.

Epoch — a complete cycle through the entire dataset.

Learning Rate — how fast weights change during gradient descent / backpropagation[14].

Activation Function — triggers the output of a node/neuron, given a set of inputs

Regularization — a "hyperparameter" to prevent overfitting

Batch Size — size of the random sample taken from the input data for each forward pass

Hidden Layer — a layer between input and output layers

Loss function — calculates the delta between predicted and actual output

Minimizing the latter of these (loss function) is the essential objective in training a neural network.

[14] We will cover Gradient Descent and Backpropagation in Chapter 5

Tools for Deep Learning

In this book, the main tools we will be using for Deep Learning are Python,[15] TensorFlow, and Keras. These tools are all open source and freely available today but it is worth noting many of the tools used in Deep Learning, including TensorFlow and Keras, have evolved out of internal development carried out by academia and Big Tech.

TensorFlow is the brainchild of Google, and is still used to power all of Google's Machine and Deep Learning on big datasets as well as other global brands such as Airbnb, Coca Cola, GE, and Twitter. It was open-sourced in 2015 and though it is a low-level language normally requiring highly specialized programmers to operate and run, like most of today's ecosystem, the underlying code has since been served through high-level "wrappers" such as Keras, developed by MIT and easily importable as a Python library.

We also use PyTorch in this book for applications of Deep Learning with Natural Language Processing. The development of PyTorch has a similar path to that of TensorFlow – it is a Facebook tool open-sourced on GitHub in 2017.

Other tools for Deep Learning include Caffe, Apache MXNet, and Theano – tend to be used where there are strong integration requirements (e.g., with NVIDIA in the case of Caffe or with Apache Kafka or Apache Spark in the case of MXNet). Theano was developed by Université de Montréal, and while it had its day, development ceased in 2017 as a result of its inability to compete with the bigger (budget) offerings from Big Tech.

Introduction to Neural Networks and DL: Hands-on Practice

Now we have set the context for Neural Networks and Deep Learning, it's worth exploring how they work. Rather than diving into code at this stage though, we will make use of great visual tool for visualizing how data is processed through a neural network in order to train a deep learning model.

[15] Although there are some labs in this book where we will run python as a script (.py format), most labs run python via either a Jupyter Notebook or Google Colab Integrated Development Environment (IDE) (i.e .ipynb format). The same labs can be run instead using Visual Studio Code or PyCharm. For Jupyter environments, the use of the RISE (Reveal.js) extension is recommended to enable code samples presented as slideshows: `https://rise.readthedocs.io/en/stable/`

TENSORFLOW PLAYGROUND

The goal of this exercise is to visualize the training process of a simplified deep learning model and to try and improve its performance by playing with some of the many tuning "levers" available to us.

1. Go to `http://playground.tensorflow.org/`
2. Have a look at the four datasets by clicking on the thumbnail icon on the RHS of the screen (under DATA),
3. Note in all cases the (supervised) datasets have labeled data: either blue or orange and the output shows these datapoints plotted on a 2D (x1, x2) grid. The inputs for our model are shown under FEATURES and are initially set to just x1 and x2 coordinates while we have two hidden layers of 4 and 2 neurons respectively.
4. Choose one dataset and press run. This will start the training process (and immediately after the evaluation process). Notice how the neural network weights get updated through each forward pass and each epoch.
5. Observe the Training and Test loss shown on the right-hand side. A good model will have both the training and test loss close to zero.
6. Stop the model training process for now. The first three datasets are relatively easy to train on. Choose the last (spiral) dataset and restart the training process.
7. As "stretch" exercises, try to improve the model performance (lower loss in less epochs) by making neural network architecture more complex – add synthetic features to the model (x1x2, power or trigonometric transforms) and/or increase the number of layers in the network. You can also change the ratio of training and test data on the LHS or increase/decrease the batch size or modify the hyperparameters along the top. Try to find a suitable configuration which achieves a loss < 0.01 in under 500 epochs.
8. Choose the last (spiral) dataset and restart the training process. For now, notice that this time the modelling process struggles to converge – the loss oscillates a lot. We will have another look at TensorFlow Playground later on in this book in Chapter 5, and try to achieve a better result training on this dataset.

Productionizing AI

Theory is one thing and delivery is another. We discussed briefly in Section 3 about the proliferation of failed AI projects - the reality is that ever since Data Science has become a "glamorous" job role backed by over-hyped job board marketing, poorly designed, and over-engineered R and Python scripts, with broken integration links have left a trail of waste across the Enterprise AI landscape.[16]

Out of this "wasteland" most organizations and businesses are attempting to redress the landscape and recognize a need to introduce a broader Design/System Thinking approach at the outset to ensure that AI solutions are built with multiuser engagement (both technical and nontechnical), end-to-end process and system-wide ecosystem, infrastructure and integration in mind.

Compute and Storage

Very few AI solutions today can be considered as "on-prem" solutions, and any true Enterprise AI solution goes way beyond the underlying machine or deep learning model. AI solutions today are inextricably tied to Cloud Computing and specific resources and services offered by Big Tech - the need for cloud stems from two key "grouped service" offerings: Compute and Storage.

Compute is essentially computer processing power - tied to computing memory, it is the ability to perform software computation and (often complex and highly parallelized) calculations. Compute is typically delivered via a Virtual Machine on Cloud.

Storage in the context of an organization's operational and strategic needs is the means by which all their data requirements are supplied, replenished, and maintained. Most Cloud providers offer File Storage and SQL/NoSQL-based options for storing both structured and unstructured data.

While original transactional DB systems required storage and compute to be as close as possible to reduce latency, faster networks and the increasing availability and scalability of database systems coupled with the need to reduce hosting costs have driven the separation of compute power and storage.

[16] "Technical Debt" is discussed further in later chapters

All businesses need two types of data: transactional and processed (batched or aggregated) data. Much of this still resides in Data Warehouses today, but because sophistication in Predictive Analytics demands complex analytical queries, businesses and organizations often decide to migrate their data to cloud as the relative increase in latency is a lower priority than the overall cost savings achieved through cloud.

The CSPs – Why No-one Can Be Successful in AI Without Investing in Amazon, Microsoft, or Google

Having established that all AI projects (or at least Enterprise AI projects) require Cloud, the question arises as to which one? Amazon Web Services gobbled up this space many years ago. Today, as can be seen in Table 1-3, there is some fragmentation, with Microsoft Azure competitive with AWS. Google Cloud Platform is also gaining market share and there are a handful of other providers such as IBM Cloud, Alibaba (mainly in China), and Heroku.

Table 1-3. *Gartner Worldwide IaaS Public Cloud Services Market Share, 2019-2020 (Millions of US Dollars)*[17]

Company	2020 Revenue	2020 Market Share (%)	2019 Revenue	2019 Market Share (%)	2019-2020 Growth (%)
Amazon	26,201	40.8	20,365	44.6	28.7
Microsoft	12,658	19.7	7,950	17.4	59.2
Alibaba	6,117	9.5	4,004	8.8	52.8
Google	3,932	6.1	2,367	5.2	66.1
Huawei	2,672	4.2	882	1.9	202.8
Others	12,706	19.8	10,115	22.1	25.6
Total	**64,286**	**100.0**	**45,684**	**100.0**	**40.7**

Vendor-lock on one particular Big Tech provider has become an issue, with many companies now trying to diversify and utilize services from multiple cloud service providers.

[17] www.gartner.com/en/newsroom/press-releases/2021-06-28-gartner-says-worldwide-iaas-public-cloud-services-market-grew-40-7-percent-in-2020

While each CSP comes with its own "Marketplace" of cloud tools, we discuss below only those that fall under the key Compute and Storage grouped services. Other grouped services such as Governance, Security, Auto-scaling, and Containerization are discussed in more detail later.

Compute Services

Azure Virtual Machines and EC2 instances, typically provisioned via a Virtual Machine on a Virtual Private Cloud on AWS, are the main options for Cloud Compute. Google Compute Engine is GCP's main offering.

Storage Services

Amazon Simple Storage Service S3 is probably AWS's most well-known storage service, and is used to securely store data and files in "buckets." Comparable services from other vendors include Azure Blob storage, and Google Cloud Storage. We will make use of these compute and storage services in some of our hands-on labs later.

Important Note Use of Cloud doesn't come free, despite the pretense to offer "Free Tiers." We will be using it frequently in this book, but be prepared for costs that could run up to $500 to follow all the labs.

Free Tiers are limited to certain resources, and these resources themselves have rather mediocre capacity limits. There are limits on usage of services, so if a customer faces costs on their cloud account, it's likely this is a result of usage on a cloud resource/ service exceeding free tier limits or the subscription has rolled over to pay-as-you-go (after one year)

It is the reader's responsibility to bear all costs related to provisioning of cloud services. We strongly advise to always stop and delete resources when finished with them. It is a source of immense frustration to this author that Big Tech will not do it automatically for you. One wonders how this practice is justifiable from the richest companies in the world.

Virtual Machines in particular incur costs when not running as they are someone else's server, and whether it's on or not, there are energy costs involved (particularly high right now with a war going on between Russia and Ukraine)

Help to manage your costs by use of sandbox environments (where available) and by deleting resources when done. The author is also happy to provide complaint letters to a CSP for onward referrals or a small donation to a charity, but ultimately it is your responsibility to bear all costs related to provisioning of cloud services.

Containerization

While the relative low cost and ease of deployment of compute and storage solutions on cloud have underpinned adoption of cloud for AI, the use of containers has become the go-to means of productionizing AI applications.

All the main cloud platforms contain containerization services – a lightweight alternative to full machine virtualization that involves encapsulating an application in a container with its own operating environment. Containerization comes with a number of benefits highly suited to building robust production-grade AI solutions including the ability to simplify and speed up the development, deployment, and applications configuration process, increased portability and server integration and scalability as well as increased productivity and federated security.

Docker and Kubernetes

Docker is the main container runtime we will utilize in this book. Docker's USP lies in its handling of dependencies, multiple (programming) language, and compilation issues when creating isolated environments to launch and deploy applications. In much the same way a "physical" container can be transported by ship, truck, or train, standardization within a Docker container means it is effectively able to run on any platform.

Although there are many similarities with Virtual Machines, as shown in Figure 1-12, Docker better supports multiple applications sharing the same underlying operating system. Docker is also fast, starting and stopping apps in a few seconds. PostgreSQL, Java, Apache, elastic, and mongoDB all run on Docker.

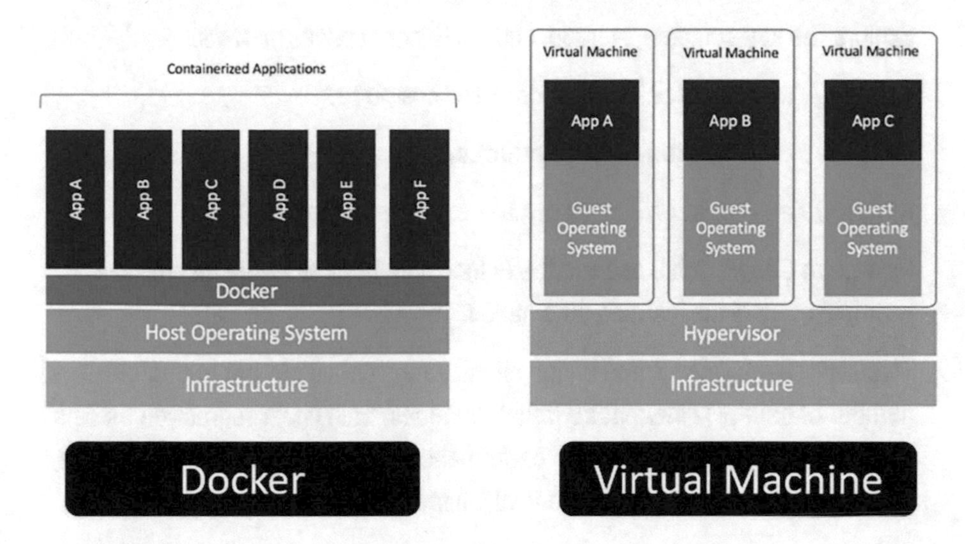

Figure 1-12. *Comparing Docker with a Virtual Machine (Source: Docker)*

Although we will not use it in this book, Kubernetes (k8s), a container management tool, is often used to orchestrate Docker "instances." As an open source platform, Kubernetes was originally designed by Google and automates deployment, management, and scaling of applications in the container.

Productionizing AI: Hands-on Practice

With that brief introduction to containers and Docker, we reach our last hands-on lab for the first chapter.

AUTOMATING THE AI PROCESS – TOOLS AND TECHNIQUES

The goal of this exercise is to get familiar with Compute, Storage, and Container solutions on cloud and to explore Machine Learning automation with Python – often the end goal of a production-grade AI solution.

1. Sign up for AWS Free Tier and open AWS Console: `https://console.aws.amazon.com`

2. Explore the key compute, storage, and container services on AWS

   https://aws.amazon.com/products/compute/

   https://aws.amazon.com/products/storage/

   https://aws.amazon.com/containers/services/

3. Now go to Google Colab and run the Python script below to see an end-to-end example of machine learning automation.

4. This script uses PyCaret and an inbuilt insurance dataset to run through a number of data pipeline/feature transformations, train linear regression models and select the best one based on performance. This same hands-on lab will be the starting point for an extensive final lab in Chapter 9 on Productionizing AI.

5. Note the script uses a static data file built into the PyCaret library but often this kind of automation with a Big Data dataset leverages cloud storage, and distributed computing to improve overall performance and data security.

Python code:

```
#ONE OFF INSTALL - comment out after running
#NB install may display an error at end of process, however this doesn't
impact the subsequent autoML modelling process
%pip install pycaret

# Import dataset from pycaret repository
from pycaret.datasets import get_data
insurance = get_data('insurance')
#insurance.head()

# NB when running below, make sure press enter to continue / accept
# Initialize environment
from pycaret.regression import *
r1 = setup(insurance, target = 'charges', session_id = 123,
           normalize = True,
           polynomial_features = True, trigonometry_features = True,
           feature_interaction=True,
           bin_numeric_features= ['age', 'bmi'])
```

```
# Train a linear regression model
lr = create_model('lr')

# save transformation pipeline and model
# NB gets saved in /content folder as pkl file
# NB model_only=True added in line with reference https://github.com/
pycaret/pycaret/issues/985
save_model(lr, model_name = 'insurance-ml', model_only=True)
```

Wrap-up

This glimpse at productionizing AI using AWS Storage and Compute instances and, separately, PyCaret hopefully gives readers a taste of things to come. While basic, the emphasis is on exposure at this stage to key tools and techniques, all of which will be elaborated in subsequent chapters.

In our next chapter, we look at the high level of AI solution implementation and best practice AI project delivery with DataOps.

CHAPTER 2

AI Best Practice and DataOps

We ran through in the first chapter the key themes for productionizing AI today. Before we proceed into an exhaustive look at data ingestion and techniques and tools for building an AI application, it's important to establish a framework for success.

That framework starts with a step-back and a "top-down" understanding of the wider context for AI, key stakeholders, business/organizational process methodologies, the importance of collaboration and stakeholder consensus, adaptability and reuse as well as best practice in delivering high-performance AI solutions. There are many best practice frameworks which can perform this function but in this book we consider the one best we consider as best placed to achieve a culture of continuous improvement in the workplace today – DataOps.

The approach taken in this chapter is focused on awareness of DataOps concepts, rather than a "deep dive" but along the way we will touch on the cornerstones of DataOps[1] including agile and how to orchestrate agile development and delivery, team and design sprint methods and collaboration.

We will also take a brief look at creating a high-performance culture, reusing materials and artifacts, version control and code automation including continuous integration (CI) and continuous deployment (CD) with Jenkins and containerization with Docker as well as test automation with Selenium and monitoring with Nagios.

In later chapters, we will "join-the-dots" when we take a look at practical implementation of a data/analytics/AI project, and adapting projects from other industries while tying implementation of best practice to DataOps techniques.

[1] See also `https://dataopsmanifesto.org/en/`

B. Walsh, *Productionizing AI*, https://doi.org/10.1007/978-1-4842-8817-7_2

Introduction to DataOps and MLOps

We start in this first section by introducing the key concept: DataOps (and MLOps) as a framework for the successful productionization of Artificial Intelligence applications.

DataOps

Let's start with the basics - DataOps is not DevOps. While DevOps is concerned with software development, Data Analytics (and therefore AI) requires this PLUS control over how data is evolving.

As data is the underlying "currency" of Data Analysts, Data Scientists, or AI Engineer roles, governance and data quality is critical if we are to generate tangible, meaningful outcomes and insights - in addition to cloning a production environment to develop an "application" or "solution," underlying infrastructure must accommodate the "continual orchestration" of changing data.

As shown in Figure 2-1 from DataKitchen a DataOps implementation stretches across the entire data pipeline - from multiple data sources, through integration, cleaning, and transformation before being consumed by (multiple) end users. The context for DataOps is that analytics and AI fail without an effective data strategy, and 89% of businesses struggle with managing data.[2]

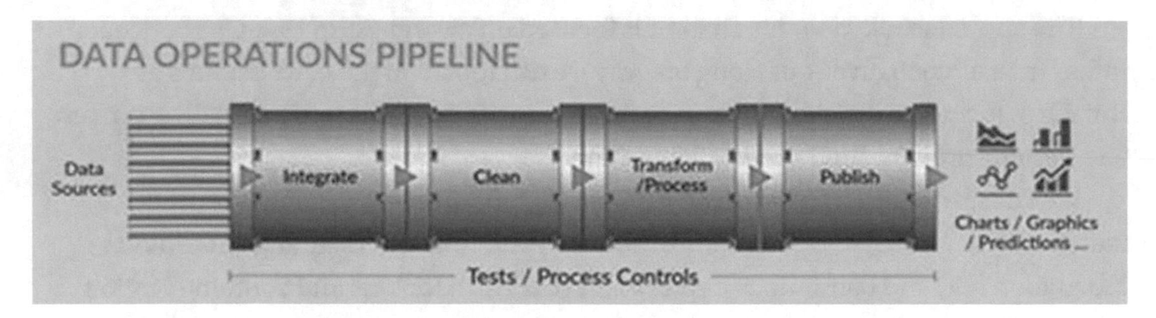

Figure 2-1. *DataOps Pipeline (DataKitchen)*

[2] Source: Experian

The Data "Factory"

DataOps is essentially the confluence of three key areas: **DevOps, Agile, and Lean,**[3] with the goal of streamlining data pipelines (more on this in Chapter 3) and improving data quality (and reliability) by decreasing the innovation and change cycle, lowering error rates in production, improving collaboration and productivity through, for example, self-service enablement. At a more granular level, data monitoring and measurement, metadata, scalable platforms, and version control are also key areas to ensure a data pipeline solution is driven by organizational goals.

Besides being a framework or methodology, DataOps is also a **culture** typically driven by a CIO/CDO or the organizational IT function. Metrics are key, both at an individual contributor level and to measure improvements in productivity and quality across projects.

The Problem with AI: From DataOps to MLOps

What about MLOps?

Even if a robust data strategy exists, less than half (Gartner: 47%, DeepLearning.ai: 22%) of all Machine Learning models go into production.[4]

We know that Machine Learning (and AI) is not just code, it's code plus data. And while code development takes place generally in a controlled environment (viz. DevOps), data, on the other hand, has **high entropy,** evolving independently from the underlying code.

MLOps is Machine Learning Operationalization and leans heavily on best practice and learnings from DataOps, this time with the focus on optimizing the production lifecycle of machine (or deep[5]) learning models rather than more generic data/analytics or AI solutions as with DataOps.

[3] specifically Statistical Process Control (SPC)

[4] Of course some may be sandbox, proof of concept (PoC), prototype, or minimum viable product (MVP) with the low-level aims being to demonstrate/test an idea and/or demonstrate core features/user journey

[5] DLOps

Besides high entropy data, the real challenge with Machine Learning isn't building the model itself. As Figure 2-2 shows, the MLOps process map/landscape from nvidia and GCP, it's integrating a machine learning system, with countless examples of poorly implemented projects resulting from a failure to adopt a "system-wide" approach:

- Inconsistent, cumbersome, (and frail) deployment
- Lack of reproducibility
- Reduced performance from skewed training and inference processes

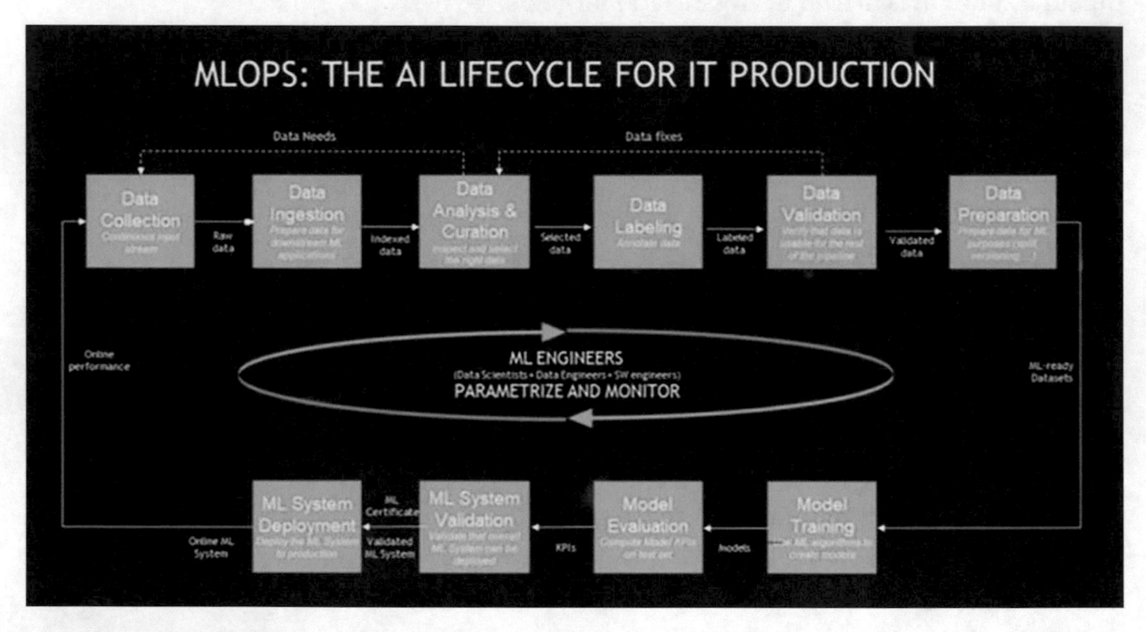

Figure 2-2. MLOps (Source: nvidia)

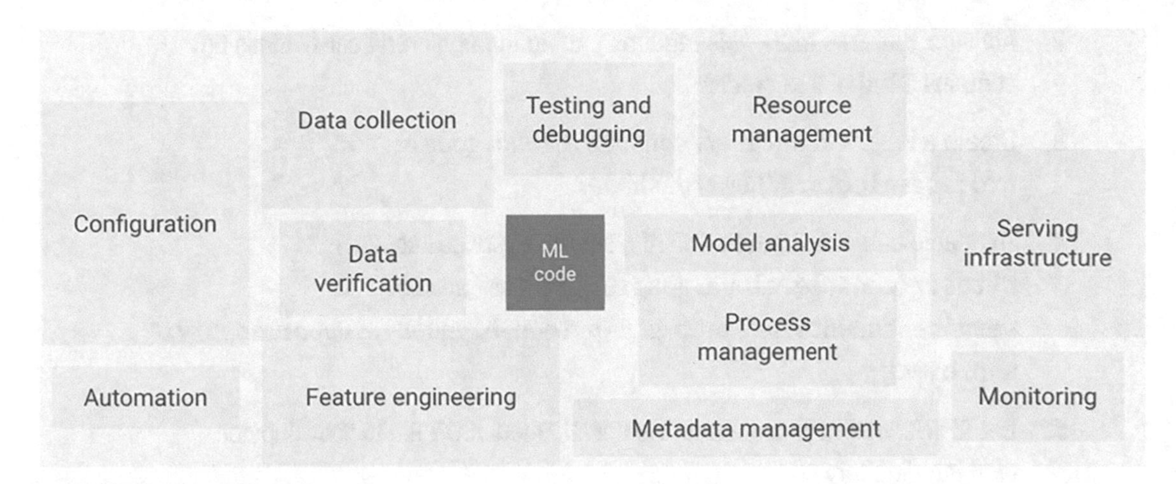

Figure 2-3. *MLOps Core Processes (Source: GCP)*

Enterprise AI

As we will see in later chapters, MLOps (and DataOps) is closely coupled with "Enterprise AI" - effectively embedding AI into an organization's company-wide strategy. MLOps and Enterprise AI are about designing a Target/To-Be architecture and implementing a robust AI infrastructure and enterprise data center while ensuring the entire workforce is aligned and trained on a company's tangible AI assets.

Enterprise AI is viewed by C-Level/Board as best practice for businesses to run AI successfully - MLOps fits that vision with many CSPs today offering MLOps as an inbuilt solution and others (such as DataRobot) adopting MLOps as their entire business model or product offering.

GCP/BigQuery: Hands-on Practice

CLOUD DATAOPS: PYTHON AND BIGQUERY

Using Python in Jupyter Notebook, the goal of this exercise is to take a first look at Google Cloud Platform and BigQuery, to help bridge the knowledge gap between standalone Data Science and a cloud-managed DataOps or MLOps solution.

1. Set up a GCP account at the link below

   `https://console.cloud.google.com`

2. Activate your free trial – this requires putting through credit card details but contains $300 of free credits[6]

3. Create a Project at `https://console.cloud.google.com/projectselector2/home/dashboard`

4. To enable authentication to GCP API, set up a Service Account `https://console.cloud.google.com/iam-admin/serviceaccounts?project=gcp-python-bigquery&supportedpurview=project`

5. Get API Service Keys and move the downloaded JSON file to your Jupyter working directory

6. Clone the Jupyter notebook from GitHub below and run through the notebook to see how the Python–BigQuery interface works

   `https://github.com/bw-cetech/apress-2.1.git`

Event Streaming with Kafka: Hands-on Practice

EVENT STREAMING WITH KAFKA

Seamless integration to data streaming architectures based on Apache Kafka is increasingly pivotal in achieving DataOps objectives. We will take a closer look at streaming (and batch data in Chapter 3) but this lab introduces readers to Kafka and how it can be used in a Big Data context.

1. Set up a Lenses portal account at

   `https://portal.lenses.io/register/`

2. After email verification, select the Lenses Demo and create a workspace

[6] Google say they won't charge unless you manually upgrade to a paid account but keep a track on usage, check API calls and free trial status (remaining credit) on the Google Console (link above) by going to Billing ➤ Overview. Remaining free trial credit is shown in the lower RHS of the screen.

3. Explore real-time data by selecting SQL Studio then sea_vessel_position_reports. Run the query below to view live vessels in motion:

```
SELECT Speed, Latitude, Longitude
FROM sea_vessel_position_reports
run query
WHERE Speed > 10
```

   Download a JSON of your results after streaming stops automatically after a couple of minutes (may be quite big c. 50 MB)

4. Exercise – run a different streaming query on one of the other datasets, for example, financial_tweets

5. By following the steps here: `https://lenses.io/box/` complete the rest of the steps below to see how a Kafka topic is set up, with Consumer and Producer and how to monitor the real-time flow

   a. Create Kafka Topic

   b. Create a Data Pipeline

   c. Carry out stream processing

   d. Consumer Monitor

   e. Monitor real-time flows

Agile

We saw in the last section that Agile is one of the three core practices intrinsic to DataOps. This next section takes a closer look at Agile in the context of AI solutionizing.

Agile Teams and Collaboration

DataOps intends to "resolve the struggle between centralization and freedom in analytics" - there has to be control, but equally companies that get ahead in an increasingly "disrupt or be disrupted" digital environment are able to foster a culture that embraces experimentation and "lab-based" innovation.

As one of the three cornerstones of DataOps, agile development and delivery is all about collaboration and adaptability to balance these seemingly competing aims.

The main part is clearly anchored on the "people" perspective. DataOps brings a diverse mix of stakeholders together on a data project with roles spanning the business/client (who define the business requirements, traditional roles: data/solution architects and data engineers, newer roles including Data Scientists and ML Engineers and IT operations or those who build and maintain the data infrastructure).

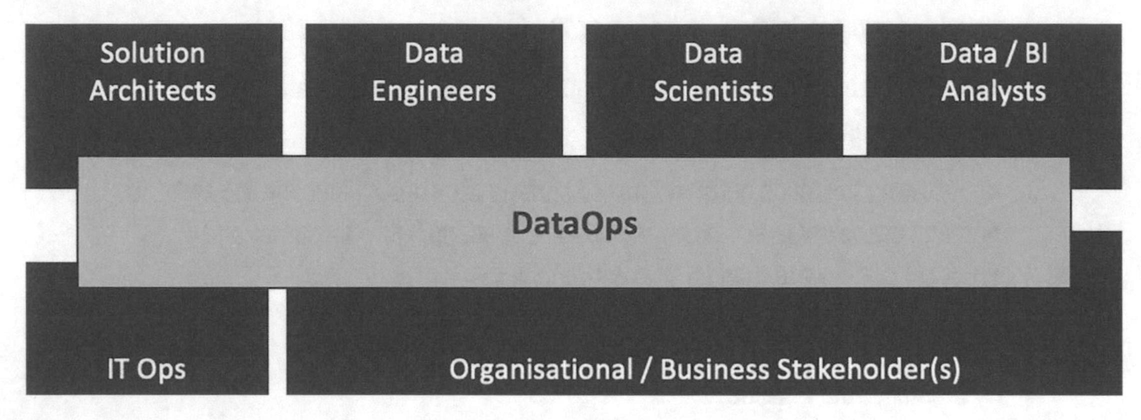

Figure 2-4. *DataOps teams (Source: Eckerson Group)*

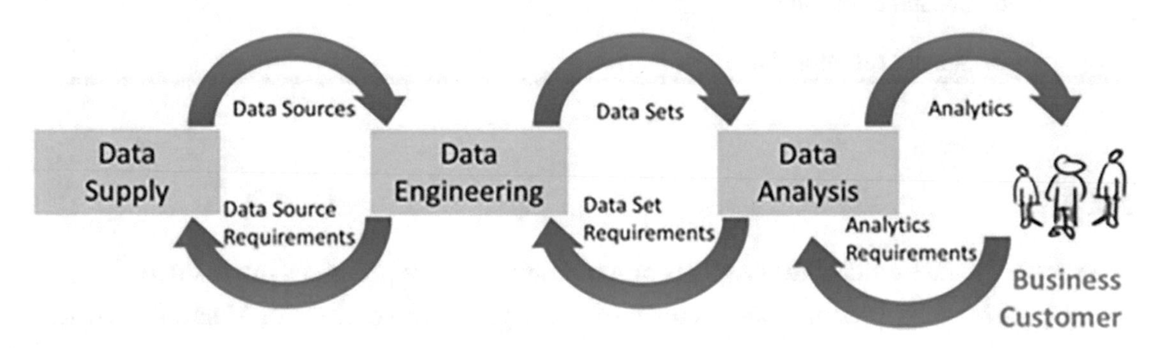

Figure 2-5. *DataOps hand-offs (`medium.com`)*

Development/Product Sprints

Prioritization of business requirements on projects makes use of techniques such as MoSCoW (i.e., **m**ust have this requirement, **s**hould have, **c**ould add if doesn't affect anything else, **w**ould like to have this feature/wishlist). Successful data projects prioritize bug fixes and features through "sprints" which walk through the pipeline process from data integration, to cleaning, transforming, and publishing and typically last 2-4 weeks in length.

While well-defined project work packages should lead to slick, agile handoffs across delivery teams, at the start, teams often rebuild organizational Data Warehousing and Analytics landscapes into an initial scaled-down sandbox (prototyping) environment. From these modest beginnings, sandboxes get scaled-up to lab environments for process orchestration and agile testing phases with continual integration (CI[7]) of code changes.

Ultimately the end goal is to dynamically deploy (CD) via a "Production Belt" and monitor the results.

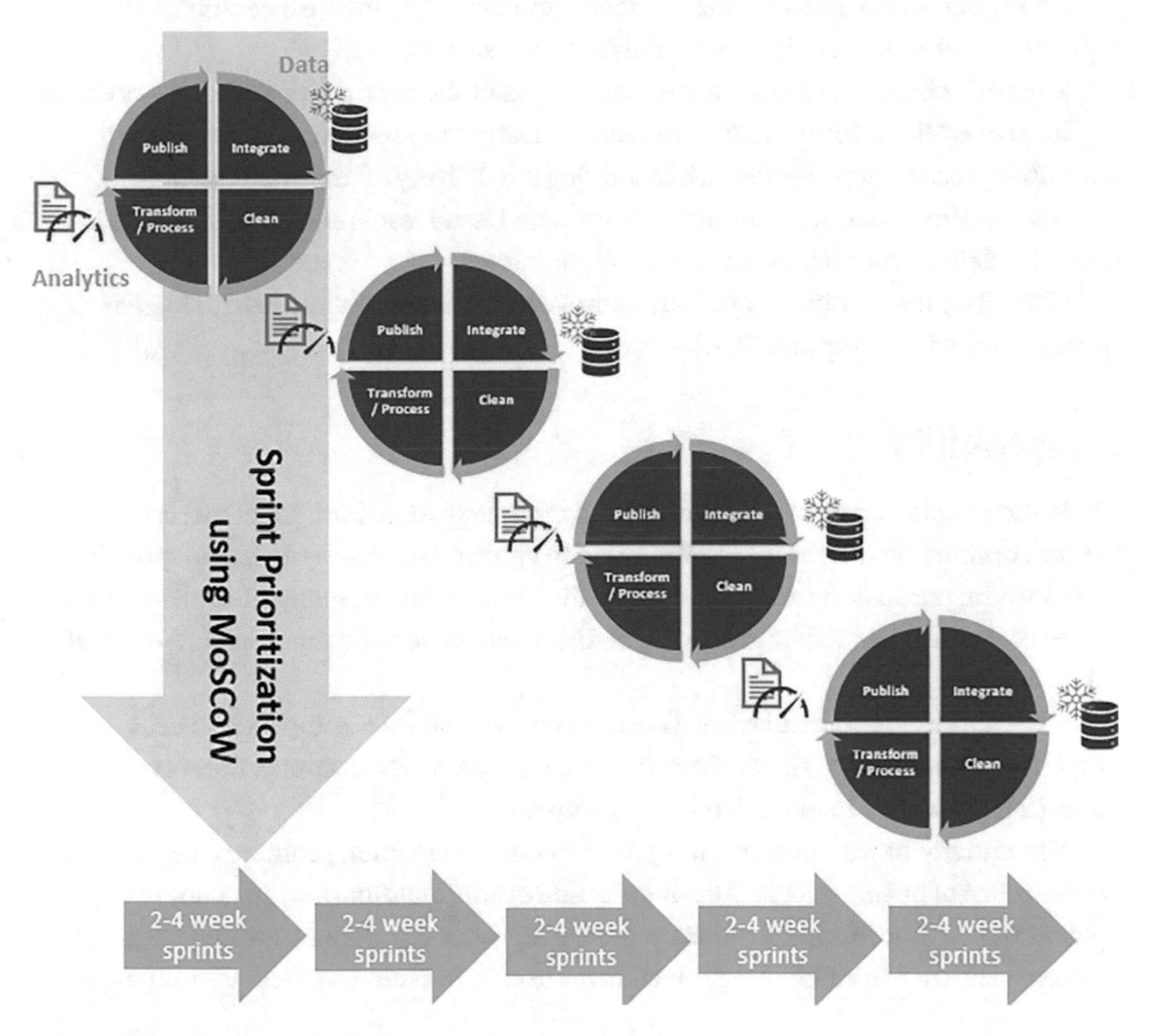

Figure 2-6. *DataOps Sprints*

[7] CI/CD will be discussed in the section below

Benefits of Agile

The DataOps test and release cycle from sandbox, scale-up through deployment is designed around agile software delivery frameworks. Data and Analytics specialists use this DataOps best practice to rapidly fix errors and implement feature requests for re-deployment into Production. The end result is a "managed delivery" which addresses the key areas below:

Changing requirements – incremental approach helps evolve a centralized data repository and supports "dynamic" analytic requirements.

Slipped schedules – iterative, stepped approach shortens the overall delivery times

Improved flexibility – allows a pipeline of feature requests turned around with continual process improvement while limiting the delivery of irrelevant features

Disappointed users – systemized, accelerated issue resolution process means less scope for deliverables not meeting user requirements

Ultimately the continual (cyclical) improvement process should lead to higher-quality product delivery and ROI.

Adaptability

While the people perspective rightfully gets prioritized, Agile (and DataOps) extends beyond optimizing teams and collaboration to applications and systems. Adaptability is important, particularly with cloud computing underpinning many AI applications today; essentially adaptability means scalability and reuse of business logic, APIs, and microservices.

Microservices in particular have become popular across the application ecosystem as companies move away from siloed code on a single server to applications as a collection of smaller, independently run components.

The underlying architectures from microservice integration promote consistent and secure reuse of business logic, API sharing, and events handling and so enable greater agility from having decentralized team ownership, elastic scalability around usage, and discrete resilience from isolating run changes from other microservices at runtime.

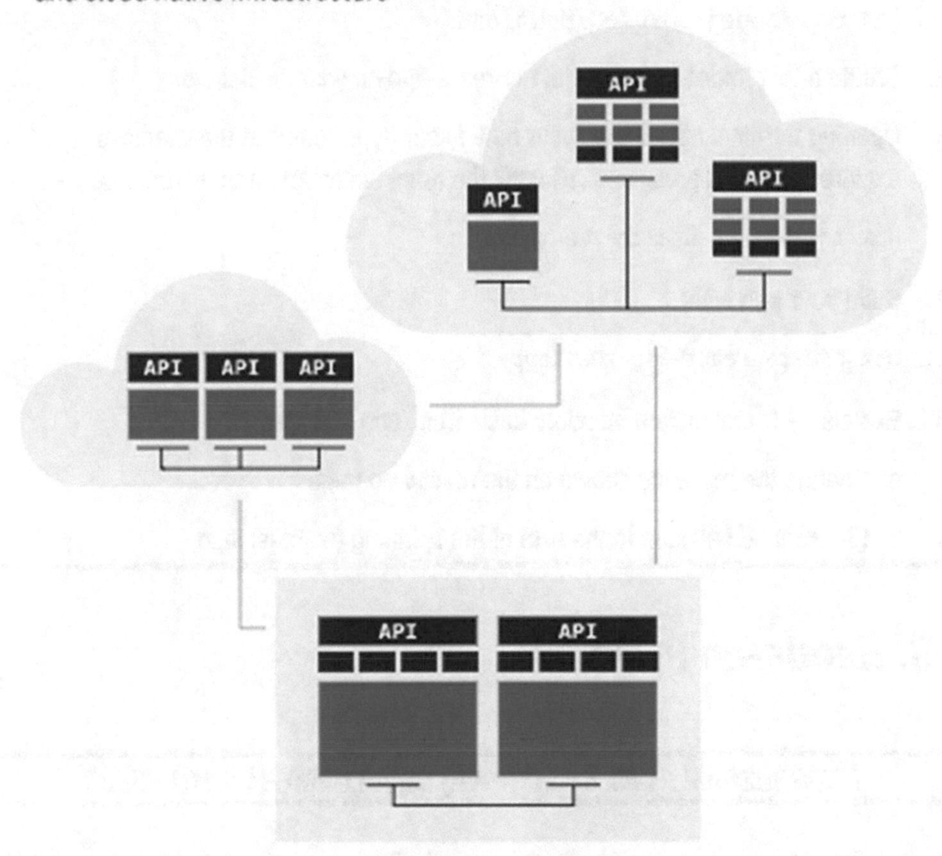

Figure 2-7. Microservices agile architecture (Source: IBM)

react.js: Hands-on Practice

FRONT-ENDING YOUR AI APPLICATIONS

React.js, along with Vue.js and Angular is one of the best and open source front-end JavaScript libraries for building an aesthetic user interface. The goal of this lab is to implement a simple react app which can be extended to front-ending AI applications:[8]

[8] We will use the same boilerplate in Chapter 7 to build a full-stack deep learning app

1. Install node and npm from the link below – both are installed from the single windows installer:

   https://nodejs.org/en/download/

2. Create a test folder, for example, my-react-app on your local drive

3. Opening terminal from within your new folder (type "cmd" in the windows explorer path and press enter), install the react boilerplate app by running:

   ```
   npm install -g create-react-app
   ```

4. Build your app with:

   ```
   npx create-react-app reactapp
   ```

5. Exercise – try to implement some basic front-end changes:

   a. Change the message shown on the react web page

   b. (Stretch) add an icon to the RHS of the spinning (react.js) logo

---

VueJS: Hands-on Practise

PROGRESSIVE AI WEB APP (PWA) DEVELOPMENT WITH VUEJS

As an alternative for building web applications, VueJS can be more suitable for front-ending smaller, less complex AI applications. While React is a library, VueJS is a Progressive JavaScript Framework. Here we walk through how to get a VueJS up and running as a template for building an AI user interface.

1. If not already installed, go ahead and install node.js (and npm) from the url in step 1 for the react.js lab above

   https://nodejs.org/en/download/

2. Open a terminal (Check the version of npm installed with

   ```
   npm -v
   ```

3. Install vue-cli (command line support for VueJS)

   ```
   npm install -g @vue/cli
   ```

4. Restart terminal and check vue-cli version

```
vue --version
```

5. Navigate to an app testing folder and create a project called "my-vue"

```
vue create my-vue
```

6. Select the Babel default source-source compiler for browser-readable .js, .html, .css (using arrow keys)
7. cd into the project folder

```
cd myvue
```

8. Run the app

```
npm run serve
```

9. Finally navigate in your broswer to the url shown in the terminal i.e. `http://localhost:8080` to see the app running

Exercise - try to update the source code to remove all the text underneath the message "Welcome to Your Vue.js App" and replace with a hyperlinked screenshot to Replika `https://replika.com/`

Code Repositories

Any cohesive team working on developing an AI solution needs to be "singing from the same hymn sheet." Code repositories "repos" are one of the key collaborative enablers for ensuring developers and data specialists are working in sync.

Git and GitHub

Version control, or source control, is the practice of tracking and managing changes to source code. In recent years version control systems (VCS) and, specifically, Distributed Version Control Systems (DVCS) have become valuable to DataOps teams. Besides intrinsic "DevOps" benefits including reduced development time and increased number of successful deployments, evolving datasets, such as widely used Covid data from John Hopkins University are increasingly widely maintained on distributed version control systems.

Git[9] is by far the most popular DVC, though there are a number of other systems, such as Beanstalk, Apache Subversion, AWS CodeCommit, and BitBucket, used for projects with specific integrations to other (usually single) CSP providers.

Besides traceability and file change history (keeping track of every code modification and every dataset change), Git eases the process of rolling back to earlier code/data states during development and has enhanced branching and merging features vital for DataOps teams working on specific application components or user stories.

The GitHub "ecosystem" comes complete with Git (effectively the command line back-end), GitHub – a cloud-based hosting service that lets you manage Git repos from a central location and GitHub Desktop – a desktop version to interact with GitHub using a GUI.

GitHub is now utilized by millions of software developers and companies worldwide with public repos in existence prior to February 2020 archived in the GitHub Arctic Code Vault – a long-term archive 250m below the permafrost of an Arctic mountain. **Forking** a repo on GitHub or **cloning** a public GitHub repo into a local directory are the main ways to access and develop preexisting source code (or update datasets) on GitHub. Both can be carried out using git commands in terminal or (more intuitively) using GitHub Desktop.

Figure 2-8. *GitHub ecosystem*

Version Control

The below image is an example of how Git performs version control on three different versions of the same (.py) file. Multiple users can select which version of the file they want to use and make changes to it independently before merging back to the single "Master" repo.

[9] Git was originally developed in 2005 by the creator of the Linux operating system kernel

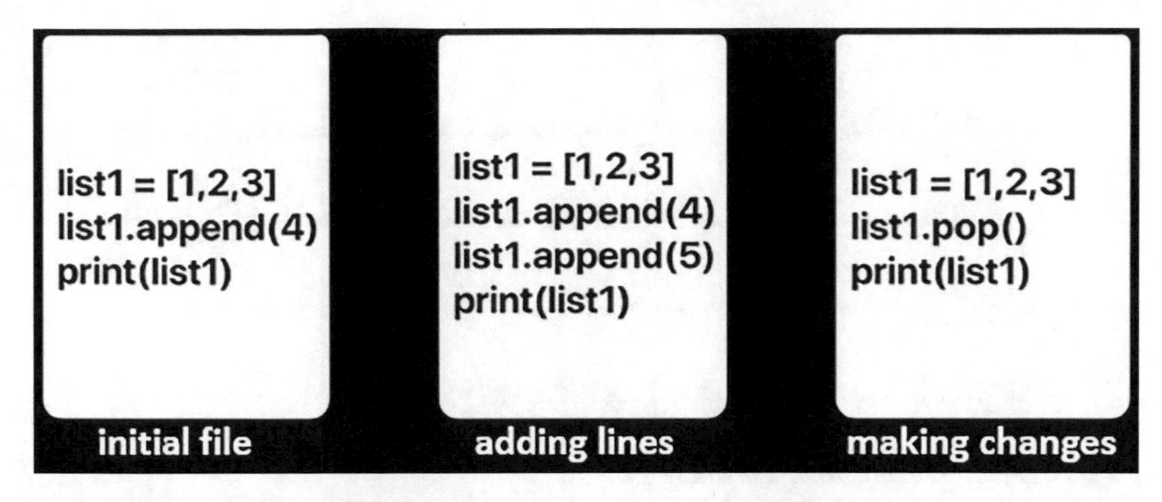

Figure 2-9. *Version Control in Git (Source: `www.freecodecamp.org`)*

Branching and Merging

Simplified "branching" is one of the main reasons why Git is by far the most widely used version control system today.

The diagram below shows branching (to a development "dev" branch) from a master repo in order to develop code further – with two other developers on the project adding feature requests. The first developer makes some minor changes and merges their code changes to the development (dev) branch, while the other developer continues working on their feature, postponing merging until later.

Once development (bug fixes and feature requests) have been completed, testing would be done on the dev branch before finally being committed (back) to the master repo.

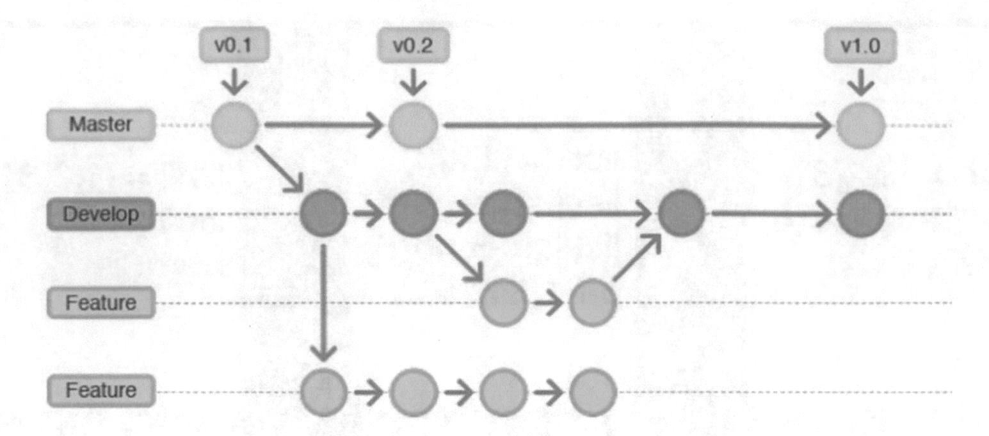

Figure 2-10. *Development branches in Git*

Git Workflows

Git workflows are the means by which changes to code and data are placed into repos. There are four fundamental "layers" through which code and data can pass: a working directory (typically a local user machine), staging area, local repository, and remote repository (typically on GitHub itself).

While a file in your working directory can be in three possible states: staged, modified, and committed, because Git is a distributed version control system as opposed to a centralized system, certain commands (such as commits) do not require communication with a remote server each time they are actioned. This is shown in the diagram below, with corresponding workflows and git commands shown underneath.

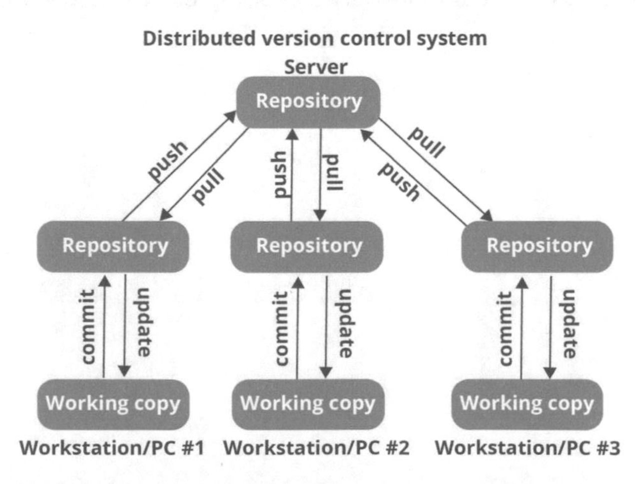

Figure 2-11. *GitHub DVCS showing commits*

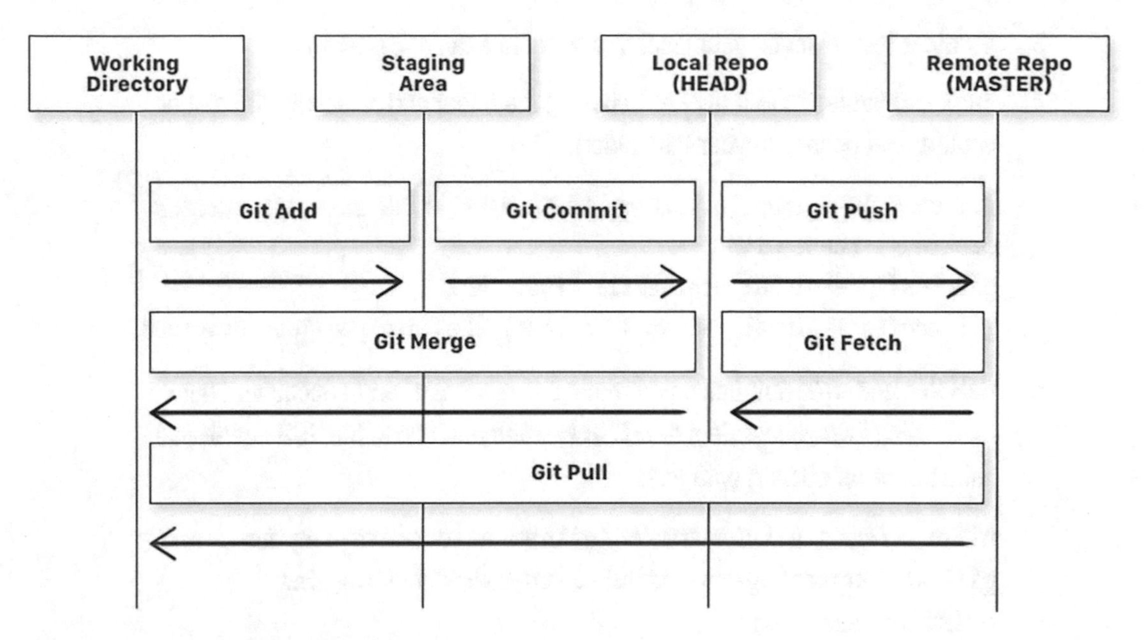

Figure 2-12. Git workflows

Git Command	Source	Destination
git add	working directory	staging area
git commit	staging area	local repository
git push	local repository	remote repository (e.g. GitHub)
git fetch	remote repository (e.g. GitHub)	local repository
git merge	local repository	working directory
git pull (fetch and merge)	remote repository (e.g. GitHub)	local repository

Figure 2-13. Git commands

GitHub and Git: Hands-on Practice

BASIC GIT

No development is done today without recourse to GitHub so let's take a look in this lab at setting up a GitHub account, installing and creating a new repo using Git:

1. Set up GitHub account – sign up at `https://github.com/`
2. Install Git from `https://gitforwindows.org/`

3. Create a test folder on your local directory, for example, git-intro

4. Introduce yourself to Git (by right-clicking the folder and selecting Git Bash or via terminal opened in your test folder):

```
git config --global user.name "USER-NAME" # NB same username as
created account with
git config --global user.email "YOUR-EMAIL"
git config --global --list # To check the info you just provided
```

5. GitHub's preferred authentication method is using SSH authentication. Check for existing keys, generate a new SSH key and add this to the SSH agent. See links below for support with this:

   `https://docs.github.com/en/github/authenticating-to-github/connecting-to-github-with-ssh/checking-for-existing-ssh-keys`

   `https://docs.github.com/en/github/authenticating-to-github/connecting-to-github-with-ssh/generating-a-new-ssh-key-and-adding-it-to-the-ssh-agent`

6. Add the new SSH key to your GitHub account

7. Create a New Repository on GitHub

8. Make sure the SSH button is selected and run the git add, commit and push commands shown on the screen by click on the copy button and pasting them to Git Bash (still opened at your local folder) or by pasting them in terminal

9. Upon refreshing your repo you should now see a README.md file in your GitHub repo

10. Exercise – try to add the react.js app source code from section 2 (Agile) to your local repo and push to GitHub

11. Exercise – clone a public repo into a different local repo (make sure you have created a new folder and are running the clone command in that folder)

NB this lab can be completed instead by installing GitHub Desktop from `https://desktop.github.com/` and using that instead of Git.

As an additional exercise with GitHub Desktop, try to fork a public repo (into your GitHub account) in addition to cloning to a local repo.

Deploying an App to GitHub Pages: Hands-on Practice

DEPLOYING AN APP TO GITHUB PAGES

GitHub isn't just about version control. This lab shows you how to deploy and host the react app from section 2 (Agile) to GitHub Pages.

NB the approach shown is via Git but a deployment using GitHub Desktop can also be carried out by following the instructions here: `https://pages.github.com/`

1. Complete the steps in the lab "react.js: Hands-on Practise: FRONT-ENDING YOUR AI APPLICATIONS" above
2. Create a Git repository
3. Copy the https url for your new repo
4. Initialize your react app directory as a local (Git) repo then:
    a. Commit
    b. Push to remote
5. Amend the "package.json" in your react app folder by:
    a. adding "homepage": "https://[YOUR-GITHUB-NAME].github.io/[YOUR-GITHUB-REPO]" (right at top of json)
    b. under 'Scripts' add:
    "predeploy": "npm run build"
    "deploy": "gh-pages -d build"
6. cd into the react app folder on terminal
7. Install GitHub Pages and deploy your app by running the command below one by one:

```
npm install gh-pages -save-dev
git init
git remote add origin [https git repo url]
git add .
git commit -m "Deploy react to GitHub pages"
```

```
npm run deploy # NB if errors running this then delete node_modules\.
cache\gh-pages folder of local app and try again
git push -u origin master
```

You should then be open the app at `https://[myusername].github.io/[my-app]` where myusername is your GitHub username and my-app is your repo name.

Continuous Integration and Continuous Delivery (CI/CD)

Having covered key team (Agile) and collaborative (GitHub) aspects of DataOps, we move on to our next section on CI/CD where the focus is on streamlining data processes in a way that facilitates, improves, and accelerates AI solution delivery.

CI/CD in DataOps

The intention of Continuous Integration and Continuous Delivery (or Continuous Deployment[10]) in software is to enforce automation through build, test, and deploy processes. Essentially this is to enable teams to release a constant flow of software updates into production to quicken release cycles, lower costs, and reduce the risks associated with development.

In the context of DataOps (as opposed to DevOps) and AI, the scope of automation has extended to data pipeline orchestration including data drift and modeling automation including the retraining process. In theory, this should mean each time a change is made to underlying code or infrastructure **and** a data change occurs (or, more realistically, each time data distributions have deviated significantly), automation kicks-in and the application is rebuilt, tested, and pushed to production.

[10] Continuous Delivery and Continuous Deployment are similar but have slightly different goals – Continuous Deployment focuses on the end-result, that is, the actual (end-point) deployment while Continuous Delivery focuses on the process, that is, the release (steps) and release strategy

Introduction to Jenkins

Jenkins is an open source automation server with hundreds of plug-ins and one of the leading CI/CD tools for DataOps. It is used by Expedia, Autodesk, UnitedHealth Group, and Boeing as a continuous delivery pipeline.

Originally built to automate testing for Java developers, but now with support for multilanguage multicode repositories, Jenkins simplifies the setup of a continuous integration or continuous delivery (CI/CD) environment. It does this via a Jenkinsfile, effectively a "pipeline" script where a declarative programming (macro-managed) model defines executable steps via a hierarchy: pipeline block > agent > stages.

A Jenkins pipeline is really a collection of jobs triggering each other in a specified sequence. For a small app, this would be, for example, three jobs: job1 build, job2 test, and job3 deploy. Jobs can also be run concurrently and for more complex pipelines, a Jenkins Pipeline Project is used where jobs are written as one whole script and the entire deployment flow is managed through Pipeline as Code. Continuous integration with Jenkins also supports GitHub automation.[11]

Blue Ocean provides a better UX for setting up Jenkins pipelines by exposing a low-code interface and low-click functional development processes, negating the need for programming a Jenkinsfile.

[11] That is, normally manual processes on GitHub such as updating code releases with notes and binary files, adding git tags to a workflow and compiling a project are all automated within Jenkins CI/CD

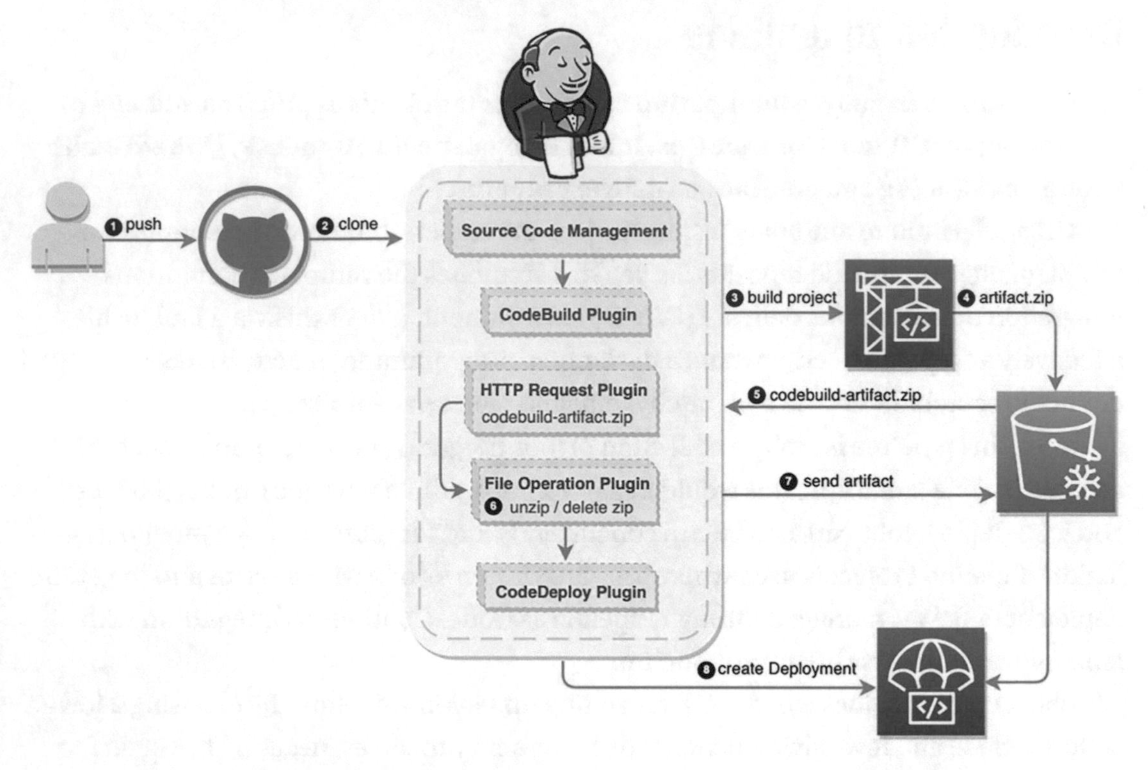

Figure 2-14. Jenkins Job pipeline with GitHub integration

Maven

Closely coupled with Jenkins (Jenkins uses Maven as its build tool), Apache Maven is a Project Management tool designed to work across the software lifecycle (performing compile, test, package, install, and deploy tasks), centrally managing project build including dependencies, reporting, and documentation.

When a Jenkins build is triggered, Maven downloads the latest code changes and updates, packages them, and performs the build. Like Jenkins, Maven works with multiple plugins, allowing users to add other bespoke tasks, but only does Continuous Delivery, not Continuous Integration (CI).[12]

[12] So no integration to merge developer code on GitHub

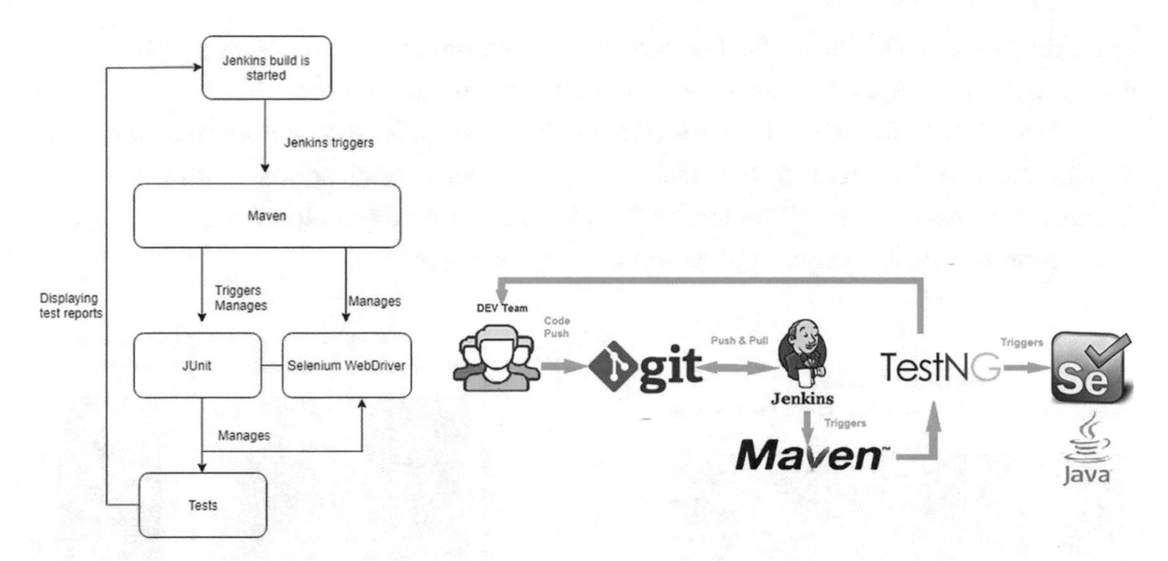

***Figure 2-15.** Maven build management*

Containerization

We finish this section with a look at containerization.

Because containers standardize deployments across multiple machines and platforms, they can naturally accelerate DataOps processes, and in particular CI/CD, where testing and debugging processes are "ringfenced" from external file dependencies.

Containerization comes with a number of additional benefits highly suited to building robust production-grade AI solutions including the ability to simplify and speed up the development, deployment, and applications configuration process, increased portability and server integration and scalability as well as increased productivity and federated security.

Docker and Kubernetes

Docker is the main container runtime we will utilize in this book. Docker's USP lies in its handling of dependencies, multiple (programming) language, and compilation issues when creating isolated environments to launch and deploy applications as containers. Although there are many similarities with Virtual Machines, as shown in the image below, Docker better supports multiple applications sharing the same underlying operating system. Docker is also fast, starting and stopping apps in a few seconds. PostgreSQL, Java, Apache, elastic, and mongoDB all run on Docker.

Although we will not use it in this book, Kubernetes (k8s), a container management tool, is often used to orchestrate Docker "instances." As an open source platform, Kubernetes was originally designed by Google and automates deployment, management, and scaling of applications in the container.

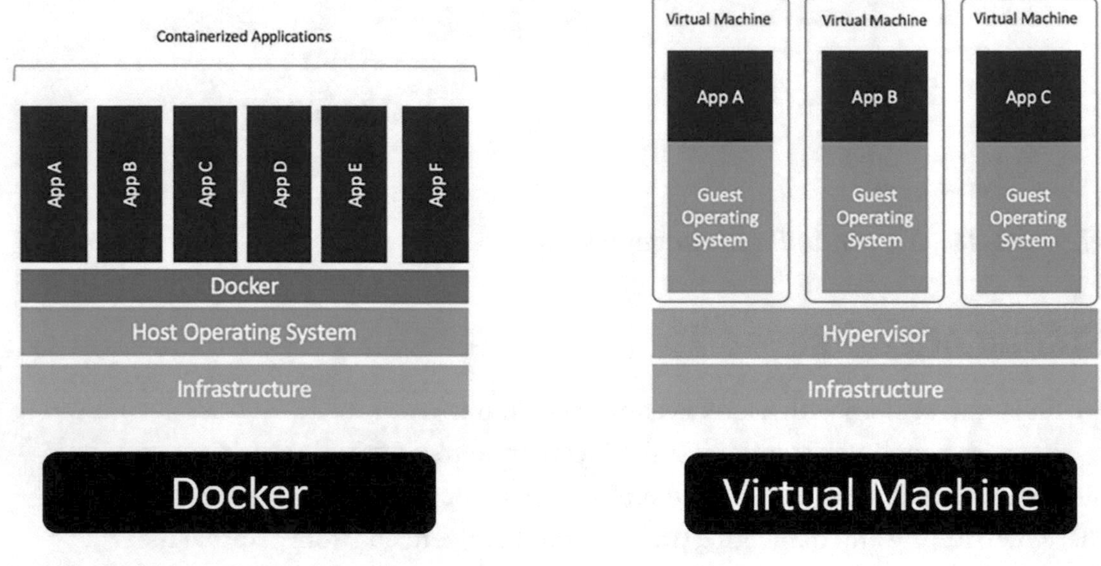

***Figure 2-16.** Docker architecture*

Play With Docker: Hands-on Practice

PLAY WITH DOCKER

Now that we have introduced containers and Docker in a CI/CD context, let's take a look at the process of containerizing a simple application with Play With Docker, which supports 4 hours of free usage:

1. Sign up to Docker at `https://hub.docker.com/`
2. After signing up and verifying your email, logon to Play with Docker at `https://labs.play-with-docker.com/`
3. When the screen opens, select "Add New Instance"

4. An embedded terminal should open on the Play With Docker screen

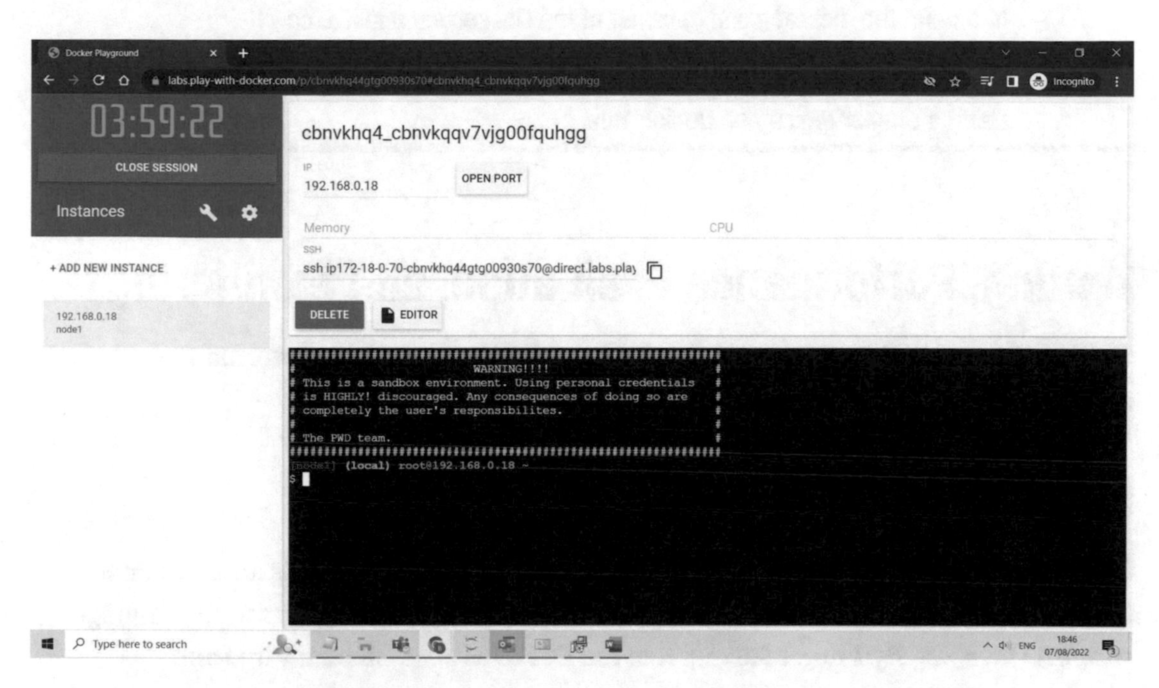

Figure 2-17. Play with Docker

5. Type the following command in your PWD terminal:

```
docker run -dp 80:80 docker/getting-started:pwd
```

This will start a container – click on port 80 to open it

NB use "ctrl + shift + V" to paste commands into the Docker command line interface (CLI)

6. The container is a tutorial for Docker – getting started.[13] Complete the tutorial to the end of the first part "Our Application" which will walk through how to build a “To-do list” app

NB when Building the App's Container Image (creating Dockerfile), use the command below:

```
touch Dockerfile
```

[13] Public url is here: `https://docs.docker.com/get-started`

7. Exercise – complete the next part of the tutorial "Updating our App" to see how to amend the message and behavior of the Dockerized application
8. Exercise – complete "Sharing our App" to see how to share Docker images, using a Docker registry on Docker Hub

Testing, Performance Evaluation, and Monitoring

Our last section in this chapter takes a look at automated testing, performance evaluation, and application monitoring in a Data/MLOps context.

Selenium

Created in 2004 to automate testing actions on web apps, Selenium is an automated framework for testing across different browsers and platforms and has multilanguage support (Java, C#, Python). Like much of the DataOps ecosystem, it's free and open source.

Not just a single tool but a software suite, Selenium is customized for specific organizational quality assurance (QA) requirements and supports important DataOps unit testing processes[14] involving data pipelines and analytics. Once unit testing is complete, continuous integration commences with QA testers able to create test cases and test suites for logically grouped test cases including data-in-transit and data-at-rest.

Running Selenium tests in Jenkins allows users to (a) run tests every time software changes and (b) deploy software to a new environment when the tests pass. Jenkins can also schedule Selenium tests to run at specific time, saving execution history and test reports.

[14] Intended to minimize the number of defects in the quality assurance testing phase

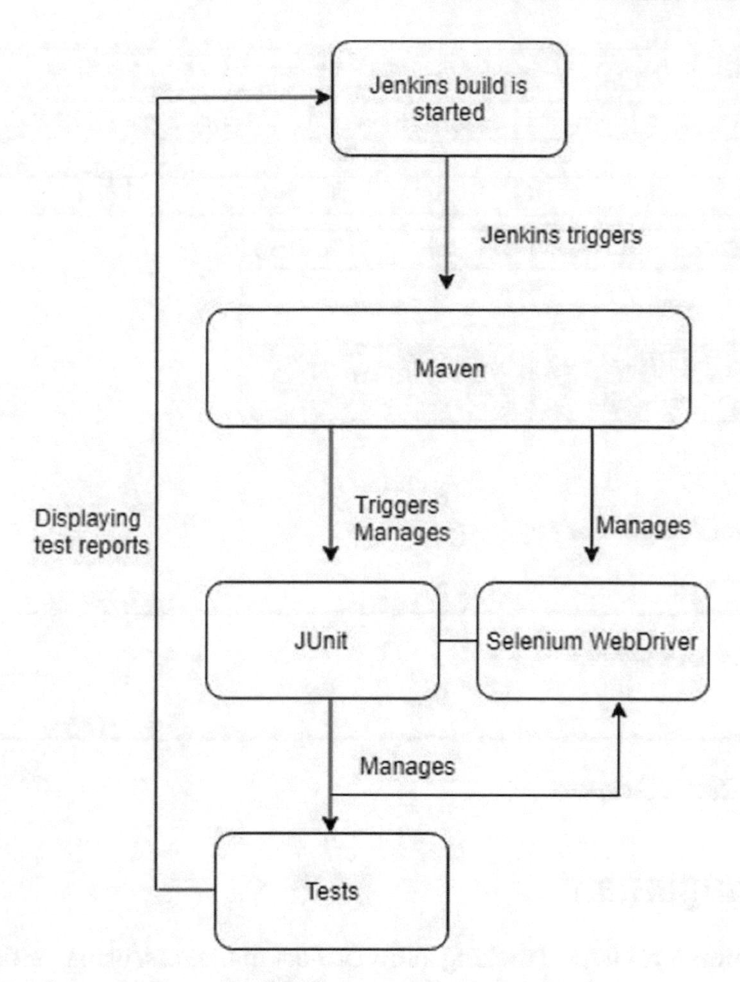

Figure 2-18. *Jenkins, Maven, and Selenium process interfaces*

TestNG

Selenium Test Scripts work with TestNG (Test Next generation) – a testing framework which addresses a reporting gap in Selenium Webdriver, generating default HTML reports after execution. These reports (shown below) identify information about test cases (such as pass, skip, or fail) and the overall status of a project.

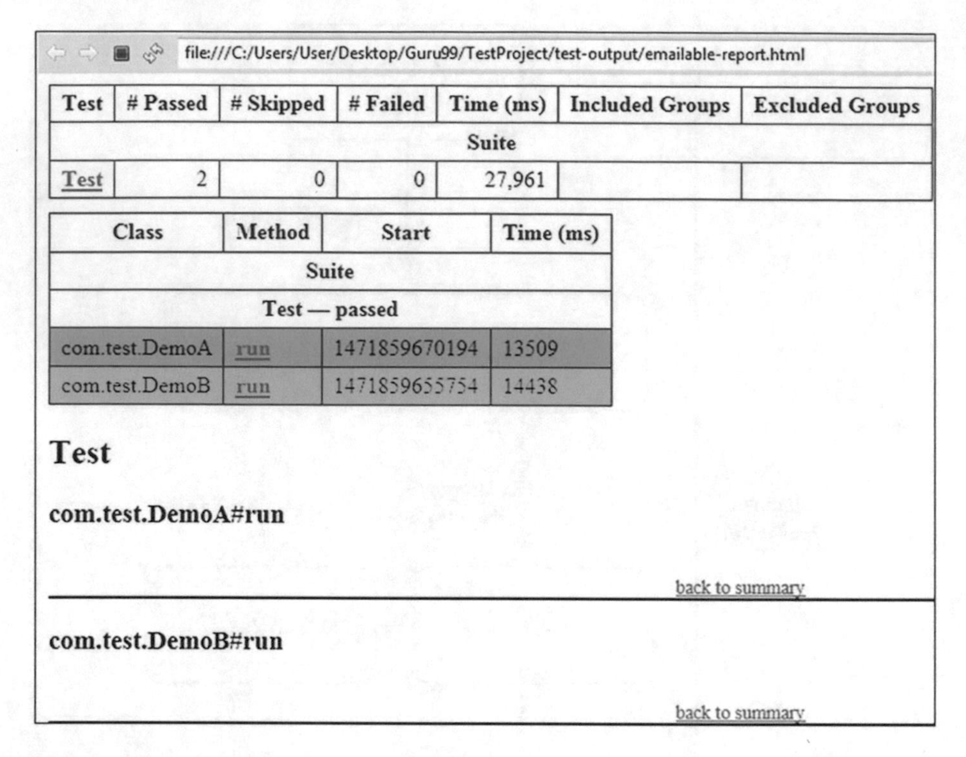

Figure 2-19. *TestNG report*

Issue Management

Issue Management and Issue Tracking allow project managers, users, or developers to **record and follow up the progress of issues on a Data project, capturing** bugs, errors, feature requests, and customer complaints. Issue tracking criteria typically include

- Level of importance
- Team members assigned
- A progress metric

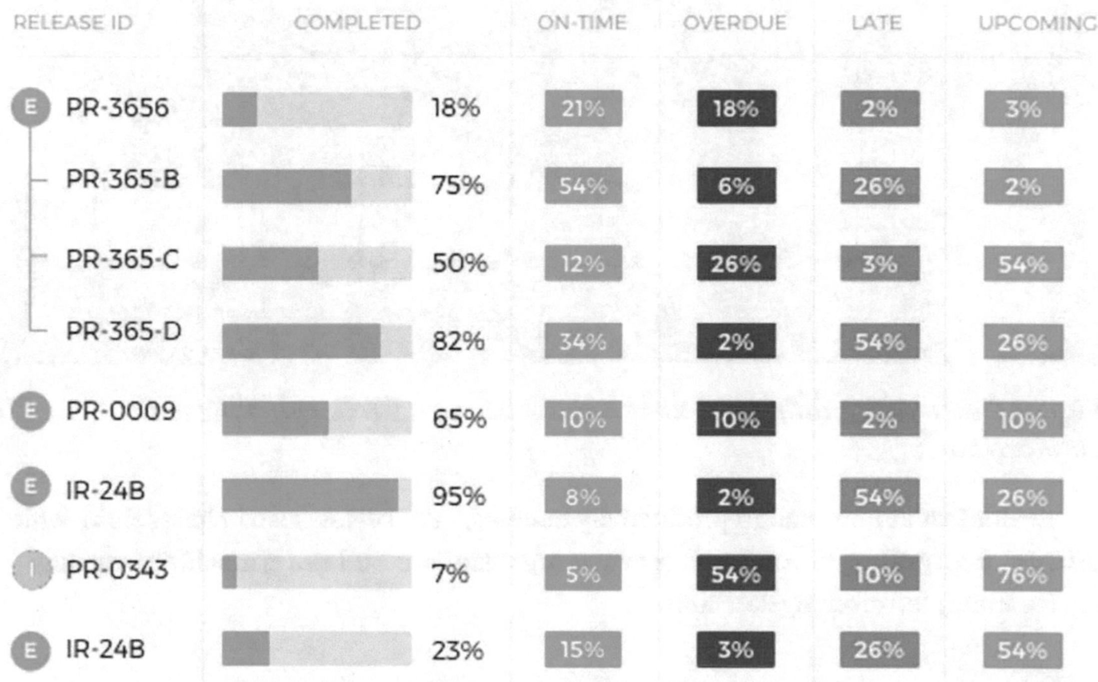

Figure 2-20. *Software delivery release – issue tracking (Plutora)*

Jira

Jira is one of the most popular issue (ticket) tracking and PM tools, although Trello, GitHub Boards, and Monday are also widely used.

Developed by Atlassian, Jira has been around a while (2002) and is an agile project management and issue/bug tracking tool. It comes with an easy-to-use dashboard and "stress-free" PM including agile delivery features such as team/user story and sprint oversight, scrum and kanban boards, roadmaps, and team performance reporting.

ServiceNow

Jira is integrated with ServiceNow – a workflow automation solution to connect people, functions, and systems across an organization. ServiceNow's product USPs are ramping up customer service and transforming workforce/employees to digital.

The below image shows how ServiceNow automates a change request workflow across a CI/CD pipeline.

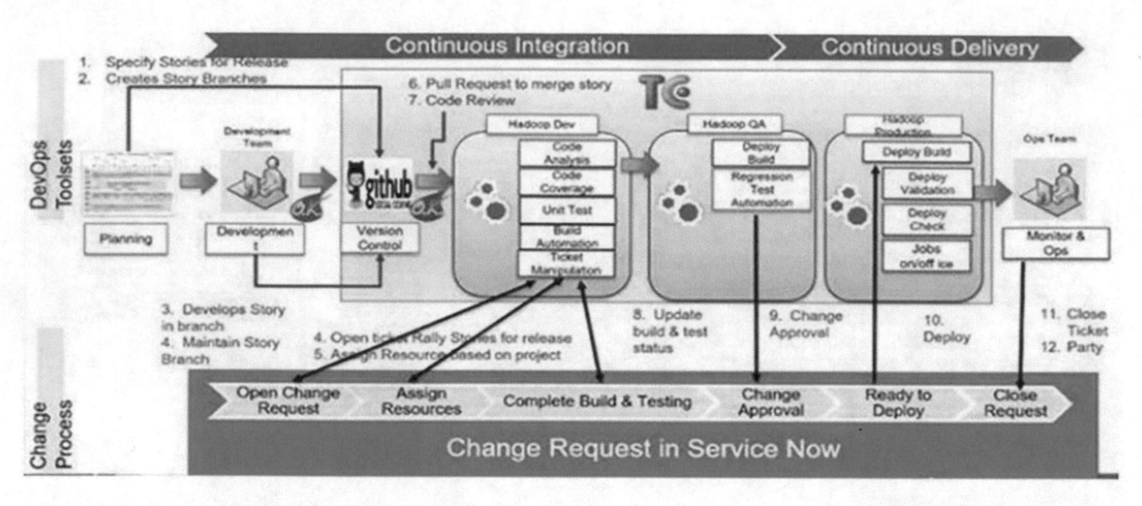

Figure 2-21. *Automating a change request across a CI/CD pipeline with ServiceNow*

Besides clear automation productivity benefits, ServiceNow also helps scale IT with "AIOps," interpreting telemetry data across organizations and using machine learning for, for example, anomaly detection.

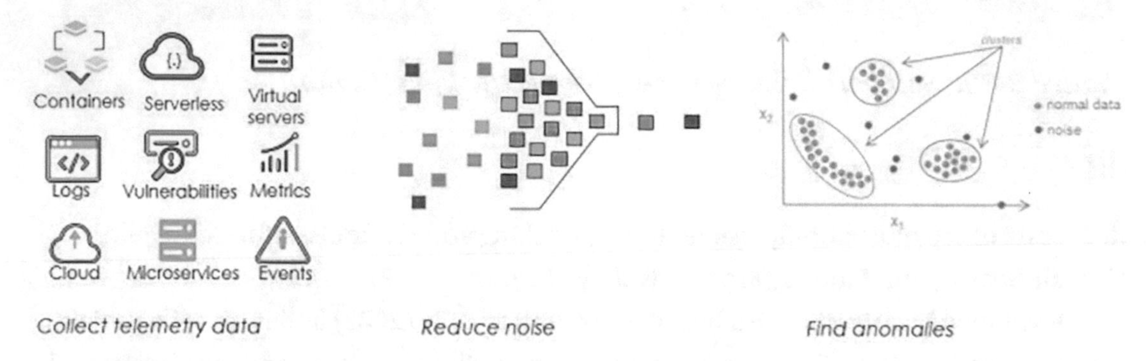

Figure 2-22. *ServiceNow AIOps - telemetry anomaly detection*

Monitoring and Alerts

Once an AI application is fully tested, and a process is implemented for managing issues, the focus turns to application **monitoring - essential** in AI due to data changes (data drift) quickly rendering model results invalid or suboptimal.

Rather than monitoring, the end goal is rather AI application "observability" – a deeper understanding of data pipelines, and data "health." The best systems have tracking, alerting, and recommendation features built into the underlying product and statistical process control (SPC) constantly monitoring and controlling the data analytics pipelines. If an anomaly occurs, analytics teams notified through automated alerts.

Nagios (looked at below) is one of the most popular application monitoring tools, but Databand's observability platform is gaining some traction in DataOps. Databand supports data engineers troubleshoot pipeline failures and data quality issues – the ambition is to achieve insight granularity, persistence, automation, ubiquity, and timeliness.

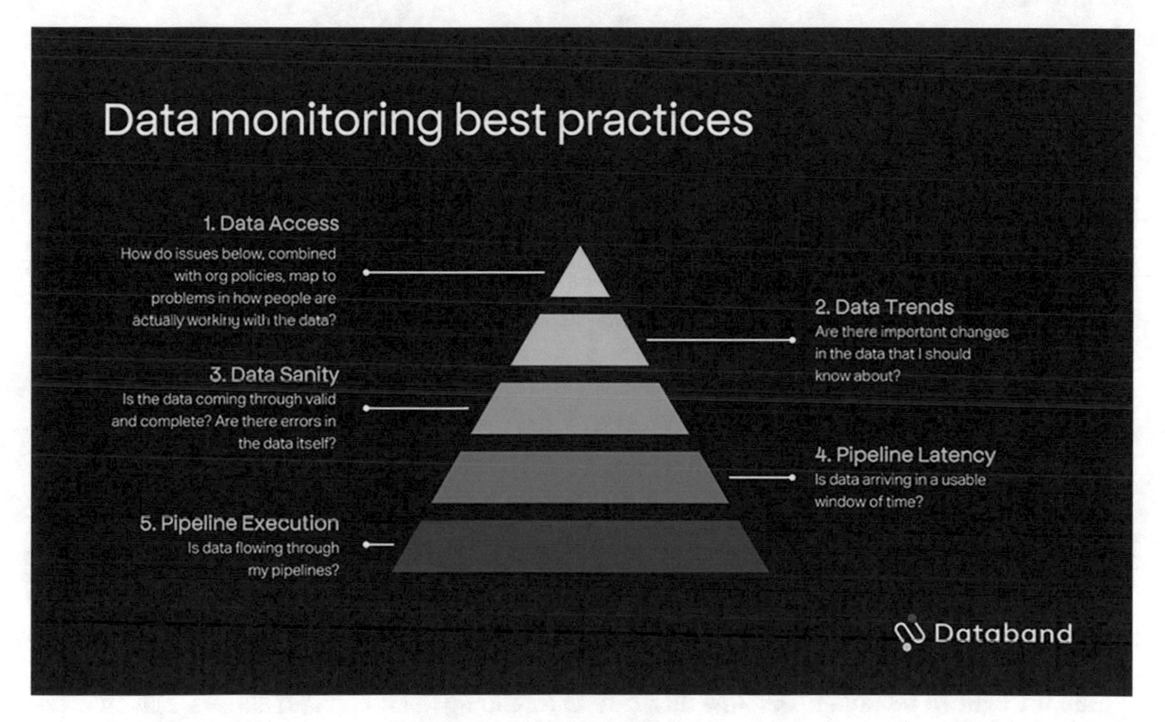

***Figure 2-23.** Databand Application Monitoring*

Nagios

Nagios monitors entire IT infrastructure to ensure systems, applications, services, and business processes are functioning properly, with technical staff alerted in the event of failure. Even older than Jira, Nagios was first launched in 1999.

The Nagios suite of tools consists of an Enterprise Server and Network Monitoring Software (Nagios XI), centralized log management, monitoring and analysis (Nagios Log Server), Infrastructure monitoring (Nagios Fusion), and Netflow Analysis with Bandwidth Utilization (Nagios Network Analyzer).

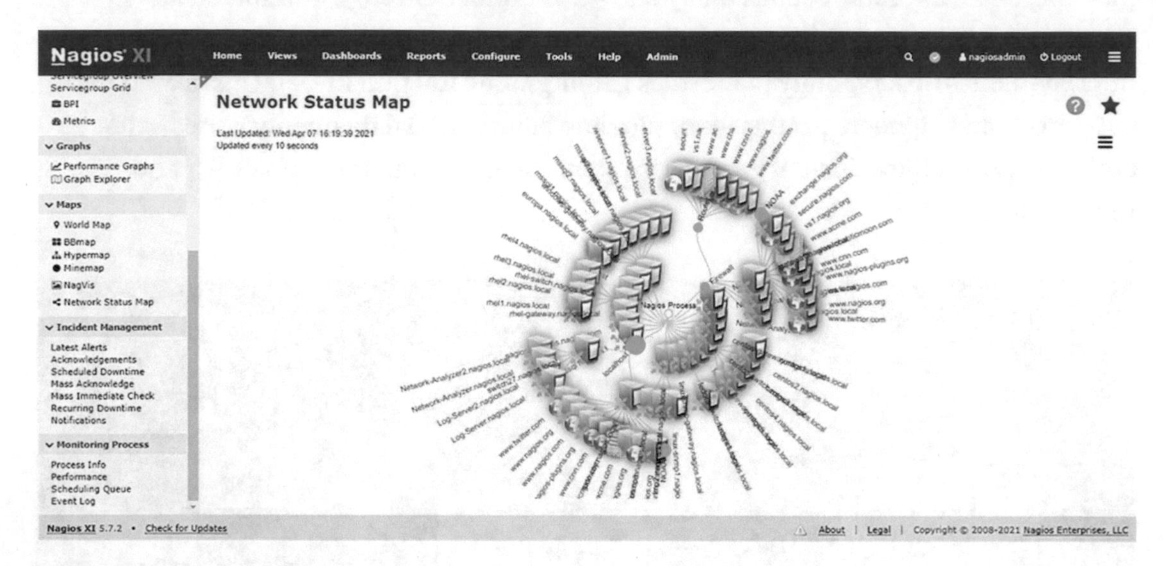

Figure 2-24. *Nagios XI*

Jenkins CI/CD and Selenium Test Scripts: Hands-on Practice

CONTINUOUS TESTING

And so to our final lab in this chapter focused on CI/CD and testing, this extensive lab introduces readers to Jenkins on Azure and running Selenium test scripts. Maven integration is also included as an exercise:

1. Create A Virtual Machine by going to the link below. This will also create a storage resource, which may incur a small monthly cost[15]

   ```
   https://portal.azure.com/#cloudshell/
   ```

[15] See Chapter 1, Important note on cloud resource usage and cost management

2. Toggle CLI settings to Bash and follow steps 3-9 in the link below

   `https://docs.microsoft.com/en-us/azure/developer/jenkins/configure-on-linux-vm`

   NB you may want to replace the Azure region (references in az group create command) to a closer Data Centre

3. Configure Jenkins by following the steps in the link above under section 4. Configure Jenkins

4. Continue to 5. Create Your First Job

   Note it may take some time at the end (10-15 mins) for the project home page to appear

5. Continue to 6. Build the sample Java app

6. Carry out Selenium Scripting by:

   a. Downloading Eclipse IDE from `www.eclipse.org/`. Eclipse is a popular IDE for Java development.

   b. Follow steps 1-7 in this link:

   `https://q-automations.com/2019/06/01/how-run-selenium-tests-in-jenkins-using-maven/`

   c. For step 8 above, add all dependencies WITHIN the project tag

   d. For step 9, the TestNG add-in is required which can be downloaded from the link: `https://marketplace.eclipse.org/content/testng-eclipse` and drag and dropped into the Eclipse workspace[16]

   e. Complete steps 10-13

7. Exercise – try to push your Selenium script to GitHub. Push source code to GitHub first and then connect from Jenkins to repo. NB to integrate GitHub to Jenkins follow Configuring GitHub and Configuring Jenkins steps here: `www.blazemeter.com/blog/how-to-integrate-your-github-repository-to-your-jenkins-project`

[16] Confirm and accept terms of license. If prompted by warning on authenticity, select ok. Restart Eclipse

8. Exercise – for Maven integration, try to follow steps under "Integrating Your Test Into Jenkins" here `https://qautomation.blog/2019/06/01/how-run-selenium-tests-in-jenkins-using-maven/`[17]

9. Exercise – to integrate all three tools: Jenkins, Maven, and Selenium, follow steps 1-12 under "Integrating Your Test Into Jenkins": `https://q-automations.com/2019/06/01/how-run-selenium-tests-in-jenkins-using-maven/` and note the following:

    a. For step 6, under Root POM put: C:\Users\[YOUR-NAME]\eclipse-workspace\SeleniumScript\target

    b. Under Goals and Options enter below:

    c. Test -Dsurefire.suitXmlFiles="$TestSuite" -Dbrowser="$BROSWER" -DURL="$APP_URL"

    d. For step 7 you need the HTML Publisher plug in. Go to Manage Jenkins > Manage Plugins, select Available tab, type HTML Publisher and click check box

    e. Download/install now and restart Jenkins

The option to Publish HTML reports should now be available under Build Settings tab (and Post Build Actions) of the Job created in step 2 of the q-automations link above.

Wrap-up

Having completed our last lab on tools and interfaces for continuous testing, we bring an end to this chapter on DataOps. From Agile development, through code repos, CI/CD, testing, and monitoring, the emphasis here has been on understanding stakeholder relationships, the end-to-end process, ecosystem of tools, and integration landscape in order to "frame" our AI solutions later in this book.

Our next chapter takes the learning from DataOps and applies best practice into one of the first (and perhaps most critical) phases of an AI project implementation: data ingestion.

---

[17] NB in step 2 if Maven Project is not shown, then go to Manage Jenkins ➤ Manage Plugin ➤ Available Tab. In the filter box enter "Maven plugin" and you will get search result as "Unleash Maven Plugin," √ enable the check-box, click on "Download now and install after restart." In the next screen click checkbox to restart Jenkins otherwise the Maven project won't show up. You can also check this link if not seeing the Maven project: `https://stackoverflow.com/questions/45205024/maven-project-option-is-not-showing-in-jenkins-under-new-item-section-latest`

CHAPTER 3

Data Ingestion for AI

Historically, before proceeding to implement any software solution, a thorough scoping exercise would capture requirements around people or stakeholders involved, business processes and tools, or an "as-is" architecture landscape. Today any resilient AI project must capture an additional, and perhaps the most important, factor, that of Data. As such, the goal of this chapter is to inform on best practice and the right (cloud) data architectures and orchestration requirements to ensure the successful delivery of an AI project.

We start this chapter by looking at the data ingestion process, examining the data types used and how approaches and storage considerations vary with scheduled/batch and streaming/transactional data, before taking a look at the cloud computing services and APIs available to AI Engineers.

We then proceed to take a look at Data Stores for AI starting with requirements gathering and the process of creating a data dictionary before moving to best practice for dealing with OLTP/OLAP data sources, and the types of data stores available to us.

Our third section looks at cloud services and tools for ingesting and querying streaming data and data storage of analytical/batch data in order service downstream enterprise analytics and BI teams. We wrap up on how to orchestrate all of this data, enterprise strategies for managing data stores, the import process into machine and deep learning models, and the growing demand for full automation of the data wrangling and model selection process in an AI project.

Introduction to Data Ingestion

So what is Data Ingestion? In simple terms, it's a means to import data, but rebranded for an age where data consumption is, well, all-consuming and where rich datasets, ideally prelabeled, can differentiate an organization's services, products, and promotions from competitors.

B. Walsh, *Productionizing AI*, https://doi.org/10.1007/978-1-4842-8817-7_3

More formally, Data Ingestion is the process by which data is moved from source to destination where it can be stored and further analyzed. Through the course of this chapter more nuanced definitions will become apparent, but we start this first section with a look at the world's current global data needs.

Data Ingestion – The Challenge Today

Today source data takes many different formats from relational or RDBMS SQL-type databases to nonrelational or NoSQL data stores, from csv and text files to API-connected or streaming data. New formats capable of compression suitable for mass data transfer have come to the fore, including avro and parquet.

IDC predicts the world's data will grow to 175 zettabytes in 2025, or 175m petabytes if you deal in those. To give readers an idea of how big this number is that's 175 trillion USB sticks of 1 GB capacity.[1] In the age of this BIG data, complexity and diversity of data presents fundamental challenges - whatever the data sources, almost always data needs to be cleaned and transformed after ingestion.

All of this data requires a vision, ideally a Data Strategy and Data Specialists with most companies today cognizant of three key data priorities:

- Speed of data ingestion/downstream processing
- Compliance and security
- Costs

The AI Ladder

Going back to our earlier failed AI projects, one of the biggest misses is in properly defining a data strategy at the outset. There is no point in undertaking an AI project without an architected solution for Data Ingestion, or put another way there is no AI (Artificial Intelligence) without IA (Information Architecture).

One of the better tools in the wider ecosystem for addressing Big Data gaps in a project is IBM's AI Ladder methodology (shown in Figure 3-1), which focuses on data ingestion best practice as a means for businesses to accelerate their AI journey. The methodology seeks to reinforce messaging around why you can't have AI without IA

[1] Source: Eland Cables

and focuses on unlocking value by **unifying data within a multicloud environment** (a cornerstone of IBM's USPs in 2022). This "Data Lake" style setup seeks to generate the following four key value levers:

- **Collect** data by making it simple and accessible wherever it resides
- **Organize** data to create a business-ready analytics foundation
- **Analyze** data to build and scale AI with trust and transparency
- **Infuse** AI throughout the enterprise and create intelligent workflows (CRPA)

We will address Data Lakes further in the following sections, but before that we address some key concepts in AI Data Ingestion.

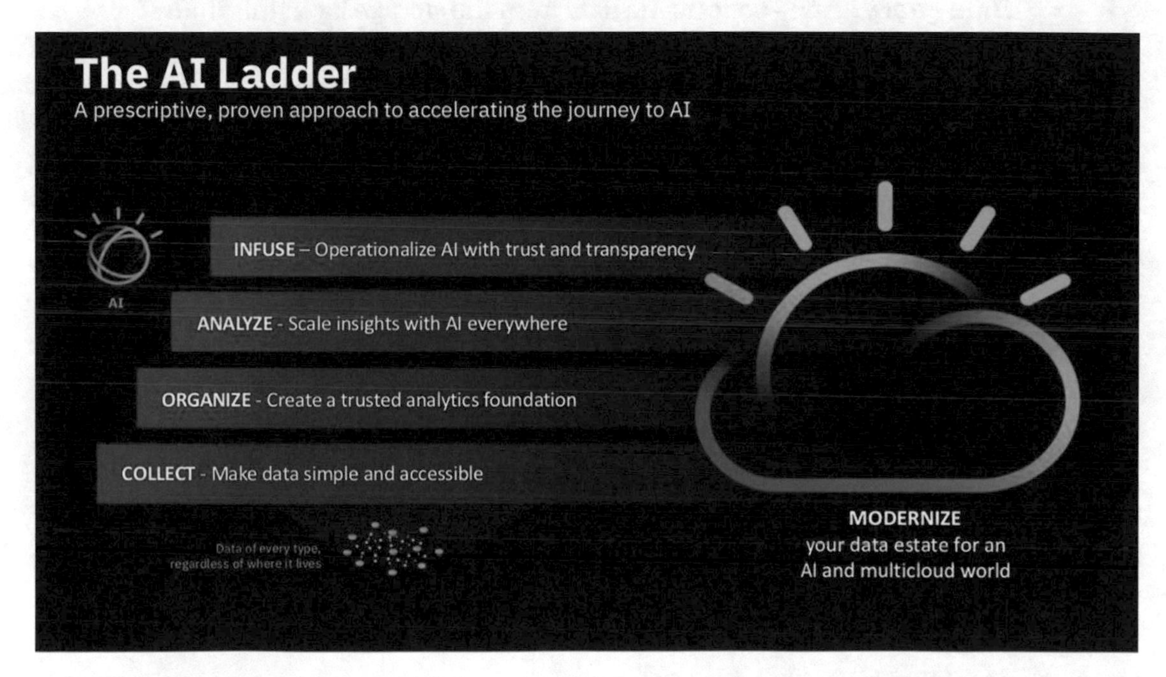

Figure 3-1. *IBM's AI Ladder*

Cloud Architectures/Cloud "Stack"

Successful AI requires an end-to-end Cloud Infrastructure and we saw in Chapter 1 that storage and compute are the main cloud-based services for Big Data handling in AI projects.

From an architectural standpoint, storage, compute, and data Ingestion are fundamental aspects of an Enterprise Solution Architecture designed to support data pipelines and the downstream analytical process. If we take Xenonstack's model, Big Data Architectures typically contain six layers:

- **Data ingestion layer** – connects to data sources, prioritizes and categorizes data
- **Data collector layer** – transports data to the rest of the data pipeline
- **Data processing layer** – processes, routes, and classifies the data
- **Data storage layer** – handles storage depending on the size of the data
- **Data query layer** – collects the data from the storage layer, this time for active analytic processing.
- **Data visualization layer** – consolidates data into a high-value presentation format

These layers are important as ultimately they define the primary concern of an end user consuming the data.

Scheduled (OLAP) vs. Streaming (OLTP) Data

The entirety of data processing in all enterprises can be viewed as simply a combination of scheduled (batch) or aggregated data vs. real-time (stream) processing of transactional data. Traditionally these two types of data were referred to as OLAP (online analytical processing) and OLTP (online transactional processing) where the primary objective was either to process data (OLTP) or analyze the data (OLAP).

Aggregated batch data. is imported (and stored, e.g., in a data warehouse) at regularly scheduled intervals – great for end-of-day, or management reporting. Stream processing, on the other hand, is a "use and lose" approach, where time-sensitive data is processed near real time and then largely discarded. Much of today's hyper-speed consumption of data, particularly via mobile/handheld devices is consumed as streaming data, for example, IoT sensor data, stock price movements, and smart meter information.

Ultimately organizations today have to carve up their data needs and architectural landscape to decide which data is handled as batch data and which as streaming data. The simplified representation in Figure 3-2 shows how most organizations deal with streaming data that needs to be stored by pushing it into an Enterprise Data Warehouse.

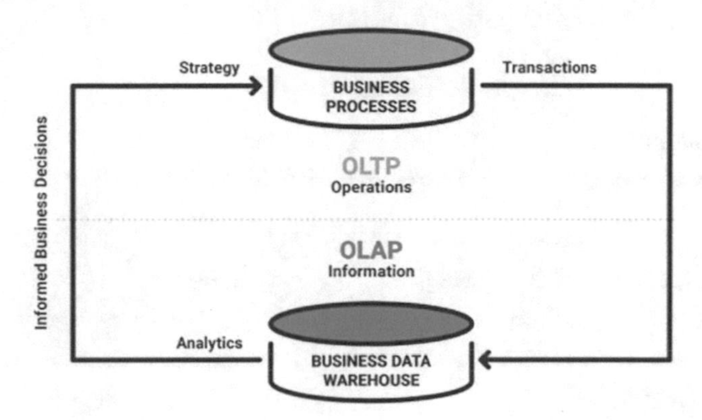

Figure 3-2. `verticaltrail.com`

Today **lambda architectures** and **micro-batching** are important concepts in storage and facilitating processing of both data ingested in batches and in real-time.

APIs

Besides remote databases and raw logs, one of the most common ways to connect to OLTP and OLAP data today is via an API (Application Programming Interface). Essentially an API is a software intermediary that allows two applications to talk to each other. The example below shows the use of Amazon API Gateway, a key component of Amazon Web Services and the means by which streaming data can be ingested from the web via REST, HTTP, or WebSocket API and AWS Kinesis into an AI application.

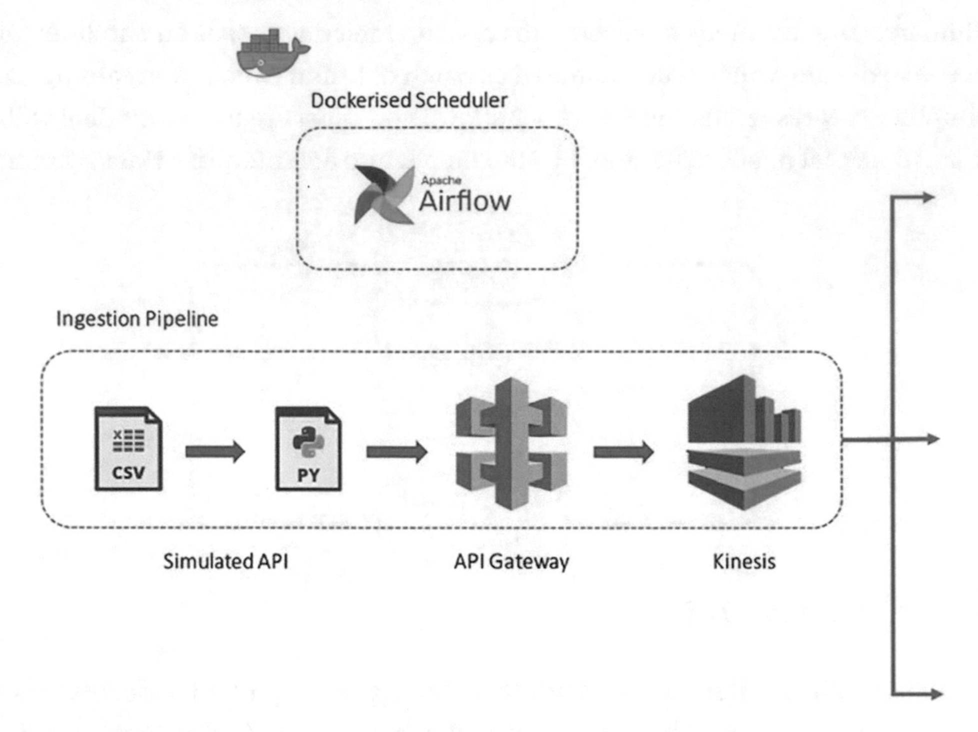

Figure 3-3. *Data Ingestion via AWS API Gateway REST API*

Data Types (Structured vs. Unstructured)

Batch vs. streaming is one perspective of Data Ingestion today, but **structured vs. unstructured** is arguably even more relevant in 2022 and mastering both of these in a Data Strategy is often the difference between an agile business able to quickly identify value-added opportunities from data.

Structured data is data that has been predefined and formatted to a set structure before being placed in data storage. This is the tabular data that most business users have been familiar with since computers have been around; csvs and excel are examples of structured data.

Unstructured data is data stored in its raw, native format and not generally processed until it is used. The explosion of interest in unstructured data in AI today stems from the ability to accumulate the data faster and its inherently feature-rich composition; pdfs and images are unstructured formats.

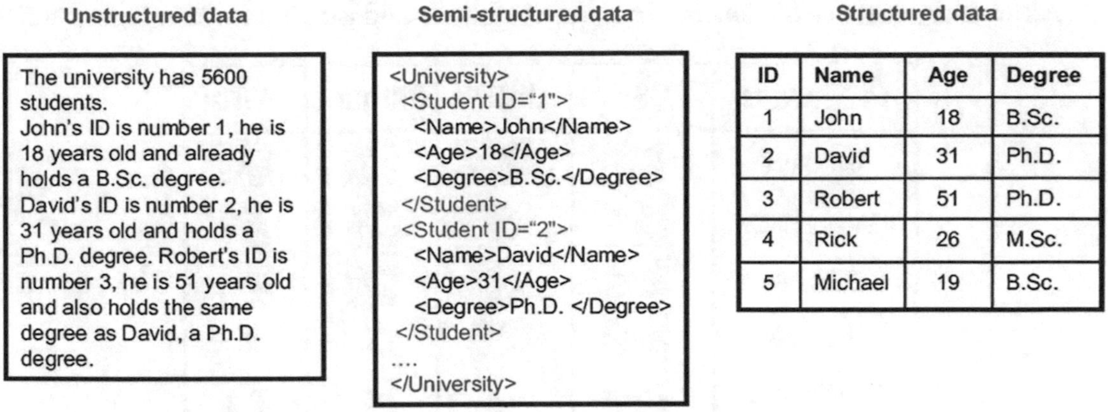

Unstructured data

The university has 5600 students.
John's ID is number 1, he is 18 years old and already holds a B.Sc. degree.
David's ID is number 2, he is 31 years old and holds a Ph.D. degree. Robert's ID is number 3, he is 51 years old and also holds the same degree as David, a Ph.D. degree.

Semi-structured data

```
<University>
  <Student ID="1">
    <Name>John</Name>
    <Age>18</Age>
    <Degree>B.Sc.</Degree>
  </Student>
  <Student ID="2">
    <Name>David</Name>
    <Age>31</Age>
    <Degree>Ph.D. </Degree>
  </Student>
  ....
</University>
```

Structured data

ID	Name	Age	Degree
1	John	18	B.Sc.
2	David	31	Ph.D.
3	Robert	51	Ph.D.
4	Rick	26	M.Sc.
5	Michael	19	B.Sc.

Figure 3-4. *Unstructured, semistructured, and structured data*

File Types

We wrap up our introduction to data ingestion with a closer look at file formats. How you store Big Data in modern data stores is critical, and any solution architect needs to consider the underlying format of the data, compression and how to leverage distributed computing/how to partition data in the fastest, most optimal way.

Traditional file formats such as .txt, .csv, and json have been around for decades and so we assume the reader is familiar with these. Newer formats such as avro, Parquet, and Apache ORC, tr.gz, and pickle formats have sprung up to mesh with rapid advances in the use of clusters (a group of remote computers) in big data processing.

Apache Parquet is an open source column-oriented data storage format of the Apache Hadoop ecosystem. A binary format that is more compressed by virtue of storing repeat data structures as columns, Parquet contains metadata about contents of the data such as column names, compression/encoding, data types, and basic stats. Compressed columnar files like Parquet, ORC, and Hadoop RCFile have lower storage requirements and are ideal for optimal performance during query execution as they read quickly (although write slowly). We will take a further look at parquet formats in a lab at the end of this section.

Avro files are a row-based binary file with a schema stored in a dictionary (specifically JSON) format. Avro files support strong schema evolution by managing added, missed, and changed fields. Because they are row-based, Avro or JSON is ideally suited for ETL (extract, transform, and load) staging layers.

A summary of these file formats compared with csv and json is shown in Figure 3-5.

Properties	CSV	JSON	Parquet	Avro
Columnar	X	X	✓	X
Compressable	✓	✓	✓	✓
Splittable	✓*	✓*	✓	✓
Readable	✓	✓	X	X
Complex data structure	X	✓	✓	✓
Schema evolution	X	X	✓	✓

@luminousmen.com

Figure 3-5. *Features of BIG data file formats (`luminousmen.com`)*

Two other important formats, Pickle and **HDF5**[2] (Hierarchical Data Format version 5) are often used in Python, generally to store trained models. Both are fast – pickle in particular, but comes with the drawback of increased storage space. HDF5 has better support for Big (heterogeneous) Data, storing data in a hierarchical (directory-like/folder) structure, with repeat data compressed in a similar way to parquet files.

Automated Data Ingestion: Hands-on Practice

Having covered a brief introduction to AI Data Ingestion, we now move on to our first hands-on practice on Data Ingestion.

[2] The file extensions .h5 and .hdf5 are synonymous

PYTHON DATA INGESTION – WEATHER DATA

Using Python in Jupyter Notebook, the goal of this exercise is to automate the ingestion of (live) semistructured weather data, then transform and extract temperatures data[3] for forecasting.

1. Clone the GitHub repo below

   `https://github.com/bw-cetech/apress-3.1.git`

2. Run through the code carrying the steps as below:

   a. Import Python Libraries

   b. Connect to MetOffice data

   c. Carry our "recursive wrangling" to extract temperature forecasts for D to D+7 (day time and night time)

   d. Tidy up and present a Pandas dataframe table of forecasted temperatures

Working with Parquet: Hands-on Practise

COMPRESSING & SCALING BIG DATA IN HEALTHCARE

We described earlier in this section how parquet files use a columnar format for data compression. We take a look in this lab at how to work with these files.

1. Clone the GitHub repo below

   `https://github.com/bw-cetech/apress-3.1b.git`

2. Download the US CDC (Centers for Disease Control and Prevention) dataset from the link below:

   `https://catalog.data.gov/dataset/social-vulnerability-index-2018-united-states-tract`

[3] Fundamental data is data that influences or drives a target variable, such as consumer demand on an underlying energy price

NB the above file is 201 MB so may take up to 10 minutes to download on a good connection (8 GB RAM laptop, c. 50 Mbps broadband download speed)

3. Run through the code in Google Colab:

   a. Import libraries to work with Parquet (here the Apache Arrow pyarrow library)

   b. Upload the CDC data downloaded above (this may take up to 30 minutes due to the file size)

   c. Perform basic EDA (data dimensions / no. of rows/columns, data types and first five rows etc.)

   d. mount Google Drive in preparation for saving the Parquet files

   e. convert Pandas DataFrame to parquet - compresses to 75 MB. NB this can be done directly rather by first converting to an Arrows table

   f. read Parquet file back

   g. repeat EDA steps on the parquet data to validate the data

4. Exercise - finally have a go at opening the parquet file in Microsoft Power BI (Get Data ➤ Parquet ➤ Connect[4]) and display the tree map shown below for total square mile area for counties in Arizona

[4] Make sure to download file first from Google Drive then paste without quotes (" ") the local path to the downloaded file

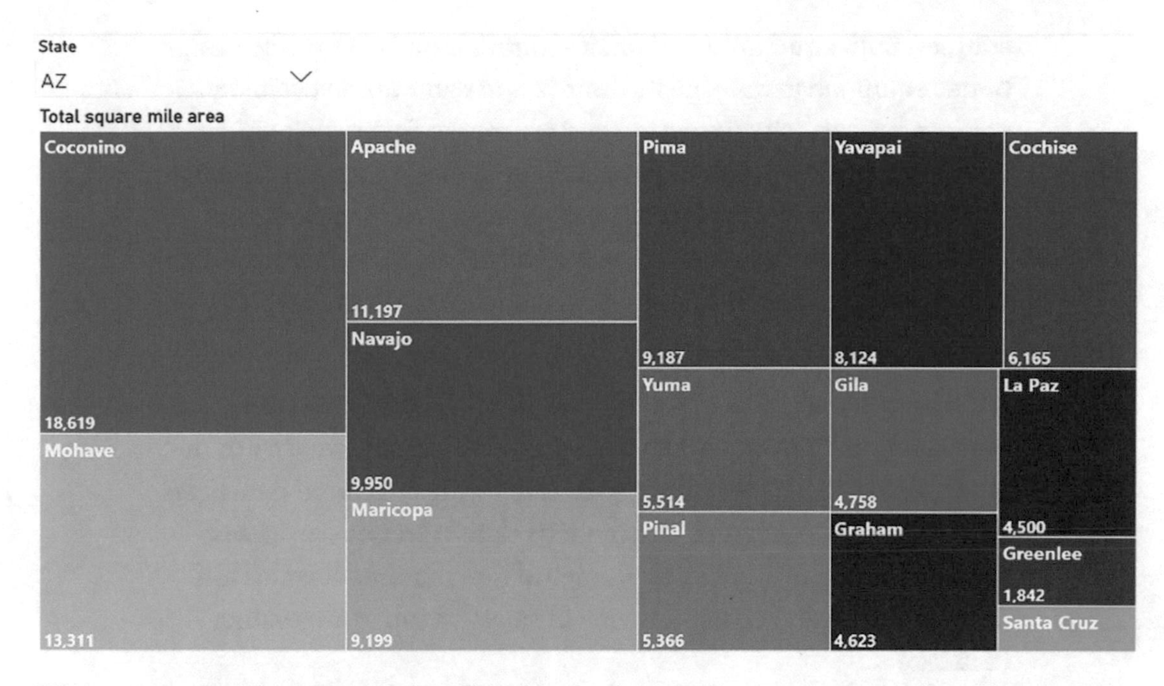

Figure 3-6. *Parquet Data visualisation in PowerBI*

Data Stores for AI

Data ingestion for an AI solution requires storage, whether transient or fixed. Our next section elaborates on what's important to consider when storing data such as defining the data requirements, what kind of data store do we need, Data Lake vs. Data Warehouse, OLAP vs. OLAP, ETL vs. ELT, SQL vs. NoSQL, and the concept of elasticity vs. scalability

Data Stores: Data Lakes and Data Warehouses

A **Data Store** is the name given to a repository for storing, managing, and distributing data across an enterprise. Usually, this refers to a production database upon which many end users are reliant, such as a CRM or ERP system.

Broadly there are three main **Data Stores** in use in most businesses today:

- **Data Lake** – around for the last ten or so years, Data Lakes are increasingly the system of choice for most data-led organizations, a Data Lake is a single storage repository for raw, multiformat data sources, both structured and unstructured. Data Lakes are by design better equipped to manage the four "v"s of data: variety, velocity, volume, and veracity (integrity) and are measured by their ability to constantly refresh multiple data streams yet optimizes performance and topology. Data Lakes do have drawbacks - cataloging data is more difficult and a perceived lack of data governance coupled with sometimes intractable data sources/models has led to the term "data swamp"

- **Data Warehouse** - Data Warehouses are clean, organized Data Stores, often designed and leveraged as a "single source of truth" for enterprise data. They are most associated with structured data types organized into a "schema" but modern Data Warehouses possess more functionality, including storage of unstructured formats and data can be organized as a schema of views on top of an existing data lake
- **Data Marts** - these are generally a filtered and or aggregated subset of a Data Warehouse. Data Marts facilitate simple, limited case querying and possess smaller schemas with focused tabular data for faster downstream analytics/BI

The above are obviously "database" style Data Stores but note that any file systems/ hard drives used in an organization are also examples of a Data Store containing additional rich datasets for AI (such as images, pdfs, and columnar formats/flat files).

Lakehouses

Many companies see the benefits of maintaining both Data Lakes and Data Warehouses served to target specific organizational needs. It is no surprise then that Lakehouses have been introduced as a new, open architecture that combines the best features of both. Examples include **Databricks Lakehouse Platform**, **GCP BigLake**, and **Snowflake** is at least partly considered a Data Lakehouse. Perhaps even more interesting, is **Delta Lake**[5] - an open source project that enables building a Lakehouse architecture on top of data lakes.

[5] https://delta.io/

Given that the "enterprise goal" is to have a Data Store system that is as flexible and performant as possible, the idea of multiple, separate data lakes, warehouses and databases might seem the most productive solution, but the cost overhead and multiple integration points create complexity and latency.

Lakehouses attempt to avoid these issues through the use of data structures and data management features akin to that of a data warehouse while at the same time possessing low-cost storage options similar to that of data lakes.

Scoping Project Data Requirements

Although there is often a choice from where to source data, businesses serious about competing in the age of digital need to "mine" data at source – legacy "ETL" approaches are inadequate.

Best practice for ingesting data in general involves:

- "Mounting" data source(s)
- Ingesting from primary source(s), not via middleware
- Seeking "atomic-level" data, not aggregated/roll-out

Besides ensuring the business is able to connect directly to raw data sources, a DataOps/Agile approach should be used to scope out the underlying business/organizational data requirements. One way of doing this is capturing requirements in a "data dictionary" – a great tool for capturing scope and later delivering both low (MVP) and high fidelity AI solutions.

A well-structured Data Dictionary helps with:

- Stakeholder alignment, buy-in, and project sign-off
- Terminology and defining data types
- Collecting and classifying data critical to the success of a Data project
- Categorizing/mapping attributes from structured and unstructured data as well as online, offline, and mobile sources
- Quick checks for primary keys for joining data
- Isolating anomalies and data flow conflicts
- Documenting experiences with new data sources
- Ongoing data maintenance

OLTP/OLAP – Determining the Best Approach

Given the data requirements on all projects boil down to OLTP and OLAP sources, what it the best approach? Historically, OLTP databases would be ingested into OLAP systems through an ETL process (more on this in the following section) but with the increasing need to source data from source, we often need to get at the "raw" form, that is, Data Lake rather than Data Warehouse access.

In general, we should identify the top three to five business priorities for each downstream analytics/BI or AI requirement to determine how to store/source our data:

- OLTP: if speed/response time is key
- OLAP: if management/strategic insights are more important

ETL vs. ELT

As mentioned in the previous section, an **ETL (extract, transform, and load)** process was historically used to get data from source into a Data Warehouse/database. In many cases, this approach is still preferred due to its inherent transformation of data into a structured, relational database, fixed-schema format and, once the overhead of transforming the data is performed, a much faster and more intuitive analysis or the preprocessed data. ETL approaches are also generally preferred for migrating data stores from on-prem to cloud.

However there is another option, sometimes better suited to the underlying design of a Data Lake and far more agile in supporting key data mining and feature engineering tasks required for Data Science and AI.

In this alternative **ELT process (extract, load, and transform),** data is extracted and directly loaded into the storage layer first without any transformation.

Because we can already query data which is already loaded into a database, an ELT process is the fastest way to expose raw data to an end user for analytics, business intelligence, or AI. However, there are often some "light transformations" required for

- Column selection (we don't need the entire data source)
- Privacy – some fields might need to be filtered/hashed, for example, Personally Identifiable Information (PII)
- Incremental extraction – only uploading new rows and accounting for changing schemas

There is also an abundance of complex transformation logic required before we can visualize the data (e.g., in a BI platform/GUI): from cleansing the data to removing duplicates or out-of-date entries; converting data from one format to another; joining and aggregating data as well as sorting and ordering data. For this reason, ELT processes are better suited for exploratory analysis and Data Science when there is (or can be) a deeper (internal) knowledge of the source datasets.

An ELT process is the preferred approach for loading data into a Data Lake; as ETL was developed when there were no data lakes, it doesn't fit well with modern requirements for direct access to raw data.

To get data into a Data lake, data is extracted from source (via an API) and (often) an extract script provided by a CSP or EL vendor. We will take a closer look at extraction scripts in our final section on Data Pipelines to follow.

SQL vs. NoSQL Databases

We could not pass by this section without tying the above concepts: Data Lake vs. Warehouse, OLTP vs. OLAP, and ETL vs. ELT to the structure and use cases for SQL and NoSQL databases.

For the most part, relational or SQL databases are used for OLAP systems and make use of SQL queries to analyze data and extract insight – SQL's simple and powerful JOIN clause to combine multiple data sources remains highly popular.

SQL databases have a rigid, structured way of storing data and typically contain two or more tables with columns and rows – they are more akin to a data warehouse than a data lake. MS SQL Server, MySQL, Oracle, IBM DB2, and PostgreSQL are all ACID compliant (Atomicity, Consistency, Isolation, and Durability) SQL databases widely used today to support AI projects.

NoSQL, or nonrelational databases, were developed in the late 2000s to improve scaling, speed of querying, and to support the frequent application changes required in DevOps. Rather than tables with fixed rows and columns, NoSQL databases are characterized by no fixed schema, with a structure based on key-value pairs (JSON documents) and nodes and edges (graph databases). Besides their schema-less advantage in handling unstructured data such as images, document files, and emails, NoSQL is ideally suited for distributed data stores with BIG (peta-scale) data storage needs. Popular examples of NoSQL databases include MongoDB, Redis, Neo4j, Apache Cassandra, and Apache HBase with all modern data lakes and modern data warehouses containing connectors to both SQL and NoSQL. Figure 3-7 compares SQL and NoSQL.

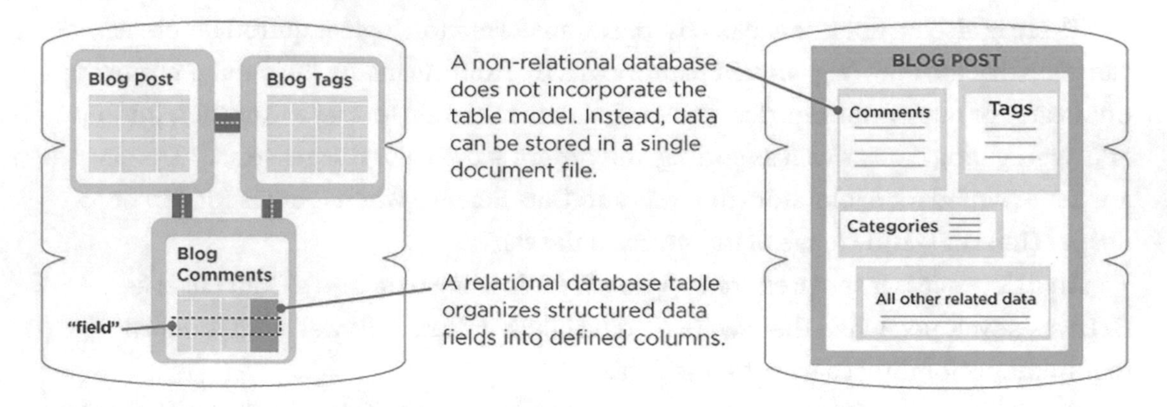

Figure 3-7. *SQL vs. NoSQL (Upwork)*

Elasticity vs. Scalability

Our final word in this section is on two important concepts tied to the underlying business and organizational drivers for the use of cloud, that of elasticity vs. scalability.

Scalable – refers to the ability to add "compute," that is, computing resources to support **static** demand and/or accommodate larger **static** load:

- Scale up – effectively means strengthening hardware
- Scale out – adding additional "nodes" (more computers or VMs)

Elasticity – refers to the ability to provide necessary resources **dynamically** for the current workload

Naturally, this brings about more sophisticated capability measure for those companies at higher (data) maturity levels – **elastic scaling,** or the ability to automatically add or remove compute or networking infrastructure based on changing application traffic patterns.

Most cloud services today are both elastic and scalable but cost models vary greatly and tying the inherent CSP capability to scale up or scale out with forecasted (or even actual) AI application usage is still mired in opacity.

Data Stores for AI: Hands-on Practice

Now that we have gone through the key concepts, let's take a look at how at two Data Stores widely used on AWS Cloud can support AI projects

AWS DATA PIPELINE

The goal of this exercise is to create a DynamoDB NoSQL table on AWS, an S3 (File Store) output bucket, and an AWS Data Pipeline to transfer data from DynamoDB to S3:

1. Follow the steps below to see how data can be transferred across data stores on AWS:

   a. Add DynamoDB NoSQL table

   b. Create S3 bucket

   c. Configure AWS Data Pipeline

   d. Launch EMR cluster with multiple EC2 instances

   e. Activate pipeline

   f. Export S3 data

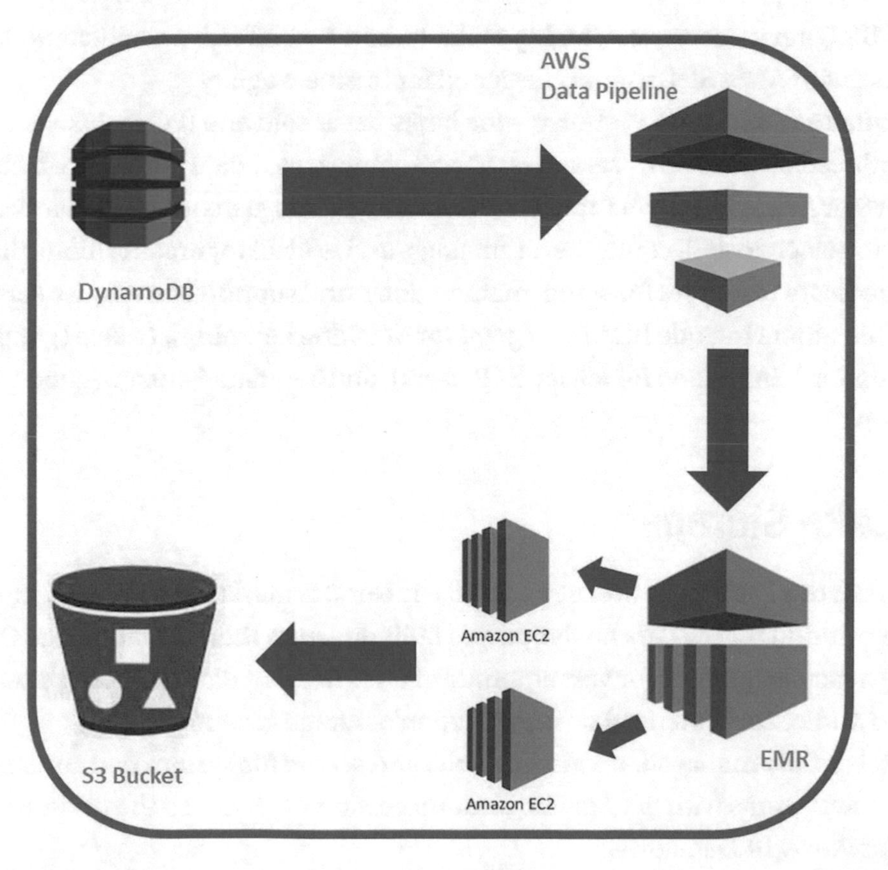

***Figure 3-8.** Data Stores for AI (Hands-on Practice) – data flow*

Cloud Services for Data Ingestion

In this third section, we take a look at cloud services and tools for ingesting and querying streaming data and data storage of analytical/batch data in order to service downstream enterprise analytics and BI teams.

Cloud (SQL) Data Warehouses

There are four main cloud data warehouses in use by companies today:

Azure: SQL Data Warehouse/Synapse Analytics – petabyte-scale Massively Parallel Processing (MPP) analytical data warehouse built on the foundation of SQL Server and run as part of the Azure Cloud Computing Platform

AWS Redshift – for querying and joining exabytes (1000*petabyte) of structured and semistructured data across data warehouses, operational databases, and data lakes using standard SQL

GCP BigQuery – serverless, highly scalable, and cost-effective multicloud data warehouse with ANSI SQL support designed for business agility

Snowflake Cloud Data Platform – the largest ever software IPO in 2020, Snowflake was introduced as a modern "serverless"/SaaS cloud-based data platform which sits on top of AWS or Azure infrastructure. It has notable advantages, such as no hardware or software to select, install, configure, or manage, and is ideal for organizations that don't want to dedicate resources for setup, maintenance, and support of in-house servers

Notable others include IBM Db2 (good for Machine Learning), Oracle (good for automation) SAP (good for legacy SAP users), and Teradata Vantage (good CSP integration).

Data Lake Storage

Microsoft Azure Data Lake Storage (ADLS) or Gen2 is marketed as a next-generation data lake solution for big data analytics. It is built on Azure Blob (Binary Large Object) storage, particularly suited for vast amounts of unstructured data including streaming video and audio, containerized on organization's storage account.

ADLS is a fully managed, elastic, scalable, and secure file system that supports HDFS semantics and works with the Apache Hadoop ecosystem. It is also the underlying storage for PowerBI Dataflows.

AWS Simple Storage Service (S3): on Amazon Web Services, S3 is the storage service of choice to build a data lake on AWS. Secure, highly scalable and durable, able to both ingest structured and unstructured data and catalog and index the data for downstream analysis, S3 is used as an underlying Data Store in many analytics projects and machine learning applications today. When block storage is required, Amazon Elastic Block Store (EBS) is used (comparable with Azure Blob Storage).

Google BigLake: BigLake has recently been launched on GCP as a new cross-platform data storage engine.

Hadoop

Apache Hadoop is the defacto framework that allows for the distributed processing of large datasets across clusters, essentially a collection of software for solving BIG data problems using a network of computers.

MapReduce is the fundamental programming model of Hadoop at the heart of Apache Hadoop for processing huge amounts of data using a Master/Slave architecture (Figure 3-9). As shown in Figure 3-10, it works by performing a (1) **mapping task** which splits data then maps to a key/value pair format and (2) **reduction task** where the mapped output is shuffled and combined into a smaller set of key/value pairs.

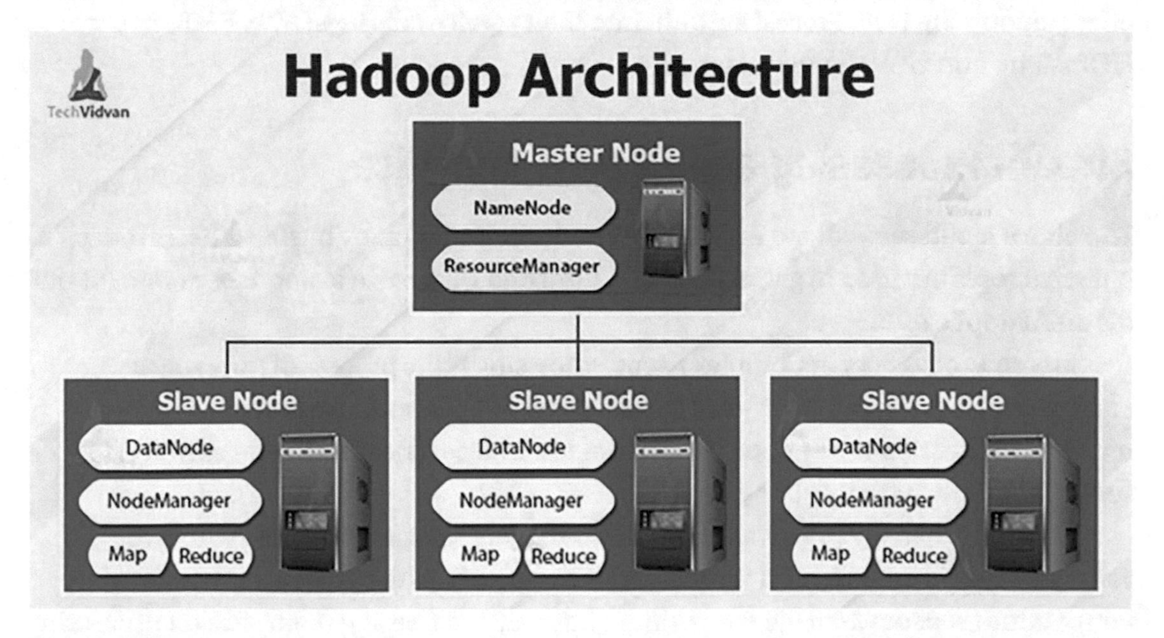

***Figure 3-9.** Hadoop Architecture (Metadata stored in NameNode). The MapReduce algorithm is shown in Figure 3-10.*

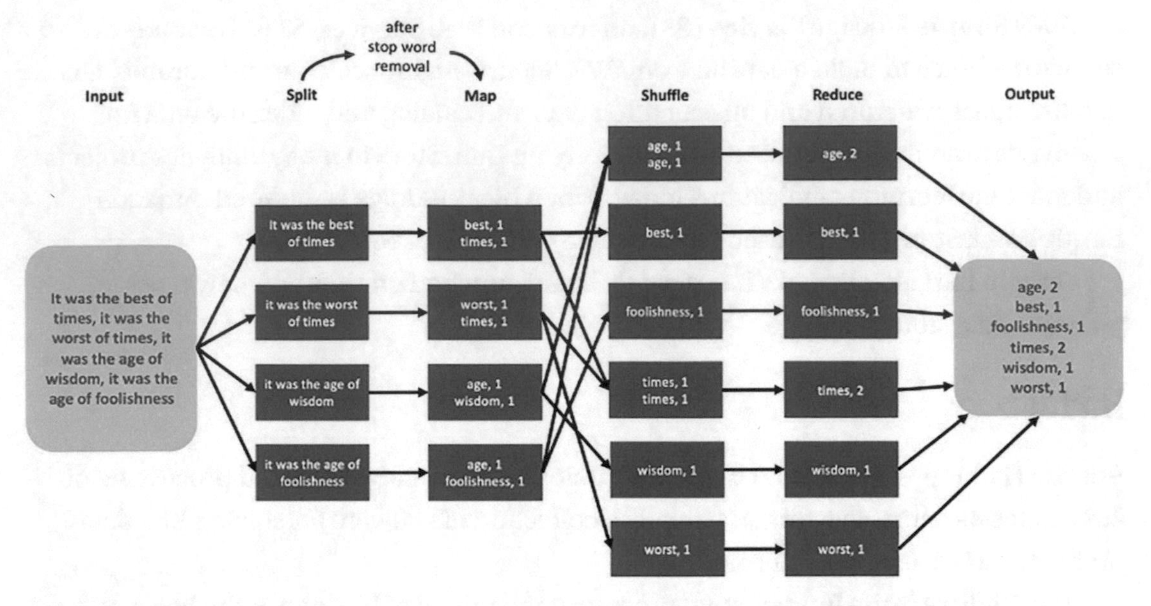

Figure 3-10. *MapReduce across a cluster with Hadoop (Source:* `guru99.com`*)*

The parallel nature of MapReduce makes it valuable for performing large-scale data analysis using multiple machines in the (cloud) cluster. Today most of the CSPs offer Hadoop Big Data Processing built into their service offerings: AWS EMR, Azure HDInsight, and GCP Cloud Dataproc.

Stream Processing and Stream Analytics

The above cloud services are the Data Stores of choice for many business users today, but what tools are used to get/extract data from one Data Store to another, or into (or out of) an AI application?

Stream Processing, or Complex Event Processing is the process of querying a continuous data stream. In an ideal world, this data flow would be real time, but in reality the detection time period varies from few milliseconds (**native streaming**) to seconds (**micro-batching**).

The goal is usually to gain as close to "immediate" insight from data where the value diminishes quickly with time. If we take a specific case of receiving an alert via a temperature sensor when the temperature has reached freezing point, stream processing would determine at what granularity the data is read (and transmitted), for example, milli or micro seconds.

Many tools which handle stream processing today come with extended analytics capabilities - **Stream Analytics**. The Apache tooling suite: Apache Storm, Sqoop, Spark, Flink and Apache Kafka, Talend, Algorithmia, and upsolver are all streaming frameworks with these inbuilt features, as is AWS Kinesis. We will take a look at three of these in our hands-on labs in this book - Apache Spark for Big Data Machine Learning, AWS Kinesis for streaming stock price data and Apache Kafka, which makes use of message brokers for handling streaming data into an AI application.

Whether it's Kafka or Kinesis, there are a huge number of AI use cases today which rely on real-time streaming technologies including algo trading, supply chain optimization, fraud detection, and sports analytics.

Simple Data Streaming: Hands-on Practise

Our first lab in this section takes a look at how to stream data in one of the mostly widely used tools out there - Microsoft Excel

STREAMING STOCK PRICES WITH MICROSOFT EXCEL & PYTHON

The goal of this exercise is to leverage Microsoft Data Streamer as one of the most basic ways to set up a streaming PoC for an AI application after scraping the latest (tech) stock prices with Python

1. Get the add-in for Microsoft Data Streamer by opening a blank excel file then selecting:

    a. File ➤ Options

    b. Add-ins

    c. COM Add-ins

    d. Go

    e. In the COM Add-ins dialogue check the box for Microsoft Data Streamer for Excel and click OK

2. Download the lab files from the GitHub repo below:

   `https://github.com/bw-cetech/apress-3.3a`

3. Stream the sample data as follows:

   a. Data Streamer tab ➤ Import Data File

   b. Point to the "df_eth.csv" file downloaded from GitHub

   c. On the "Data In" sheet ➤ Highlight cells B7:C22

   d. Press alt + F1 to create a chart – this will be blank for now

   e. Data Streamer tab ➤ Play Data

   You should need be able to see the streamed data as a timeline in the chart

4. Exercise: making use of the downloaded python script "StockPriceScraper.ipynb" in Jupyter Notebook, export live stock prices (reader can choose which tech stock they want to export) from Jan-21 to D-1 as well as the latest Ethereum EOD prices. Stream the stock prices in an excel chart as described above

Cloud services for Data Ingestion: Hands-on Practise

Lets also take a look in this next lab at streaming data through one of the main cloud services for data ingestion.

STREAMING WITH PYTHON & AWS KINESIS

The goal of this exercise is to build a python script, create an access key in AWS, an IAM Policy and IAM User then connect to Kinesis and stream stock price data.

1. Create a new (blank) python script (e.g. in Jupyter notebook, Google Colab) and add the Python libraries for AWS (boto3) shown in the sample notebook here:

   `https://github.com/bw-cetech/apress-3.3`

2. Create an access key in AWS by following steps 1-4 here: `https://docs.aws.amazon.com/IAM/latest/UserGuide/id_credentials_access-keys.html`

3. Create a Kinesis stream by following steps 1-7 below: https://docs.aws.amazon.com/streams/latest/dev/tutorial-stock-data-kplkcl-create-stream.html
4. Create an IAM policy by following the steps here: https://docs.aws.amazon.com/streams/latest/dev/tutorial-stock-data-kplkcl-iam.html
5. Create an IAM user by following this link:

   https://docs.aws.amazon.com/streams/latest/dev/tutorial-stock-data-kplkcl-iam.html
6. Add the remaining config steps shown in the above sample Python notebook to view the streaming data.
7. Try the following as "stretch" exercises:

   a) once your data is streaming, view metrics in AWS CloudWatch

   b) try to get DynamoDB to consume the data

   c) modify your code to API into live stock price data on yahoo / quandl

Data Pipeline Orchestration – Best Practice

So far in our hands-on labs in this chapter we have looked at establishing a live connection to weather data, made use of cloud tooling (AWS Data Pipeline) to push data from source to sink (DynamoDB to an S3 bucket), and viewed a data flow through Kinesis in Python with the boto3 (AWS) library.

Our wrap-up of this chapter takes a closer look at best practice for flowing data into and out of an AI application, expanding on these hands-on labs to examine how best to architect AI solutions dependent on OLTP or OLAP data, or a mixture of both.

Storage Considerations

At the commencement of an AI project storage considerations should be on the list of priority considerations. Can we get way with a simple database or data mart, or is integration (as in most enterprise projects) required with a data Warehouse or data lake?

Cost is always important as is flexibility/agility, particularly around unstructured data sources. In both cases data lakes are usually the preferred option but data warehouse companies are improving the consumer cloud experience and making it easier to try, buy, and expand data warehouses with little to no administrative overhead.

If security is a priority, an AI project may favor integration with a legacy data warehouse with a mature security model. Data warehousing may also have better support for machine learning and AI – generally they have faster "reads" of up-to-the-minute data than in a data lake. Data Warehouses also reduce duplication, have schema-overlay to improve data quality, fit reusable features and functions, and are better at preautomating transformations.

Eighty percent of the job of a Data Scientist is in data prep and wrangling so the target "end user" can heavily influence the approach – if research and exploratory data analysis is more aligned with the company's vision then a data lake will provide a wider scope for data mining while if automation and explanatory data analysis is key then integration should where possible be to a scaled-down database or data mart.

Data Ingestion Schedules

After deciding on a data store, effective data ingestion should follow a process of

- Prioritizing data sources
- Validating individual files
- Routing data items to the correct destination

As described in section 2 above, documenting the parameters for Data Ingestion helps to inform the process and continual improvement:

- **Data velocity** – how often is the source data updated?
- **Data size** – what is the storage size of each relative data source?
- **Data frequency** –how frequently do we NEED to access the data? Can we do a batch/analytical upload or should it be streamed/ real time?
- **Data format** – is all the data structured/in tabular format or do we have semistructured (.json, .css, etc.) or unstructured (images, audios, videos)?

A rigorous take on the above can lead to a more performant data ingestion process and accessible data lake similar to that used at Just Eat (using Apache Airflow) through the Ingestion, Transformation, Learn, Egress, and Orchestration cycle.

Serverless Computing

One of the most popular serverless components on cloud is AWS Lambda. Lambda's event-driven, serverless compute architecture allows end users to focus more time on rapidly building data and analytics pipelines by virtue of its independence from infrastructure management and a pay-per-use pricing model.

While a source and sink data store are obviously not isolated from this process, serverless computing can simplify the often complex alert-driven events configured for data ingestion.

End-of-day Processes

Many corporate companies undertake end-of-day processes (EOD). Particularly in banking and retail sectors, for example, outbound billing and reconciling tills after stores have closed, these are crucial operational workflows.

Typically an EOD process involves (a) updating, verifying, and posting daily sales information and (b) aggregation of raw transactions into meaningful business data. The automation requirements here coupled with job scheduling and workflow automation are ideally suited for batch processing and data pipeline orchestration. These automated workflows can involve other (non-daily scheduled) batch processing, such as quarterly or annual reporting.

Data Import for Machine and Deep Learning

The aim of many AI implementation projects today is to achieve production-grade AI models and there are some obvious advantages in automating the entire data ingestion process/data pipeline through an EOD overnight batch process:

- Collect data
- Send it through an enterprise message bus
- Process it (e.g., re-train a model) to provide precalculated results
- Provide guidance for next day's operations

However, the above process doesn't fully capture requirements for ML/DL applications, because features and predictions are time-sensitive. As examples, Netflix's recommendation engines, Uber's arrival time estimation, LinkedIn's connections suggestions, Airbnb's search engines all require training, or at least inference (predictions) in real time.

Data Ingestion for ML/DL needs to consider both online model analytics (real time, operational decision making) and offline data discovery (learning and analysis on historical aggregated data) as shown in Figure 3-11.

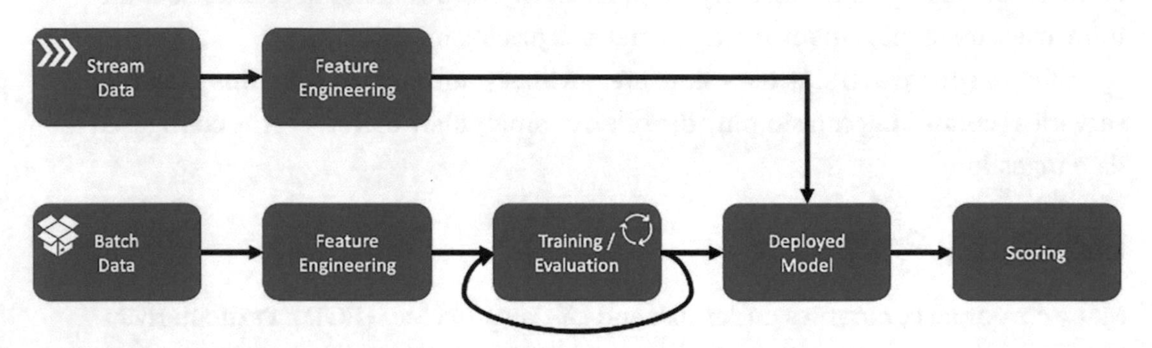

Figure 3-11. *towardsdatascience.com*

Building a Delivery Pipeline

So how do we best build a data pipeline? Every project is different, but we can use best endeavors at the start to guide us on the right path.

We end this section and this chapter by breaking the orchestration process down into a documented approach:

Establish the business context – building new analytics pipelines in traditional analytics architectures typically requires extensive coordination across business, data engineering, and Data Science and analytics teams to first negotiate requirements, schema, infrastructure capacity needs, and workload management

Define the specific problem – business users, Data Scientists, and analysts are demanding easy, frictionless, self-service options to build end-to-end data pipelines because it's hard and inefficient to predefine constantly changing schemas and spend time negotiating capacity slots

Look for a "frictionless" solution – for example, a serverless data lake architecture enables agile and self-service data onboarding and analytics for all data consumer roles across a company on shared infrastructure

Write a detailed solution – it's helpful to consider data lake–centric analytics architectures as a stack of six logical layers, each of which has multiple components:

- Ingestion
- Storage with three separate "zones"
    - Raw zone
    - Cleaned
    - Curated
- Cataloging and search layer
- Processing layer
- Consumption layer
- Security and governance layer

As there is no "one size fits all," we show below four examples of robust data ingestion architectures and pipelines built to better service, speed-up, and scale AI and analytics solutions.

Example: XenonStack

XenonStack's Big Data Ingestion Architecture consists of six key layers: Ingestion, Collector, Processing, Storage, Query, and Visualization.

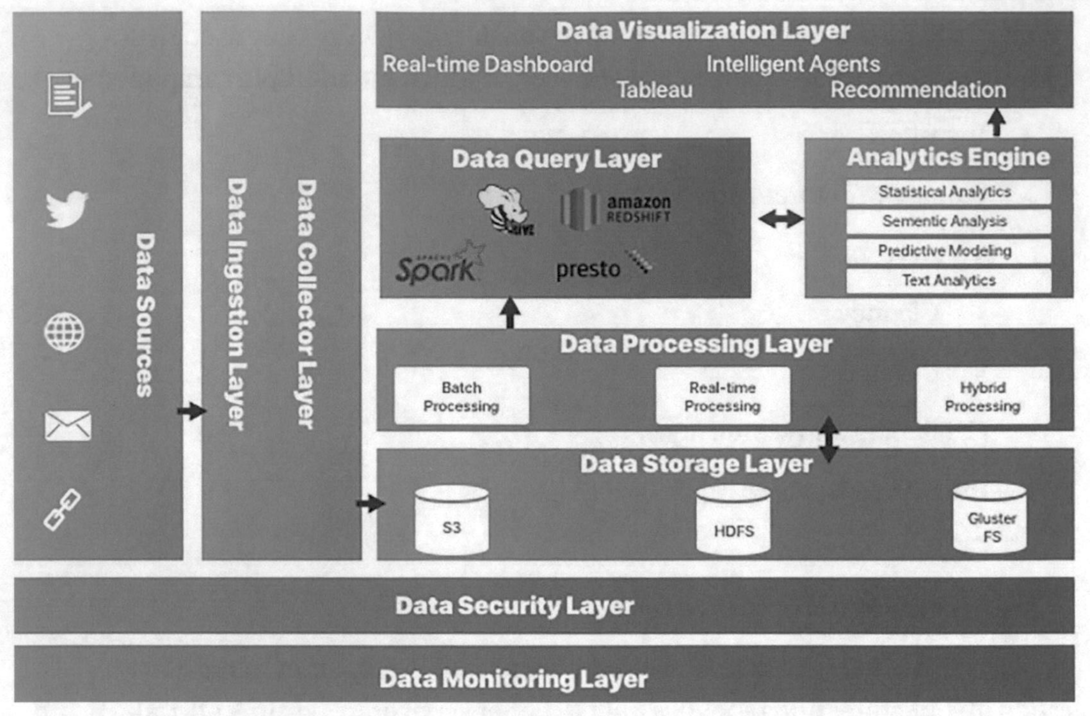

Figure 3-12. *XenonStack Big Data Ingestion Architecture*

Example: Red Hat/IBM

Red Hat's Ingest Data Pipeline with object storage and Kafka streaming.

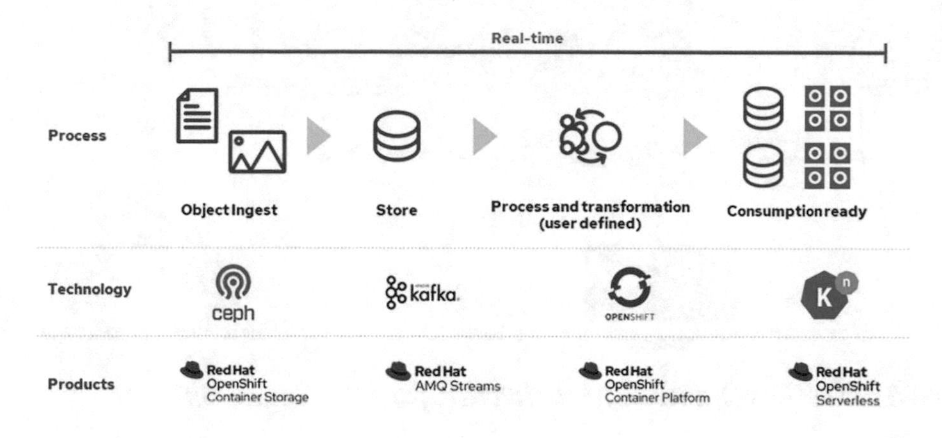

Figure 3-13. *Red Hat Data Pipeline Ingestion*

Example: AWS Serverless Architecture

AWS Serverless Architecture – Lambda is the key (serverless) component but interfaces to AWS Glue for running large workloads with Scala or Python. A larger Architecture Diagram for generic Data Ingestion on AWS is also shown underneath this one, involving a more expansive provisioning of AWS resources.

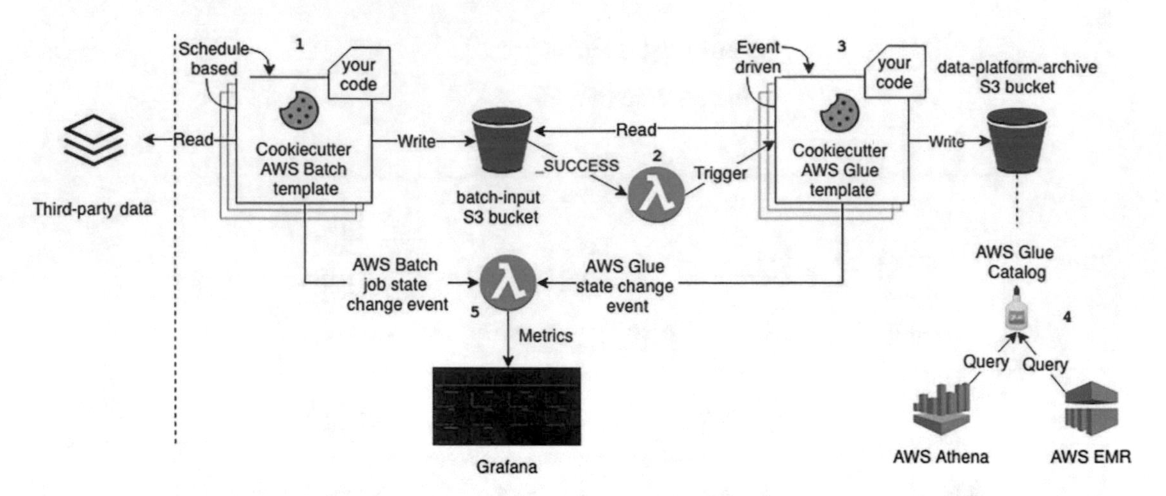

***Figure 3-14.** AWS Serverless (Lambda) Architecture for Data Ingestion*

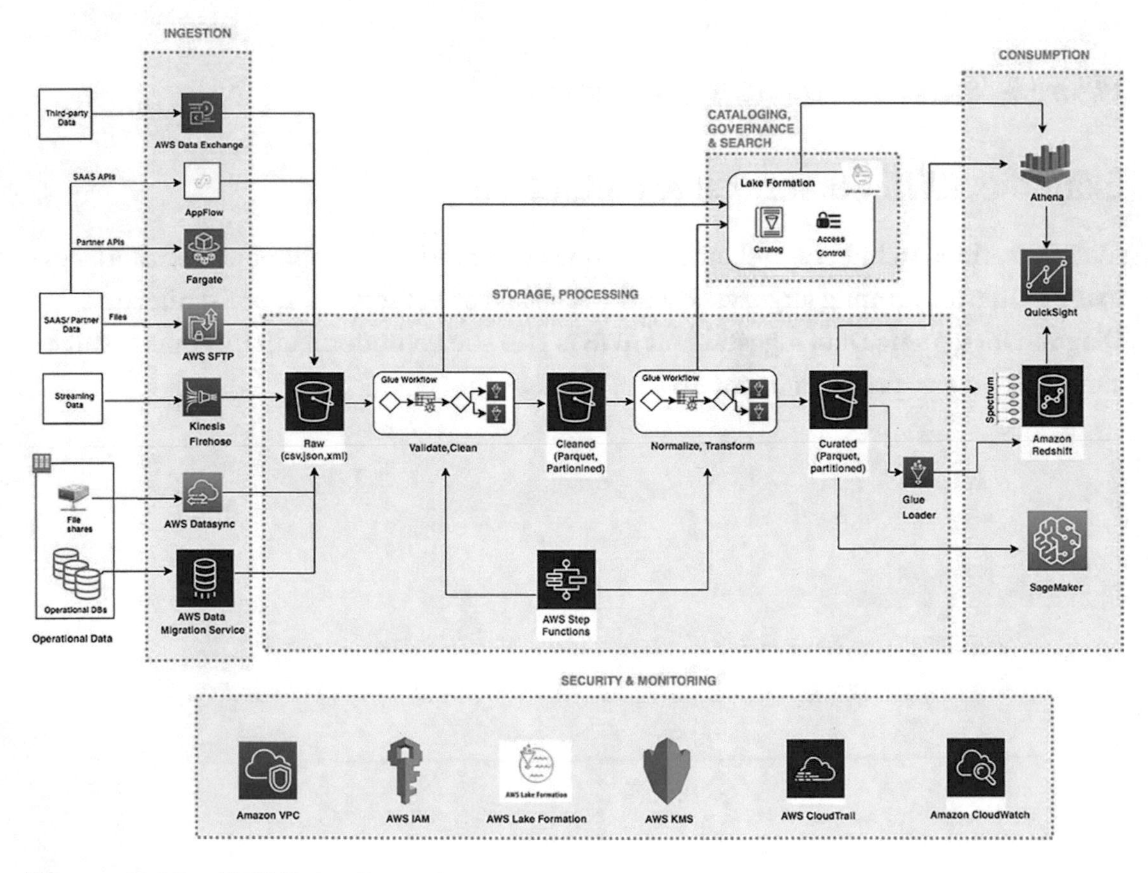

***Figure 3-15.** AWS Data Ingestion*

Example: Databricks with Apache Spark

This Databricks example shows how both Apache Kafka and Spark interface for streaming analytics and AI/BI reporting. Databricks with Apache Spark is the subject of our final hands-on lab in this chapter.

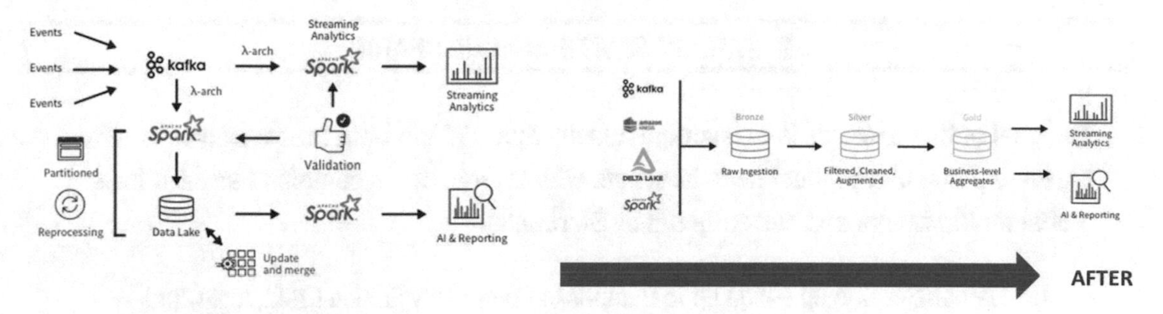

Figure 3-16. *Databricks with Apache Spark*

Example: Snowflake Workload Management

Finally, Figure 3-17 shows Snowflake's architecture designed to ensure data pipeline performance is aligned with workload management and competition of resources across a corporate environment:

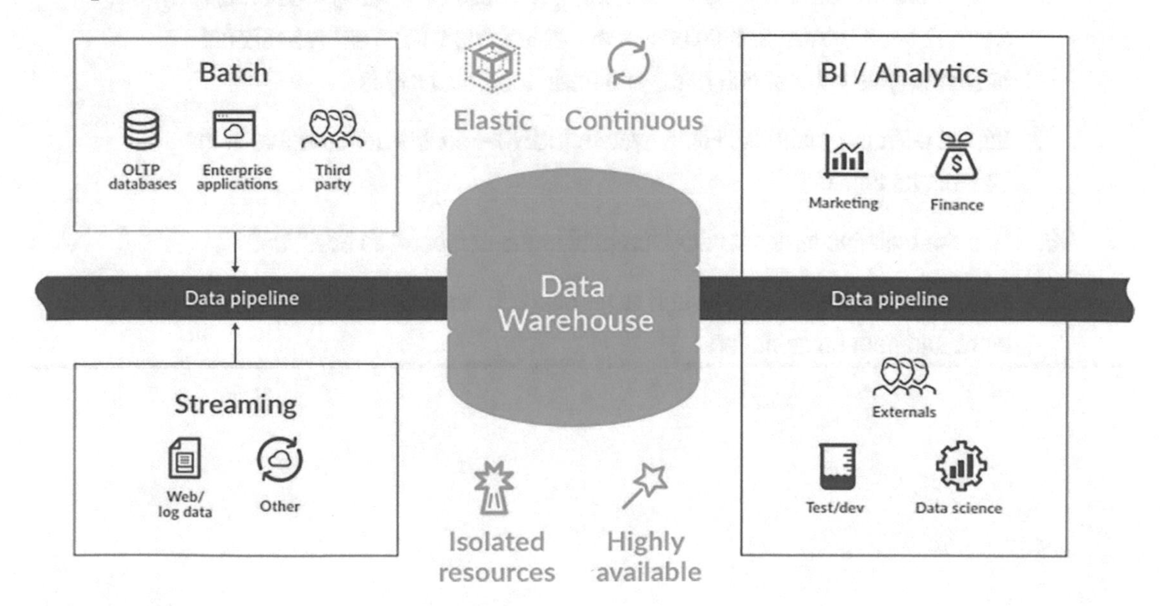

Figure 3-17. *Snowflake workload management*

Data Pipeline Orchestration: Hands-on Practice

So with these well-architected data pipelines in mind, our final hands-on lab below takes a look at how to implement one ourselves, leveraging one of the industry-leading tools for Big Data Processing to scale our AI solution.

DATABRICKS WITH APACHE SPARK

The goal of this exercise is to leverage Apache Spark's[6] Big Data Processing capability to run regional and product level forecasts with fbprophet – a common use case for a global multinational and currently run by Starbucks.

1. Databricks sign up – sign up to Databricks Community Edition (DCE), login, and spin up a cluster
2. Import notebook by pasting into a browser: `https://databricks.com/wp-content/uploads/notebooks/fine-grained-demand-forecasting-spark-3.html`, selecting "Import Notebook" and copy url
3. Attach to cluster and import libraries
4. Download the data by going to `www.kaggle.com/c/demand-forecasting-kernels-only/data`. Scroll down and select "download all" (you should accept Kaggle competition rules) and unzip the train.csv file
5. Import Data to Databricks Filing System (DBFS) – enable admin rights in the Databricks console
6. Run the baseline forecast with fbprophet in the notebook as described
7. Using a Data Pipeline, leverage Apache Spark to scale the forecast to every store and item combination

[6] Apache Spark is written in (statically typed) Scala to compile faster but we will be using Python in this hands-on lab

Wrap-up

Our last lab brings to an end this chapter where we have introduced Data Ingestion in the context of AIaaS, looking at the challenges today in dealing with the deluge of data available to companies, how to store this data, how to process the data using cloud services, and how to orchestrate the various different data sources, ideally through a seamless, automated data pipeline.

Our next chapter takes a whistle-stop tour through Machine Learning. As most of the core concepts should already be known to the reader, our main interest is on machine learning as a bridge to a more in-depth look at Deep Learning (in the subsequent chapter). However, we will also lay out a best practice roadmap for successfully implementing Machine Learning, covering critical data wrangling, training, testing, benchmarking, implementing, and deployment phases.

CHAPTER 4

Machine Learning on Cloud

In terms of (Gartner) Hype Cycle transitions, Machine Learning has long since passed its "peak of inflated expectations," but it remains the core AI technique in use in most businesses and organizations today.

Before embarking on an extended look at Deep Learning, we carry out in this chapter a quick refresher on Machine Learning with reference to applications on cloud. As mentioned in Chapter 1, it is expected readers have some grounding in Machine Learning already, so will assume a basic understanding of supervised and unsupervised machine learning exists.

As such this fourth chapter is an accelerated run through of the mechanics of Machine Learning, covering critical processes from Data Import through EDA and Data Wrangling (cleaning, encoding, normalizing, and scaling) as well as the model training process. We will look at both unsupervised (clustering) techniques and supervised classification and regression as well as time series approaches before interpreting results and comparing performance across multiple algorithms.

We wrap up with the inference process and deploying a model to cloud. Following a more expansive look at Neural Networks and Deep Learning in AI in the next chapter, we will revisit machine learning again in Chapter 6, specifically usage of increasingly important NoLo code UIs and AutoML tools for AI: Azure Machine Learning and IBM Cloud Pak for Data.

B. Walsh, *Productionizing AI*, https://doi.org/10.1007/978-1-4842-8817-7_4

ML Fundamentals

As mentioned in Chapter 1, Machine Learning is a technique enabling computers to make inferences from complex data. High-level definitions for the main types are given below, focused on the inherent difference between these machine learning approaches:

Supervised – training on data points where the desired "target" or labeled output is known

Unsupervised – no labeled outputs available but machine learning is used to identify patterns in data

Semisupervised – an initial unsupervised ML approach applied to a large amount of unlabeled data followed by supervised machine learning on labeled data

Reinforcement learning – training a machine learning model by maximizing a reward/score

Supervised Machine Learning

At a basic level, there are two types of Machine Learning: Supervised and Unsupervised Machine Learning. Reinforcement Learning is sometimes considered a third type, although can equally be considered a type of Unsupervised Learning.

Classification and Regression

What distinguishes Supervised Machine Learning from Unsupervised Machine Learning is the prevalence of "labeled" or "ground truth" data, that is, a specific target field or variable that we wish to train a model to predict.

There are two main types of Supervised Machine Learning:[1] classification and regression. Whereas the label or target variable in classification is discrete (usually binary, but sometimes multiclass), the label in a Supervised Regression problem is continuous. The objective of supervised classification is to find a decision boundary which splits the (training/test) dataset into separate "classes" while for supervised regression the aim is to find a "best-fit" line through the data – a straight line for linear regression or a curve for nonlinear regression as shown in Figure 4-1.

[1] Or three if Time Series Forecasting is considered as distinct from Regression. We will look at this separately below

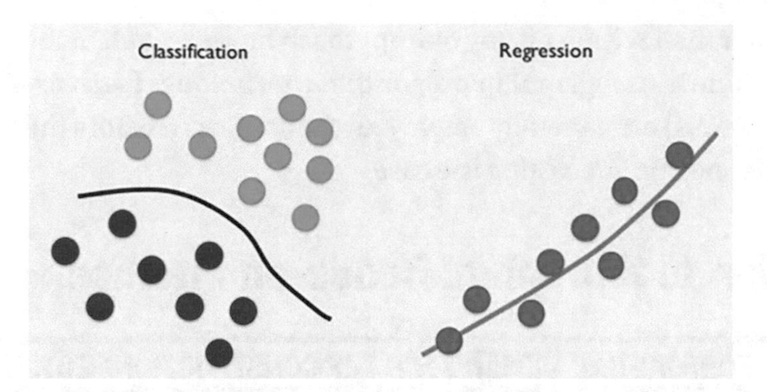

Figure 4-1. *Supervised classification vs. regression*

A key industry machine learning application using a classification technique is determining whether a customer is likely to churn (or not), while forecasting customer revenue is an example of a regression technique. In both of these cases, the **features** used to predict or forecast the target variable are typically customer attributes, normally both transactional and demographic but can also include behavioral or attitudinal data such as time spent on a web page or sentiment from social media engagement.

Time Series Forecasting

As mentioned in Chapter 1, the use of AI in forecasting is trending right now. We will take a look at the specific application of neural networks and deep learning in the next chapter as well as AutoAI approaches in Chapter 6[2] but there are plenty of machine learning applications as well which expand on traditional regression techniques: Autoregression (AR), Moving Average (MA), Autoregressive Moving Average (ARMA), Autoregressive Integrated Moving Average (ARIMA), and Seasonal Autoregressive Integrated Moving-Average (SARIMA).

Facebook's open-sourced algorithm fbprophet for time series forecasting is particularly popular.[3] The algorithm automates forecasting on time series data with nonlinear trends, seasonality, and holiday effects, capturing the four key time series components: secular trends, seasonal variations, cyclical variations, and irregular

[2] AutoAI in IBM Cloud Pak for Data for instance automates time series forecasting

[3] LinkedIn's open source Python forecasting library Greykite was only released in May 2021 but may have speed and accuracy advantages: `https://engineering.linkedin.com/blog/2021/greykite--a-flexible--intuitive--and-fast-forecasting-library`

variations. Rather than relying on any one approach however, and in line with common practice in machine learning, multiple algorithmic techniques (such as ARIMA, fbprophet, and RNNs) are generally employed and compared before the final model selection for the specific forecasting use case.

Introduction to fbprophet: Hands-on Practice

FORECASTING DEMAND FOR THE CONSTRUCTION SECTOR

Using Python in Jupyter Notebook, the goal of this exercise is to introduce fbprophet as an accelerator for forecasting. In this specific case we forecast UK quarterly housing demand, but adaptability of the code sample supports easy swapping out of the data for other sectoral/industry forecasts.

1. Clone the GitHub repo `https://github.com/bw-cetech/apress-4.2.git`
2. Running the notebook in Colab,[4] walk-through the code sample
3. The code will
    a. Import fbprophet (now renamed prophet)
    b. Import construction data from the Office of National Statistics
    c. Wrangle the data into the required format (showing quarterly time series for UK Public Housing Output)
    d. Derive fbprophet forecasting components from the data
    e. Output a ten-year forecast
4. Exercise (stretch): repeat the exercise to forecast at monthly level
5. Exercise (stretch): automate the data import to read in the latest data (i.e., pick up the latest quarter)

[4] No install of fbprophet is needed in Colab

Unsupervised Machine Learning

Unsupervised machine learning is all about identifying hidden patterns in data, specifically when there is no "labeled" data to tell us otherwise. Clustering is the predictive modeling technique but we also take a look in this section at unsupervised ML as means to reduce "Big Data" datasets.

Clustering

With unsupervised machine learning, we have no labeled data or "ground truth," so the goal of predictive modeling in this context is instead to identify an unseen pattern on the underlying data. These patterns are identified as "clusters" often over multiple dimensions, where intracluster distances (the distance between datapoints within a cluster) are minimized and intercluster distances (distance between separate clusters) are maximized.

Because of its inherent ability to find hidden patterns in BIG datasets running into hundreds, thousands, or even more parameters/features, unsupervised clustering is ideally suited to mine customer data buried in a CRM platform (or multiple systems) and for **anomaly detection**. Given a specific number of groupings (clusters), an unsupervised machine learning approach can throw up customer segments which have some degree of commonality (high spend, low to mid income, located in a specific region, etc.). Equally after training an unsupervised machine learning model on the data, anomalies in the data stand out as isolated clusters, as shown in Figure 4-2.

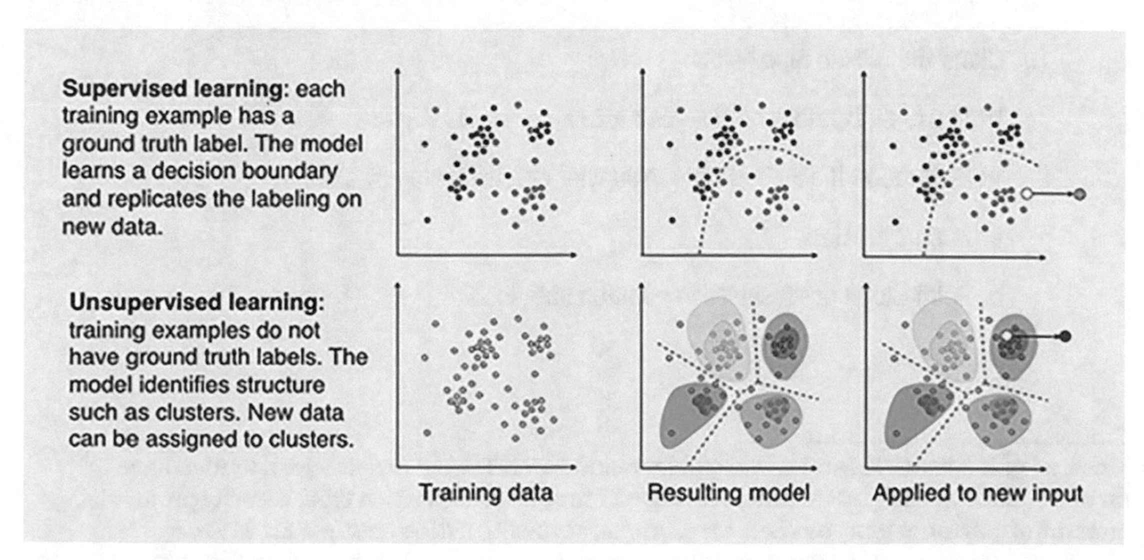

Figure 4-2. *Supervised vs. unsupervised ML (Source: ResearchGate)*

Dimensionality Reduction

Dimensionality Reduction is also considered an unsupervised technique. Here the idea is that a machine learning algorithm is used to simplify (reduce) the data, often from thousands of underlying features into tens of features.

The process of reducing the number of dimensions in your data can of course be carried out in both a manual way (by dropping unwanted features) or automated via an algorithm. The main unsupervised technique used is Principal Component Analysis, where the data is "collapsed" into a form that statistically resembles the original data.

While this approach can greatly reduce datasets and improve runtime/performance, it is not strictly a Machine Learning modeling technique in the normal sense as the outcome in this case is another dataset, albeit compressed, rather than a trained model.

Unsupervised Machine Learning (Clustering): Hands-on Practice

K-MEANS APPLIED TO AERIAL LIDAR POINT CLOUD DATASET

Using Python in Jupyter Notebook, the goal of this exercise is to apply unsupervised machine learning technique (using the K-Means algorithm[5]) to a LIDAR dataset – here an unlabeled "point cloud"[6] representing an airport terminal:

1. Clone the GitHub repo below:

   `https://github.com/bw-cetech/apress-4.3.git`

2. Walk-through the Python code in Jupyter notebook step by step:

   a. Import the data

   b. Point Cloud quick selection – flatten data to 2D

[5] Instead of K-Means, K-Modes is useful for feature-rich Telco datasets which tend to have an abundance of categorical variables: dependents, Internet service type, security protocol, streaming, payment type, etc. See, for example, `https://medium.com/geekculture/the-k-modes-as-clustering-algorithm-for-categorical-data-type-bcde8f95efd7`

[6] A set of data points in space

    c. Point Cloud Filtering

    d. K-Means Clustering Implementation

3. Exercise – display an "elbow plot" to compare different values of k – is k=2 optimal?

4. Exercise (stretch) – try to apply the same technique to the car dataset provided in the GitHub link above

Semisupervised Machine Learning

Not strictly a technique in its own right, semisupervised machine learning essentially involves two processes: unsupervised machine learning followed by supervised (classification or regression). As such, semisupervised machine learning uses both unlabeled and labeled data for training - typically a large amount of unlabeled data and a small amount of labeled data.

A large amount of unsupervised data is used as unlabeled data is less expensive and takes less effort to acquire - semisupervised learning is useful when the cost associated with labeling (often involving annotation by subject matter experts) is too high to allow for a fully labeled training process.

For the above reasons, semisupervised machine learning is often adopted by businesses and organizations looking to scale an AI strategy, but there are also applications such as identifying a person's face on a (low-quality) webcam, which rely on only a few labeled training images[7] and are able to achieve high performance.

Machine Learning Implementation

Having covered the key concepts above, let's take a look at the process of implementing a machine learning model. Although machine learning is never quite as "linear" as we would like, as shown below, the process essentially progresses from Data Import through Exploratory Data Analysis (EDA), Data Wrangling involving a number of subprocesses described below, Modeling and Performance Benchmarking and Deployment.

[7] So therefore "semi-supervised" in the sense of partial labeling, rather than an unsupervised-supervised two-step process.

Design Thinking and Data Mining and Data Import have been covered in Chapters 2 and 3. We will cover each of the subprocesses from EDA to Performance Benchmarking in the sections that follow with Deployment covered in Chapter 7 on Application Development.

Figure 4-3. *Machine Learning process*

Exploratory Data Analysis (EDA)

Exploratory Data Analysis is the process carried out typically once we have imported our data sources, but "ground work" on the process can also occur prior to import. The main activity is to understand the data, firstly by performing basic analysis such as looking at the size of the dataset(s) (number of data files, number of rows and columns in each, etc.), as well as looking at the top and bottom rows,[8] data types, and high-level statistics.

Following this basic EDA, a secondary process involves "going a level deeper" and checking, for example, the number of missing values, the specific values in each column and the frequency of these values. We may also look at specific column/field names in the case of larger datasets (e.g., 20+ columns) to get a "feel" for which fields/variables are likely to be good predictors, and which might be "target" variables.

Although we are initially just looking to output/display key metrics from these EDA tasks, invariably we progress to use of graphical analysis to visualize the data – first single feature plots (histograms for distributions and boxplots for outliers, for example) then multiple feature plots for looking at how features vary against a target variable (e.g., via a pairplot) or for multifeature analysis, for example, via correlation plots (Pearson's correlation for continuous variables, or Spearman rank correlation for categorical variables[9]).

[8] Python .head and .tail methods

[9] Correlation values measured between -1 and 1 with the absolute value determining the intensity of the relationship – negative values indicating an inverse relationship. Spearman rank correlation is just Pearson's correlation on the ranked values of the data.

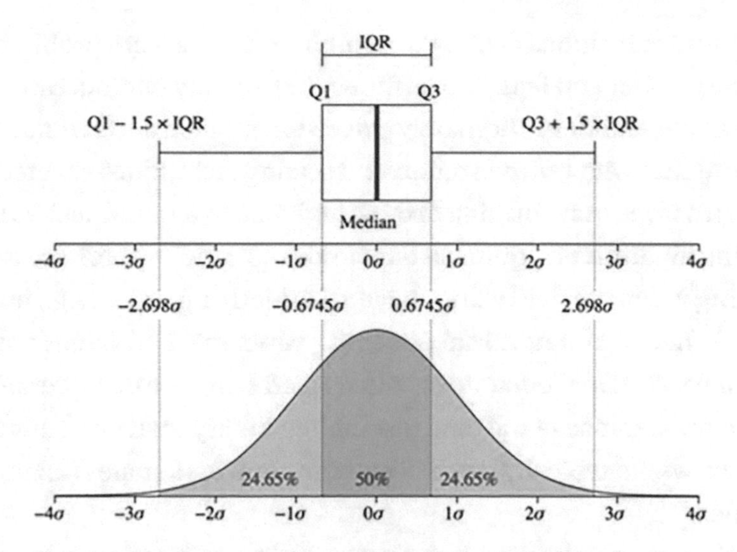

Figure 4-4. *Boxplot interpretation for dealing with outliers*

Data Wrangling

While EDA involves **passively observing** the data,[10] Data Wrangling is focused on **actively changing** the data in some way. There are a number of subprocesses here, described below as a desired sequence of steps, although many are iterative and, depending on the dataset, there can be many loopbacks to an earlier sub-process

Data cleaning: primarily dealing with **incomplete data**, that is, missing values, but formatting or dealing with **invalid data** such as negative ages or salaries, **inaccurate data** (incorrect application of a formula at source, e.g., profit), removing duplicates, and treating outliers are additional tasks that may be required. Dealing with missing values involves taking a decision as to whether the data can be (a) dropped outright (e.g., drop rows when < 1% values missing in a column or drop column when sparsity > 97%), (b) replaced with a proxy value: mean for normal distributions, median for skewed or mode for categorical variables, or (c) interpolated when more sophistication/performance tuning is required.

[10] Although "passive," good Data Science includes documenting findings along the way, typically using comments or markdown process in a Python notebook

Data cleaning also includes dealing with outliers. Outliers are problematic in machine learning as they can lead to overfitting, but equally outliers can be important to building sophisticated models. Ultimately a decision is taken about removing outliers, usually in conjunction with normalization and scaling techniques covered below:[11]

Encoding - involves transforming text (string) data to a numerical format. Most machine learning algorithms require all data to be numeric[12] - there are two approaches: **ordinal encoding** when the field values have an underlying order (e.g., movie ratings - good 2, average 1, bad 0) and **nominal encoding** when the field values have no intrinsic order (e.g., gender). Nominal encoding is often referred to as **one hot encoding** due to the transformed appearance of the data resembling binary machine language (in the case of gender we would typically create 2 columns one for females (taking 0 or 1 values) and one for males.[13]

Normalization - normalization is the process of transforming a skewed field/ variable distribution into a normal distribution. Log transforms are normally used for right-skewed distributions while power transforms are used for left-skewed.

Standardization (Scaling) - refers to the process where data is brought under the same scale, with treatment different for data that is normally distributed (StandardScaler in Python) or skewed (MinMaxScaler or RobustScaler).

Feature Engineering

Data Wrangling is considered by some to include Feature Engineering - the process of selecting which features to use in a predictive model (i.e., in the machine learning training process). This process is highly iterative, occurring before and after each iteration of the machine learning training process.

[11] A good "line in the sand" for an outlier is any value outside [-1.5 x IQR, 1.5 x IQR], however depending on whether this is too broad, a capping method (outside [5th, 95th] percentile) or too narrow (three or more standard deviation away from mean) may be used instead. Clustering as a preprocess (see earlier section on Unsupervised Machine Learning) is also an option.

[12] CART (classification and regression tree) methods are an exception - decision tree and random forest

[13] Normally implemented with just one column/field, for example, for Females, with Males taken as the complement of the Females column

Feature Engineering includes manual processes such as dropping variables which are not likely to be good predictors in an ML model (such as Control IDs) or adding new derived features such as calculating the number of months that have elapsed since a customer made their last purchase.

Typically feature engineering will also include a postprocess to looking at a correlation plot during EDA. Correlation plots (whether Pearson or Spearman) tell us two things: to what degree features are correlated to the target variable (and hence which are likely to be strong predictors), as well as how well features are correlated to each other.

Care must be taken in the latter case to rule out **multicollinearity** – a variable dependency which can skew model performance and create model bias. Calculating the Variable Inflation Factor (VIF) for continuous variables and carrying out chi-square test(s) for categorical variables can be carried out in those cases where we suspect variable dependencies.

Feature Engineering also includes automated techniques such as KBest (selecting the K best features which explain the variance of the target variable) or Recursive Feature Elimination (RFE) which eliminates features after fitting a model until a predefined number of features is reached.

Shuffling and Data Partitioning/Splitting

Like Feature Engineering, shuffling and splitting the data into training and test sets can be considered as part of Data Wrangling. Typically we split the data 70/80% for training and 30/20% for testing. This is implemented in Python (sklearn library) using the train_test_split function. By default, data is shuffled so that the row selection/split occurs randomly (i.e., to prevent nonrandom assignment of rows and the possibility of model bias).

A further level of randomization is typically employed to render the probability of nonrandom assignment almost negligible. This technique, **KFolds cross validation**, splits the training set further into a training (subset) and validation set.

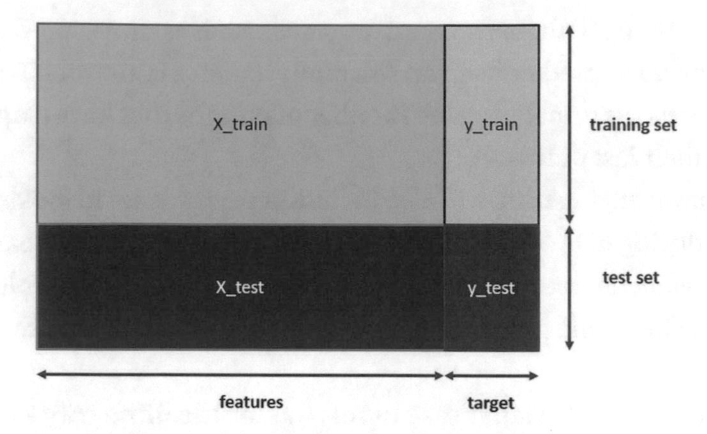

Figure 4-5. *Data partitioning for ML modeling*

Sampling

In the next two sections, we will take a look at the algorithmic process and performance benchmarking. As it is highly iterative, data wrangling also encompasses a number of tasks not typically performed until several model runs have been undertaken. Normalization and scaling are two such techniques mentioned above, which are not always necessary to get a model "up and running" but rather to fine tune once there is a stable model process. Sampling, specifically over or undersampling, is another when it's applied to imbalanced datasets.

Most datasets have inherent imbalance[14] - take fraud detection where the number of fraudulent transactions is typically minuscule in comparison to the number of nonfraud transactions. The same applies in cybersecurity - a DDoS attack can be like finding a needle in a haystack of benign network activity.

When we imbalanced data, random undersampling refers to the process of only sampling enough of the majority class so that we have approximately the same volume of both majority and minority class samples. For random oversampling, we instead synthesize data (duplicates from the minority class to achieve the same effect).

Under or oversampling can often make the difference between a good and badly performing model although undersampling runs the risk of losing information in the data that may be valuable to the model while oversampling can lead to overfitting.

---

[14] or worse, feature "bias" where multiple features have statistical bias in the dataset or modelling process

We will take a look at a specific application of undersampling, SMOTE (Synthetic Minority Oversampling Technique) in Chapter 8.

End-to-End Wrangling: Hands-on Practice

CREDIT RISK MODELING PREP USING THE KAGGLE API KEY

Using Python in Google Colab, the goal of this exercise is to apply EDA and Data Wrangling techniques to a common banking challenge – detecting and predicting customer credit risk:

1. Clone the GitHub repo `https://github.com/bw-cetech/-apress-4.4.git`
2. Referencing the TOC on the LHS in Colab walk-through the processes below:
    a. Library import
    b. Config
    c. Direct connection to the dataset using a Kaggle API key
    d. EDA
    e. Data Wrangling
    f. Feature Selection
3. Proceed to the modeling part of the notebook, carrying out a baseline run then a scenario where specific features are scaled
4. Exercise – try to improve model performance by carrying out more runs and changing the data/enhancing the feature engineering process
5. Exercise – do the same but this time carry out a parallel run across multiple algorithms

Algorithmic Modelling

After splitting our data into training and test sets (or training, validation, and test sets), we are ready to "fit" a model. The algorithmic technique takes the training data and fits an algorithm to the model.[15] Once the model has been trained, using the .predict() function in sklearn, we put the test data through the trained algorithm in order to benchmark model predictions/forecasts (see section below).

This book is a practical look at building AI applications, not a theoretical (and often irrelevant[16]) discussion on the mechanics of machine or deep learning algorithms. That said, there are plenty of labs in this chapter and others showing how underlying algorithms, both machine and deep learning, are applied and plenty of relevant best practice on how to fine-tune and get the most out of the modeling process in later chapters.

For reference going forward in this book, we summarize below the main machine learning algorithms used and their strengths and weaknesses. This is followed by a hands-on lab in the next section where we train multiple machine learning (and one deep learning) algorithm on three different datasets and compare performance.

---

[15] k times in the case of k folds cross-validation, with each iteration also tested against the validation set

[16] See e.g. Chapter 1 discussion on "technical debt"

Table 4-1. *Supervised Classification Algos*

ALGORITHM	STRENGTHS	WEAKNESSES
NAÏVE BAYES	Simple model based on conditional probability of classes. Easy to implement	Assumes independence of all features – rarely the case
LOGISTIC REGRESSION	Straightforward, easy to understand – outputs are probabilistic	Underperforms with multiple or nonlinear decision boundaries
CLASSIFICATION TREES (CART)/RANDOM FOREST/ GRADIENT BOOSTED TREES/ CATBOOST[17]	Can learn nonlinear relationships, robust to outliers	Prone to overfitting when unconstrained (alleviated by constraining/using ensembles)
SUPPORT VECTOR MACHINES	Can model nonlinear decision boundaries; fairly robust against outliers in high-dimensional space	Memory intensive, hard to tune, don't scale well to larger datasets
MULTILAYER NEURAL NETWORKS (DEEP LEARNING)	Performs very well when classifying audio, text, and image data	Require very large amounts of data to train, not general-purpose

[17] CatBoost is an algorithm for gradient boosting on decision trees (tree ensembles). It's simple install, superior model exploration, and performance tracking (global and locally supported SHAP value analysis means it has become one of the more performant ML algorithms). See also e.g. `https://towardsdatascience.com/why-you-should-learn-catboost-now-390fb3895f76`

Table 4-2. Supervised Regression Algos

ALGORITHM	STRENGTHS	WEAKNESSES
LINEAR REGRESSION	Straightforward, easy to understand	Performs poorly with nonlinear relationships
LASSO, RIDGE, AND ELASTIC-NET	Regularization – penalizes large coefficients to avoid overfitting	Complex hyperparameter tuning
REGRESSION TREES (CART)/RANDOM FOREST/ GRADIENT BOOSTED TREES/XGBOOST/ LIGHTGBM/CATBOOST	Can learn nonlinear relationships, robust to outliers	Prone to overfitting when unconstrained (alleviated by constraining/using ensembles)
K NEAREST NEIGHBORS	Simple search for most similar training observation method	Memory intensive, perform poorly on high-dimensional data
MULTILAYER NEURAL NETWORKS (DEEP LEARNING)	Can learn extremely complex patterns, efficient learners of high-dimensional data, excel at computer vision, speech recognition	Computationally intensive, high degree of expertise to tune. Not suitable for general purpose as require BIG data to train on

For unsupervised machine learning, K Means is the main algorithm, but there are several variants including hierarchical clustering. Neural networks (specifically autoencoders) can also be used for unsupervised machine/deep learning,[18] and Principal Component Analysis (discussed earlier) is essentially also an unsupervised machine learning algorithm.

[18] Autoencoders are discussed in Chapter 5

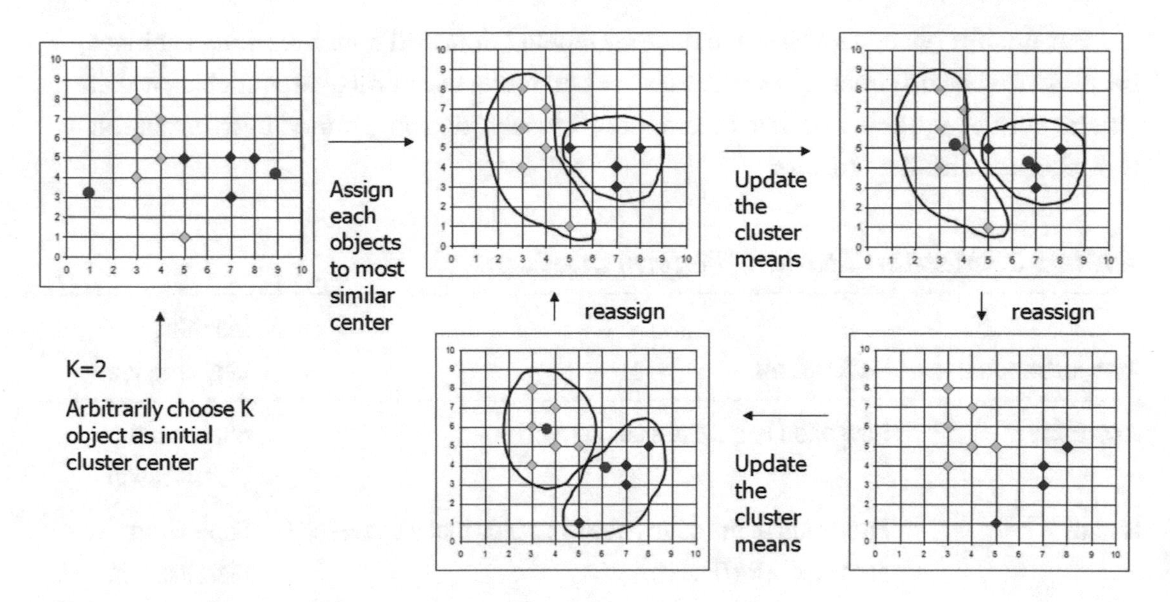

***Figure 4-6.** KMeans algorithm – operational mechanism*

Performance Benchmarking

Building an AI app requires constant training and testing – knowing how to benchmark performance and which measures to use is a key overhead. It's always a good idea to compare model output with a baseline.

For supervised classification models, this can be a model which randomly assigns classes to each datapoint in the test set, or a model which uses all available features (unaltered except for data cleansing associated with runtime errors, e.g., missing values removed, text data encoded, etc.) – essentially a model with no "feature engineering." For supervised regression models, a **persistence** forecast is often used as a baseline, where the forecast just "persists" (i.e., repeats) a pattern of data from the test set, for example, day minus 1, day minus 7, or year minus 1, etc. As with classification models, it's a good idea to run a baseline with all features to compare later runs where more sophisticated feature engineering has been carried out.

The above baseline is useful for supervised machine learning "up and running" but after 5-10 model iterations more sophistication is required and, as shown in the table below, a number of measures are used to determine how well a model is performing.

For unsupervised learning, we have no labeled data and benchmarking against a baseline makes no sense. However here we can make use of measures such as within cluster sum of squares to get an idea of how "closely" data is grouped together in the model output model clusters.

Table 4-3. *Machine Learning Performance Metrics*

Metric/Benchmark	Description	Machine Learning type
Accuracy	How often is the classifier correct?	Supervised Classification
Recall	Proportion of the actual positive cases that we correctly predicted: TP/(TP+FN)	Supervised Classification
Precision	Proportion of correct predictions: TP/(TP+FP)	Supervised Classification
f-measure	Weighted average of the true positive rate (recall) and precision: 2 * recall * precision / (recall + precision)	Supervised Classification
Confusion Matrix	Number of correct and incorrect predictions compared to the actual outcomes	Supervised Classification
Classification Report	Displays precision, recall, F1, and support scores for the model	Supervised Classification
ROC Curve/AUC	Trade-off between sensitivity and specificity	Supervised Classification
Root mean Squared Error (RMSE)	Square root of mean error between predictions and actual values	Supervised Regression
R2 (R-squared)	A measure of how well a model fits predictions to the actual values	Supervised Regression
Variance	Error from sensitivity to small changes as a result of overfitting	Supervised Regression

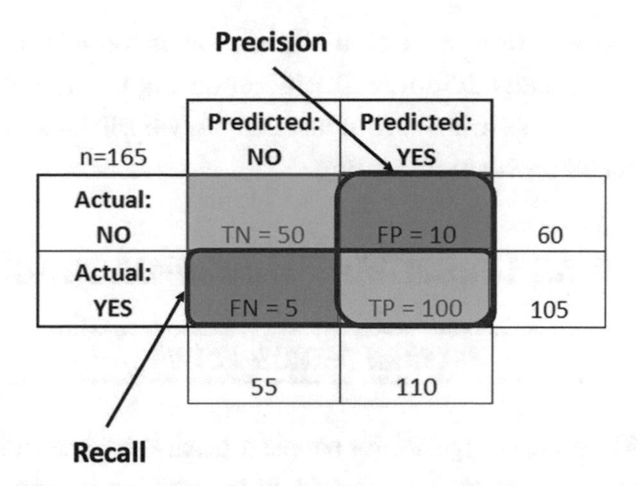

Figure 4-7. *Confusion Matrix with recall = 100 / (100 + 5) = 0.95 and precision = 100 / (100 + 10) = 0.91*

Continual Improvement

Continual improvement is of course part of performance benchmarking - rather than relying on static metrics and blind trial and error approaches to improve results, best practice should be adopted to ensure over time we can surpass tolerances on model performance.

Almost always revisiting the data should be the first port of call and critiquing the approach used for key data cleansing tasks such as imputing proxy values for missing values - has too much valuable (other) feature information been lost when we removed missing data from another feature, is it appropriate to just replace missing values with the mean value, etc.? Some of these same concerns around loss of information and/or model bias extend to the sampling approaches already described for imbalanced datasets.

The question of data extends further to the data acquisition process - while more data doesn't always mean better results, a larger sample is statistically more likely to yield better performance. More data also doesn't necessarily mean more records or rows of data - new features, often derived from existing ones (such as "no. of days since last purchase," "freq of purchase," etc.) can have a huge impact.

Whatever the granular approach taken, only after an exhaustive reexamining of the data should we look to further algorithmic testing (parallel or otherwise) and hyperparameter tuning.

And while the above focus is on continually improving model training and test results, model resiliency is an additional consideration. A great model today isn't always fit for purpose in one month's time – in our last chapter we will look at data drift and automated retraining for problem mitigation.

Machine Learning Classifiers: Hands-on Practice

SKLEARN RETROSPECTIVE

Scikit-learn (or sklearn) is the "go-to" for simple predictive analysis and modeling – in this lab we compare machine learning algorithms in scikit-learn against three different datasets.

Not just "entry-level" ML, sklearn is widely used in production by well-known brands including J.P.Morgan, Spotify, and booking.com. Built on NumPy, SciPy, and matplotlib, and besides the main algorithmic processes in machine learning, sklearn comes with some built-in datasets, preprocessing, feature engineering, and model selection functions/methods:

1. Clone the Python notebook from: `https://github.com/bw-cetech/apress-4.9.git`
2. As described in the notebook:
    a. Import the three dummy datasets
    b. Scale
    c. Carry out a parallel run and fit the data to the ten different algorithms
3. Compare visually how the multiple algorithms fit to the training data
4. Finally, have a go at the exercises:
    a. Extract the model scores (raw values in a list displayed on the screen)
    b. Isolate the multilayer perceptron (MLP) for the "make_circles" dataset only. Plot this with the input data and make it bigger on the screen

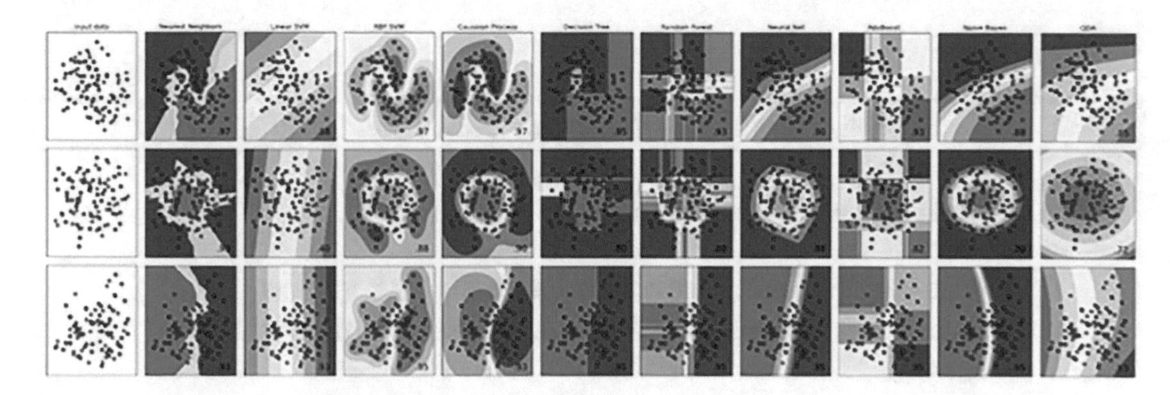

Figure 4-8. *Comparison of ten machine learning algorithms against three different datasets*

Model Selection, Deployment, and Inference

The above performance metrics are the basis for initial and final model selection - a process which is more commonly carried out in 2022 using automated techniques such as Bayesian Optimization and AutoML/AutoAI. This is the subject of Chapter 6 on AutoAI and we will leave a discussion of the process until then.

Machine (and Deep) Learning deployment is the focus of Chapters 7 and 9 - essentially it is the process of deploying a model into production, where **inference** is carried out on new data - running new data through the model to obtain a prediction/ forecast.

Inference is performed in our last hands-on lab in this chapter and in essence is no different from the test prediction step described in the Algorithmic Modeling section, only in this case we are exposing new data to the trained model, instead of a sample datapoint from the test set.

Inference: Hands-on Practice

AZURE MACHINE LEARNING – INFERENCE API TESTING

Using Python in Jupyter Notebook, the goal of this exercise is to…

1. In Azure Machine Learning Studio, follow the steps below to train and evaluate a machine learning model on predicting whether income is above or below $50k:

   `https://gallery.azure.ai/Experiment/3fe213e3ae6244c5ac84a73e1b451dc4`

2. Now follow the steps below to set up a web service and perform a simple API inference test:

   a. Making sure model has already run, set up web service

   b. Select predictive web service (recommended)

   c. Run

   d. Deploy web service

   e. Click on TEST (blue button) beside REQUEST/RESPONSE to do a simple test

   f. SHOW with some default values (age = 12, income = 45) customer predicted to have LOW INCOME (shown in small text at the bottom of the screen)

   g. Exercise – after deploying the Web Service and using the config settings shown under "Consume," add the Azure ML add-in to MS excel and entering sample data, call the API from excel

Reinforcement Learning

Reinforcement learning involves real-time machine (or deep) learning with an agent/environment mechanism which either penalizes or rewards iterations of a model based on real-time feedback from the surrounding environment (how accurate is the model).

There are three primary components to the algorithm, with the aim being to discover through trial and error which actions maximize the expected reward over a given amount of time:

- The agent (the learner or decision maker)
- The environment (everything the agent interacts with)
- Actions (what the agent can do)

While the scope of this book is mainly focused on mainstream business and organizational applications, advances in reinforcement learning are in general where there is considerable hype in the media – essentially this is the underlying technique that drives "industrial-scale" applications such as Google's Search Engine, autonomous vehicles, robotics, and gaming.

Wrap-up

Whether Reinforcement Learning is Machine or Deep Learning may be a moot point but yawning skill gaps means AI ambitions for most companies today are in establishing in-house capabilities in more prosaic Machine Learning solutions.

Deep Learning represents a step-up in organizational AI maturity and having reached the end of this chapter's accelerated run through on Machine Learning, we move onto how cloud data-fed neural networks have become mainstream and their specific application in Deep Learning solutions in our next chapter.

CHAPTER 5

Neural Networks and Deep Learning

At the heart of most of today's "flagship" or "hyped-up" applications of Artificial Intelligence is Deep Learning, and, specifically the use and predictive power of Artificial Neural Networks, or ANNs.

How these neural networks consume vast amounts of data in order to operate and ultimately achieve, often astonishingly sophisticated predictions is the subject of this chapter.

More specifically, the main takeaways of this chapter are knowing where to start with AI, from the historical context and the business/user journey from machine to deep learning to underlying Big Data/Cloud architectures necessary to productionize Deep Learning. To get there we will cover underlying stochastic process theory and how it relates to neural networks as well as the different types of Neural Networks used in Deep Learning. We will look at the deep learning mode lifecycle, techniques to model and infer from Big Data as well as critical diagnostics around the model training and testing process and the many levers we have at our disposal, such as activation functions, pooling layers, and dropout to achieve better performance.

Bringing everything together, ultimately the goal here is how to establish the right (best practice) artificial neural network architecture, data orchestration pipeline, and infrastructure for Deep Learning. We will cover many AI tools throughout the rest of this book and in our hands-on labs, but the main focus in this chapter will be around TensorFlow, Keras, and PyTorch.

B. Walsh, *Productionizing AI*, https://doi.org/10.1007/978-1-4842-8817-7_5

Introduction to Deep Learning

In our first section we start with an introduction, covering key concepts, historical milestones, the AI hype cycle, and the neural network architecture underpinning deep learning.

What Is Deep Learning

Before we embark on specifics of neural networks, we should address the relationship between these (often misunderstood) algorithmic structures for solving complex problems and the relatively recent concept of "deep" learning.

The common consensus today is that Deep Learning is essentially a subfield of machine learning based on artificial neural networks, while both are subfields of Artificial Intelligence. So, like machine learning, deep learning involves automating a machine to perform a (predictive) task, but this time the underlying algorithm is far more complex, and highly nonlinear. The name "deep" comes from the hidden layers within a neural network where much of the optimization and convergence process takes place.

Neural network development goes back to the 1950s, but today we have fast enough computers and data handling capacity to actually train large neural networks. It is this, together with inherent performance sophistication, that has really led to their explosion in popularity as a tool in businesses and organizations worldwide trying to get a handle on both revenue growth opportunities and cost optimization. The sky is essentially the limit – as evidenced by Andrew Ng, the co-founder of Google Brain; traditional algorithms performance is limited as data scales but performance in Deep Learning continues to improve as the volume of training data increases (Figure 5-1).

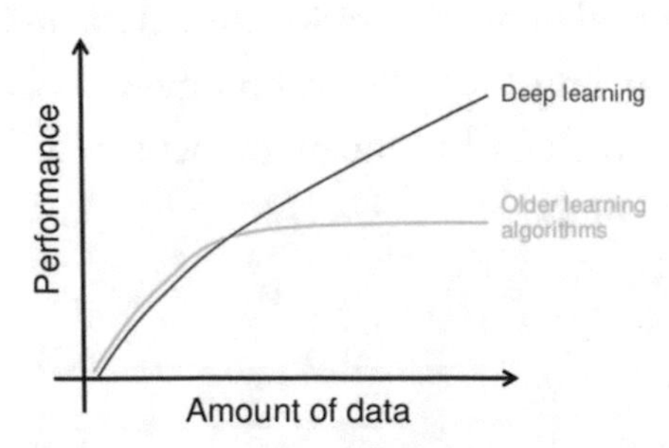

Figure 5-1. `https://machinelearningmastery.com/`

Most of the "hyped-up" AI applications discussed in the media today are Deep Learning solutions: from Driverless Cars (Autonomous Vehicles) to Search Engines, Computer Vision/Image Recognition, Chatbots, and Portfolio Optimization and Forecasting.

Deep Learning – Why Now?

While most people would consider the 1950s as a key milestone on the development journey of deep learning (by virtue of the birth of the modern computer), it is really only in the last 20-25 years that things have exploded. Much of this would not have been possible without the rapid improvement in semiconductor technology which has underpinned processing power in today's PCs and laptops. Below are some of the key dates during this period which have contributed to the mass-perception that Deep Learning is a key differentiator for digital-first businesses and organizations.

- 1997 IBM's Deep Blue defeats Garry Kasparov at Chess
- 2011 IBM Watson wins the quiz show "Jeopardy!"
- 2012 ImageNet competition - AlexNet Deep Neural Networks result in significant reduction in error in visual object recognition
- 2015 Facebook puts deep learning technology - called DeepFace - into operations to automatically tag and identify Facebook users in photographs
- 2016 Google DeepMind algorithm AlphaGo masters the complex board game Go and beats Lee Sedol, one of the world's best professional Go players

These events have been widely publicized by the media and when taken together with the increasing scalability of industry deep learning solutions, a snowball effect has essentially led to a broader market commissioning of Deep Learning implementation projects.

AI and Deep Learning Hype Cycle

While the above significant milestones for Deep Learning have given a "green light" to companies looking to generate a return on investment from an adoption of deep learning technologies and solutions, there remains considerable hype around AI as a whole as frequently illustrated in Gartner Hype Cycles such as the one below.

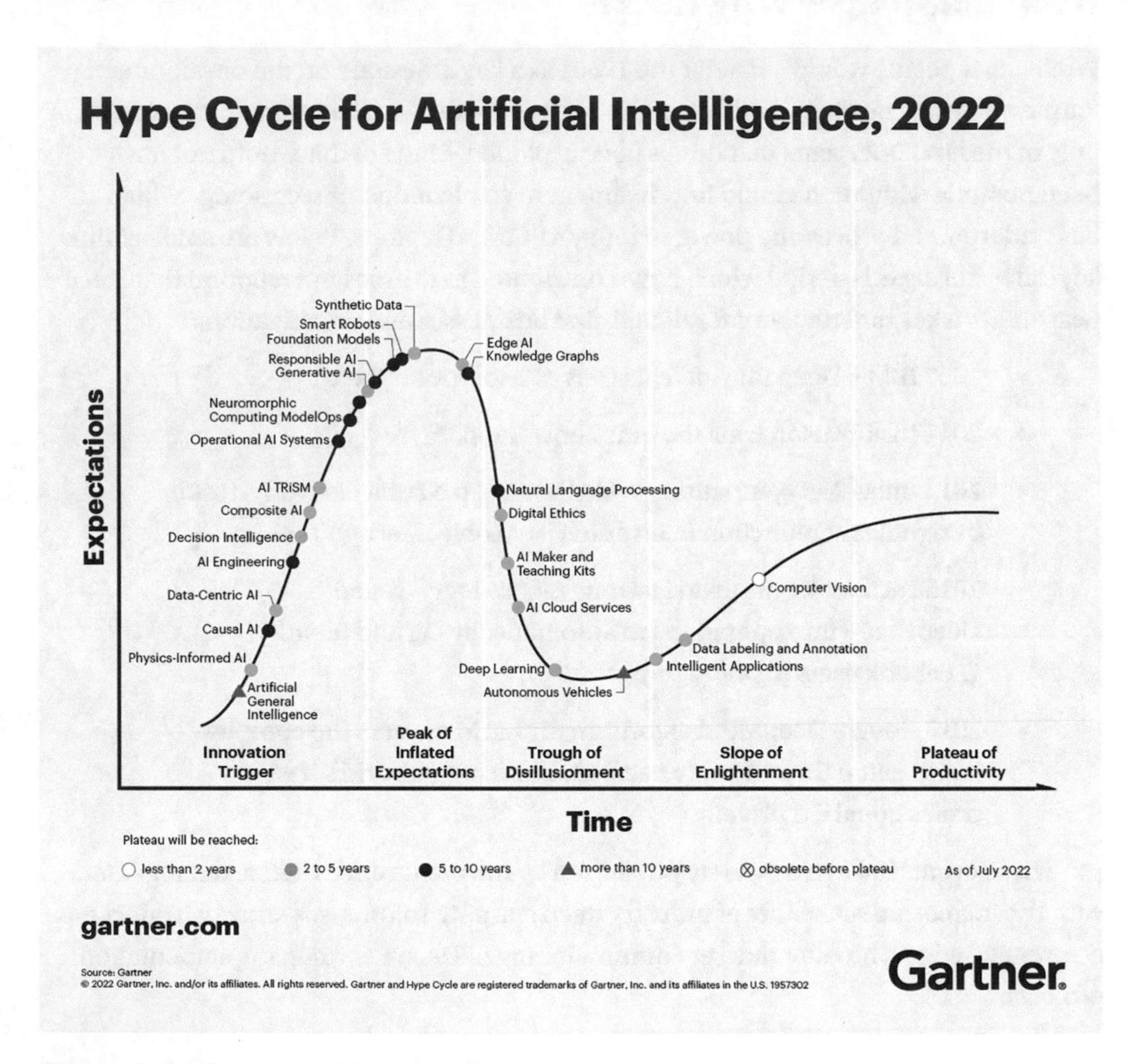

Figure 5-2. *Gartner AI Hype Cycle*

This hype among businesses is part ignorance, part science fiction. AI in the job-world means "Augmented" Intelligence, rather than Artificial Intelligence. Artificial Intelligence is too often confused with the "Artificial General Intelligence" of the movies and all its scary apocalyptic visions.[1]

Real "Augmented" intelligence is today generally perceived as delivering on its potential and the tangible benefits for businesses and organizations are a reality, driven by an existential need to accelerate digitalization during Covid such as:

- Provision of chatbot support during the pandemic
- Deep learning applied to healthcare diagnostics
- Computer vision supported social-distancing
- Machine Learning for modelling the effects of reopening economies

In 2022 and beyond, the focus has turned to the "democratization" of AI and Deep Learning post-Covid, shifting projects from expert/niche knowledge and a mountain of "technical debt" to achieving buy-in (and understanding) across a wider ecosystem of key stakeholders (all employees, customers). Similarly, the "industrialization" of AI has moved front and center, necessitating a push for reusability, scalability, and safety of AI solutions.[2]

High-Level Architectures

As described above, Deep Learning extends Machine Learning to the use of neural networks to solve hard and very large Big Data business problems. The computational overload is often achieved with parallel processing over a cluster or Graphical Processing Units (GPUs) to accelerate the internal calculations.

As we will see later on in this chapter, the heavy computation takes place inside an artificial neural network of ANN. These ANNs underpin Deep Learning and are inspired by the structure and function of the brain. A simple diagrammatic representation is shown as follows:

- Two hidden layers of four nodes (or neurons)
- Four inputs
- One output

[1] No-one really wants to see that happening, although plenty is being done in research labs around the world.

[2] These two trends are central to Chapter 7 on AI Application Development

Most ANNs and therefore Deep Learning models can trace their architecture back to this fundamental structure. To get a flavor of how these ANNs actually operate and how we can use them to perform predictions we will now proceed to our first hands-on lab in this chapter.

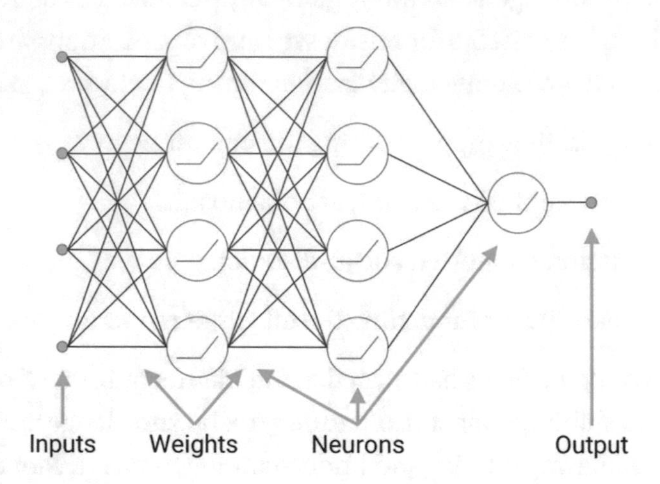

Figure 5-3. *Typical ANN architecture*

TensorFlow Playground: Hands-on Practice

No Deep Learning today can be undertaken without some knowledge of TensorFlow, Google's now open-sourced library for AI. We will go into much more detail on this tool later and the code itself (via Keras, Python's interface to TensorFlow), but we start here with a "big picture" look visually at what is happening when the tool is used to train an Artificial Neural Network (and thus Deep Learning model)

VISUAL DEEP LEARNING

The goal of this exercise is to get familiar with an Artificial Neural Network, how data is imported, and how it can be trained to predict an output:

1. Go to `https://playground.tensorflow.org`
2. Notice the data configuration on the LHS of the screen – there are four sample datasets to choose from, each of which is defined by a set of x1 and x2 coordinates

3. The neural network architecture is shown in the middle from left to right, that is, Features (transformations of the x1 and x2 coordinates in the dataset) through the network's Hidden Layers and Output
4. A number of hyperparameters (levers for fine-tuning the modeling process) are shown along the top – we will cover these later in this chapter
5. Notice that the Problem type is set to "classification" – we are trying to predict whether data belongs to the blue or orange group
6. Taking the first (concentric circles) dataset try to play around with (a) number of features, (b) number of hidden layers, and (c) number of neurons in each layer and observe how quickly the model converges (loss close to zero). Observe that the more hidden layers and neurons we have, the quicker the convergence
7. Exercise: try to train the model on the most complex / non-linear dataset (the spiral pattern) and see how long it takes (how many epochs) to achieve a loss < 0.05
8. Stretch Exercise: find a suitable configuration which achieves a loss < 0.01 in under 500 epochs

Stochastic Processes

Before we proceed further and examine the different types of neural networks leveraged for actual industry used cases today, we first need to take a look at some fundamental probability theory to understand the intricate processes happening in our ANNs.

Generative vs. Discriminative

A key concept in understanding how deep learning models work is the difference between a discriminative and a generative model.

Discriminative and generative models are used to perform both machine and deep learning. What distinguishes the two is that while discriminative models **make predictions based on conditional probability p(y|x)**, seek a decision boundary between classes and are used in supervised learning, generative models **predict conditional probability from a joint probability distribution p(x,y)** by explicitly modeling the **actual distribution** of each class and are used for unsupervised feature learning.

To understand the concept, an analogy is when considering how to classify speech as one of two languages. We can either use the difference within the linguistic models (discriminative) or learn each language (generative).

Because neural networks attempt to learn which features x will map to y, and therefore tries to discriminate among inputs, they use a discriminative model.

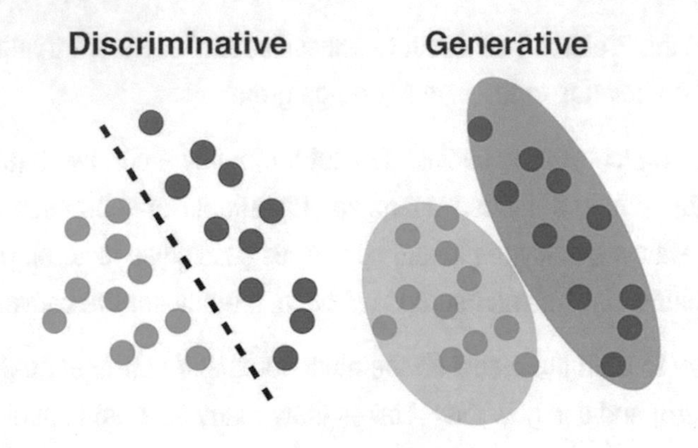

Random Walks

Another important concept for understanding neural networks is a random walk.

A random walk is a stochastic or random process describing a path that consists of a succession of random steps on a variable. A two-dimensional random walk can be visualized if we were to consider a person tossing a coin twice every time they are standing at a crossroads in order to determine where they can go ahead, turn back go left or right:

- heads, heads -> forward
- heads, tails -> right
- tails, heads -> left
- tails, tails -> backward

We will create this 2D random walk shortly in the hands-on lab for this section.

The important aspect here is that each of the outcomes from each step of a random walk have equal probability. Random walks are important for, for example, image segmentation, where they are used to determine the labels (i.e., "object" or "background") associated with each pixel. They are also used to sample massive online graphs such as online social networks

Markov Chains and Markov Processes

A Markov chain is a stochastic model describing a sequence of possible events in which the probability of each event depends only on the state of the previous event.

The defining characteristic of a Markov chain is that no matter how the process arrived at its present state, the possible future states are fixed. In effect, Markov chains are memoryless – the conditional distribution of future states of the process given present and past states depends only on the present state and not at all on the past states.

Consider another gambling analogy is helpful here – betting on the outcome of a sequence of independent fair coin tosses:

- If head, she/he gains one dollar.
- If tail, she/ he loses one dollar.

The stopping criteria is (a) if she/he reaches a fortune of N dollars or (b) if his purse is ever empty.

Systems modeled as a Markov process give rise to hidden Markov models (HMMs) – statistical models similar to the hidden layers/states in neural networks/deep learning.

To see how they work and how they can be applied to real-world phenomena, take a look at a Markov chain in action: `https://setosa.io/ev/markov-chains/`

Other Stochastic Processes: Martingales

The concept of Martingales is important in time series analysis, quantitative finance, algorithmic trading, and the betting industry.

Originally a class of betting strategies in 18th-century France, a gambler (a) wins their stake if a coin toss comes up heads, or loses if tails and (b) doubles their bet after every loss. The idea is that if a gamblers wealth and availability is infinite, the probability of eventually flipping heads approaches 1.

The statistical power of Deep Learning models, particularly Recurrent Neural Networks, bears similarities to ARMA/ARIMA time series models underpinned by Martingale theory.

Implementing a Random Walk in Python: Hands-on Practice

NEURAL NETWORK FUNDAMENTALS – STOCHASTIC THEORY

Using Python in Jupyter Notebook, the goal of this exercise is to implement a 2D random walk with Python:

1. Clone the GitHub repo below:

   `https://github.com/bw-cetech/apress-5.2.git`

2. Walk through the code sample:

   a. Import Python Libraries

   b. Define number of steps in your random walk and lists to capture the outcomes

   c. Try the exercise to create a 2D random walk using a For loop and four If statements

   d. Plot the outcomes and observe the fractal

   e. As a stretch exercise try to implement a random walk in 3D !

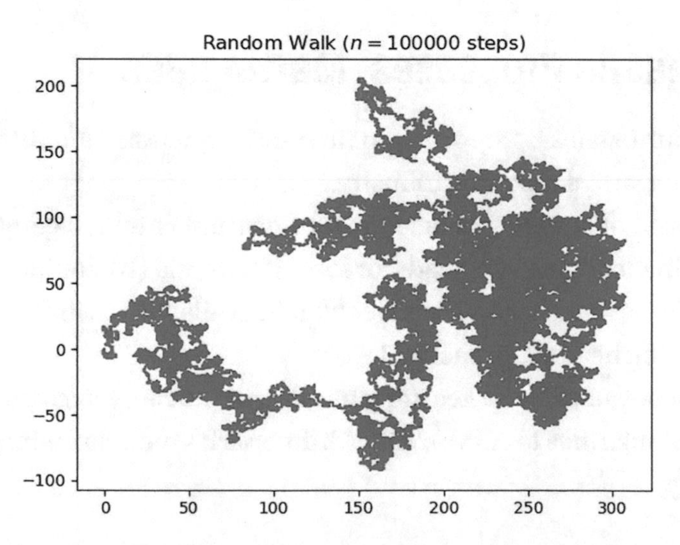

Figure 5-4. *2D Random Walk in Python*

Introduction to Neural Networks

Taking the above probability theory into our main focus – its application to neural networks, we now take a deeper dive in this section into what constitutes a neural network, looking at the different types of neural networks and what distinguishes them.

Artificial Neural Networks (ANNs)

As we will see below, Artificial Neural Networks, or ANNs, are biologically inspired networks able to extract hierarchical abstract features from underlying data. Essentially the generic name for any type of neural network used to solve a predictive problem, Artificial Neural Networks consist of an input layer, a hidden layer, and an output layer, taking as inputs various data points, feeding that data through the network and outputting some numerical value. All the main AI applications in use today including speech recognition, image recognition, machine translation, and forecasting make use of a type of Artificial Neural Network

As we will see, the number of ANN architectures and algorithms used in Deep learning is wide and varied. We will cover the main ones below, starting with the "vanilla" versions: the Simple Perceptron and the Multilayer Perceptron before proceeding to Convolutional Neural Networks and Recurrent Neural Networks – the two main types for AI applications in use today, from speech and image recognition to machine translation and forecasting.

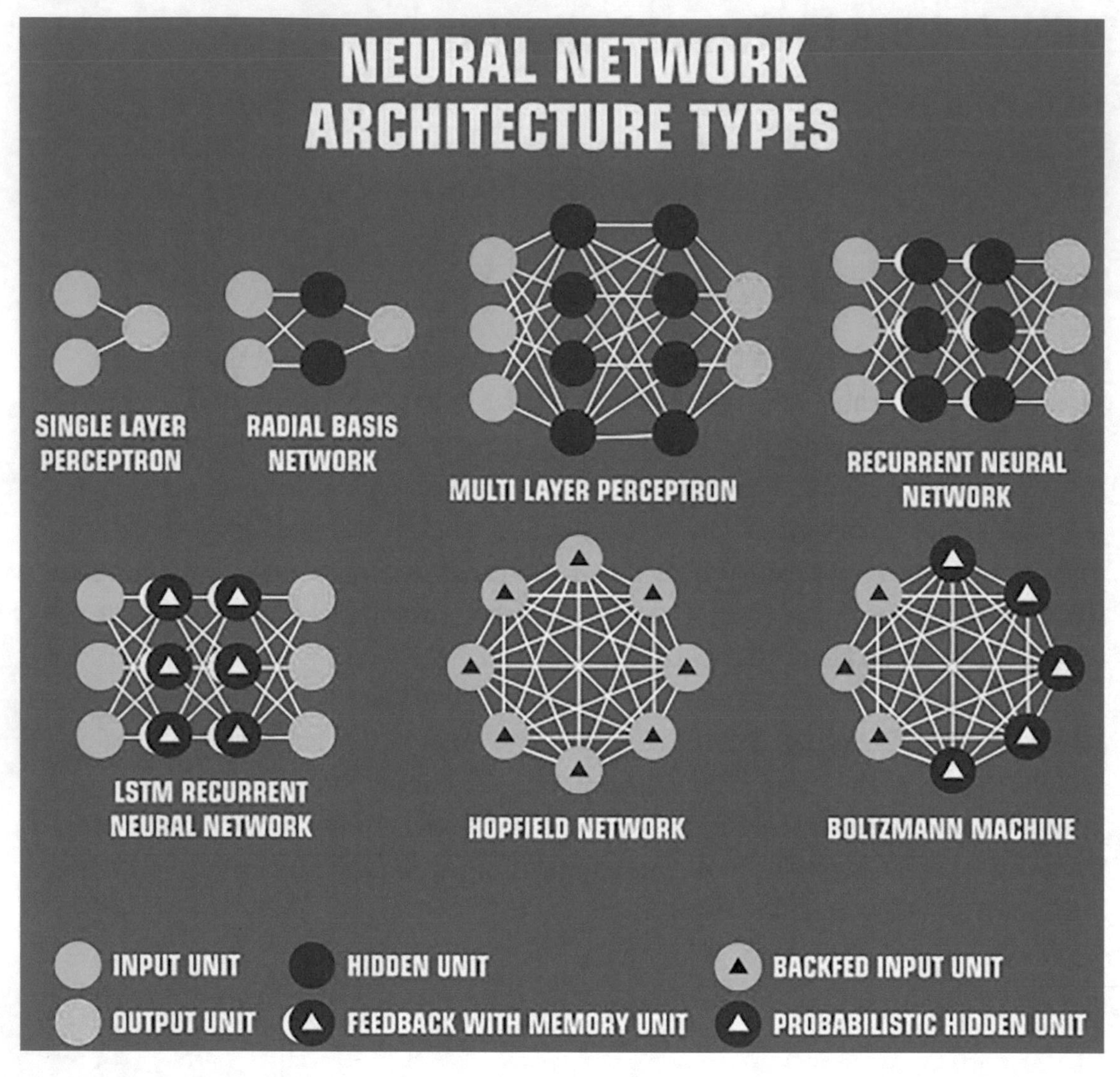

Figure 5-5. *Types of ANN*

The Simple Perceptron

The Simple (or Single Layer) Perceptron is the building block of all Artificial Neural Networks. Essentially it is a simplified model of the biological neurons in our brain.

Neurons in our brain have multiple inputs from other neurons and based on these inputs the neuron either fires off or it doesn't. A simple perceptron is much simpler – a single neuron which takes a number of inputs and learns a linear function to produce an output. An activation function (Heaviside step function) helps the simple perceptron classify the binary output (on or off).

If we structure these perceptrons in a multilayered network and enable each neuron to learn nonlinear functions using inputs/neurons from prior layers, we get a **multilayer perceptron (MLP)** and something that more closely resembles our brain.

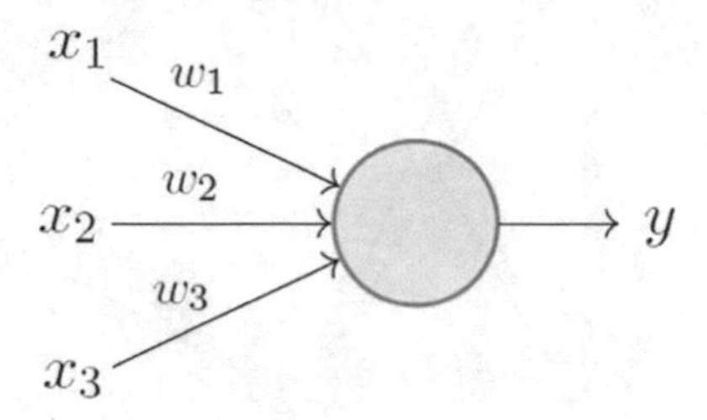

Figure 5-6. *Simple perceptron*

Multilayer Perceptron (MLP)

Formally, a Multilayer Perceptron (or MLP) is a class of feedforward ANN, essentially a neural network connecting multiple layers in a **directed** graph (or **finite acyclic graph**). Because it has multiple "hidden" layers of neurons, a Multilayer Perceptron is a type of **Deep Neural Network.** It is these hidden layers that distinguish between a machine learning model and a deep learning model.

The "direction" stems from the **forward pass** mechanism where input data is fed into the network, passing through the nodes (neurons) in only one direction (i.e., left to right) until we arrive at our output layer.

In contrast to a simple perceptron, each node in a Multilayer Perceptron, apart from the input nodes, has a **nonlinear activation** function, typically sigmoid, tanh, or ReLU. We will discuss these in the section below.

The "feedforward" nature of an MLP should not be confused with another important process which occurs after a forward pass, that of **backpropagation**. Backpropagation is a supervised learning algorithm for updating the weights in the network (based on the error between the prediction and the actual labeled data) after a forward pass.

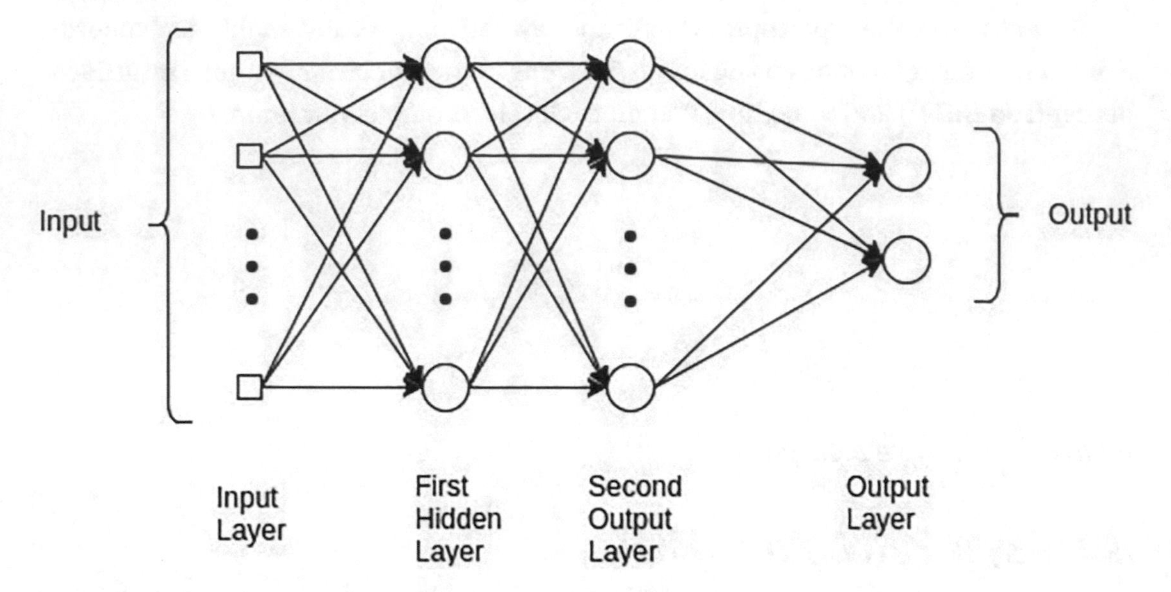

Figure 5-7. *Multilayer Perceptron as a class of feedforward ANN*

Convolutional Neural Networks (CNN)

Multilayer perceptrons take vector inputs and while capable of harder problems such as image classification (image pixels can be reduced to a single long vector of input data) are not themselves sufficient to carry out the vast majority of predictive challenges faced today.

In particular for Image Classification problems, since the seismic improvements presented at the ImageNet competition in 2012, Convolutional Neural Networks (CNNs) have become the preferred approach, with the image data fed in as tensors rather than vectors.

These CNNs are specialized neural network models designed for working with two-dimensional image data where learning spatial relationships are critical.

And in contrast to an MLP, where the number of weights in a neural network rapidly becomes unmanageable (because they are fully connected), a CNN will utilize "convolutions" and techniques such as pooling and dropout to reduce the number of weights.

A convolution is a mathematical way of combining two signals to form a third signal and in neural networks this is the simple application of a filter (i.e., **a set of weights**) to an input that results in an activation. The popularity of CNNs stems from their ability to automatically learn a large number of filters in parallel during the model training process and thereby **extract hierarchical patterns from input data** (such as identifying a person or building in a composite image).

Across the entire network, CNNs rely on several convolutional layers, repeatedly applying the same filter to the input data, resulting in a **feature map** of activations indicating the locations and strength of a detected feature in an image.

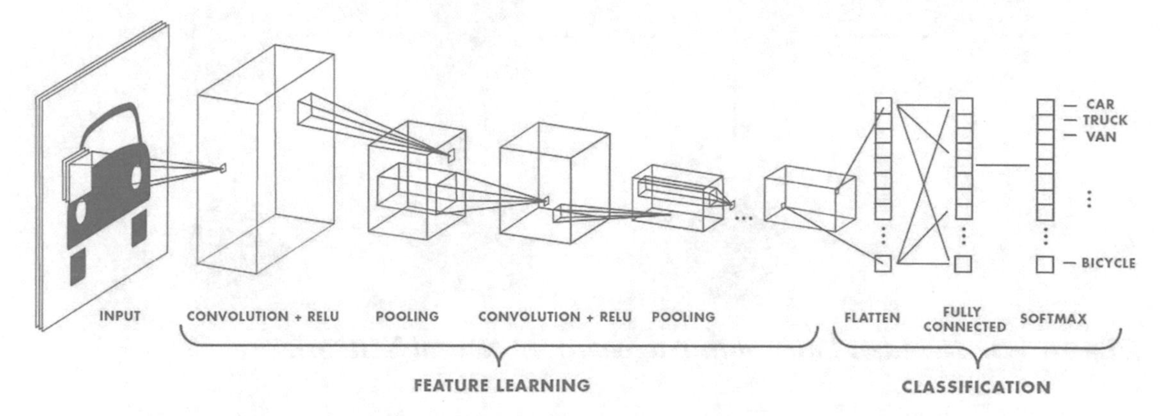

***Figure 5-8.** Example CNN used for Image Classification*

Recurrent Neural Networks (RNN)

While Convolutional Neural Networks are the artificial neural network of choice for Image Classification, Recurrent Neural Networks (RNNs) are the class of ANNs commonly used for ordinal or temporal problems, including Natural Language Processing, language translation, speech recognition, and forecasting.

Like feedforward networks such as CNNs, RNNs utilize the input/training data to learn with the connections between nodes forming a directed graph but this time along a temporal sequence. This allows RNNs to address an inherent sequential data weakness in feedforward networks, that is, when the order of the inputs matters.

As such, RNNs are distinguished by their "memory" as they take information from prior inputs to influence the current input layer and output. Effectively a **hidden state** is used to carry pertinent information from one input item in the series to others. Or put another way, and as shown in the diagram below, the outputs of recurrent neural networks depend on the prior elements within the sequence.

Recurrent Neural Networks are the ANN architecture behind popular applications such as Alexa, Siri, voice search, and Google Translate. High-accuracy, industrial-scale recurrent neural networks (RNNs) are also increasingly able to outperform traditional methods in **forecasting**, particularly specialized Long Short-Term Memory networks (LSTMs).

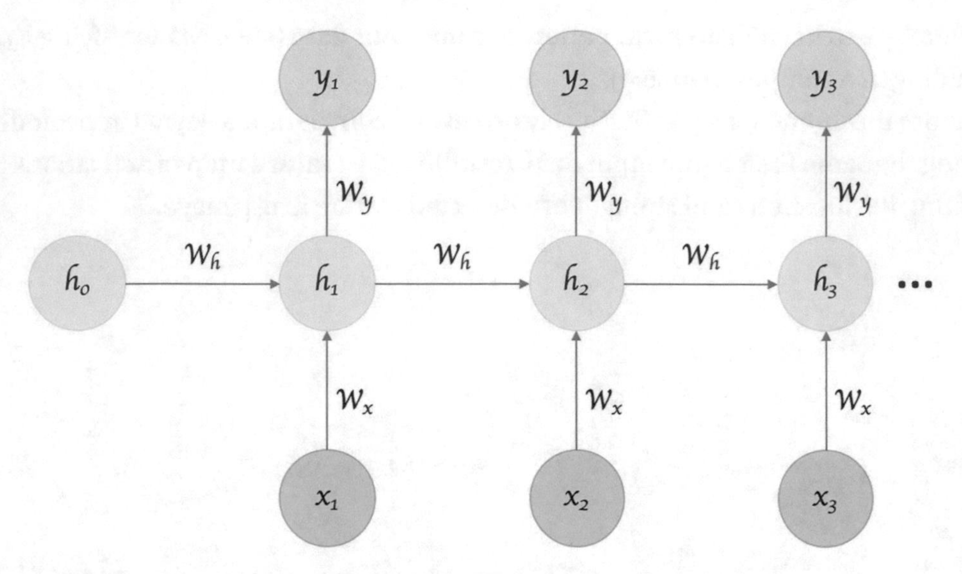

Figure 5-9. *"Hidden state" within a Recurrent Neural Network*

Long Short-Term Memory (LSTM) networks

LSTMs are a type of recurrent neural network capable of learning order dependence in sequence prediction problems.[3] LSTMs are distinguishable from more general Recurrent Neural Networks as they are able to store past information that is important, and forget the information that is not.

The internal "gates" of a Long Short-Term Memory network in effect enable LSTMs to learn much longer-term dependencies than a vanilla RNN. Whereas standard RNNs fail to learn in the presence of time lags greater than 5 to 10 discrete time steps, LSTM can learn to bridge minimal time lags in excess of 1000 discrete time steps by enforcing constant error flow.

LSTMs are commonly used for (time-series) forecasting but also in speech recognition and translation, by virtue of their ability to learn more sophisticated grammar relationships.

[3] Gated Recurrent Units (GRUs) also have this feature, but are less complex (two gates: reset and update) as opposed to three for an LSTM (input, output, forget). GRUs generally use less memory and can be faster than LSTMs but are less accurate with longer sequences.

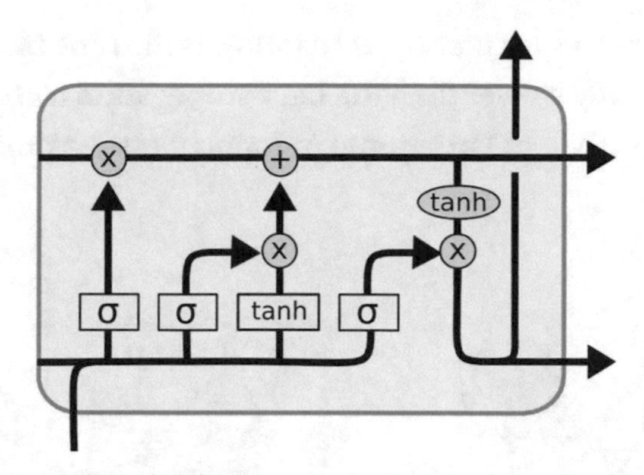

Figure 5-10. *LSTM Architecture*

Other Types of Neural Networks

Convolutional and Recurrent Neural Networks are used as deep learning architectures to train on Supervised Learning problems, that is, where we have prelabeled data (such as classified images or end-of-day stock prices for forecasting). This is the focus of our first hands-on exercise at the end of this section.

Before we go to that, we will take a quick look at other architectures which are generally applied to solving Unsupervised Deep Learning problems. These include **Restricted Boltzmann Machines (RBMs)**, **Deep Belief Nctworks (DBNs)**, and **Deep Boltzmann Machines (DBMs)** as well as **Autoencoders, Variational Autoencoders,** and **Generative Adversarial Networks**. Autoencoders are addressed in our second hands-on lab in this section.

Restricted Boltzmann Machines (RBMs)

RBMs are generative, stochastic two-layered artificial neural networks which learn a probability distribution over a set of inputs. There are only two types of neurons, hidden (h in diagram) and visible (v in diagram), all of which are connected to each other. There are no **output nodes.**

Whereas Boltzmann Machines have connections between input nodes, Restricted Boltzmann Machines are a special class with restricted connections between the visible and the hidden units. This allows for more efficient training using gradient-based contrastive divergence algorithms.

In general, RBMs are less frequently used today as the time needed to calculate probabilities is significantly slower than the backpropagation algorithm. Generative Adversarial Networks (GANs) or Variational Autoencoders are preferred, both of which are discussed below.

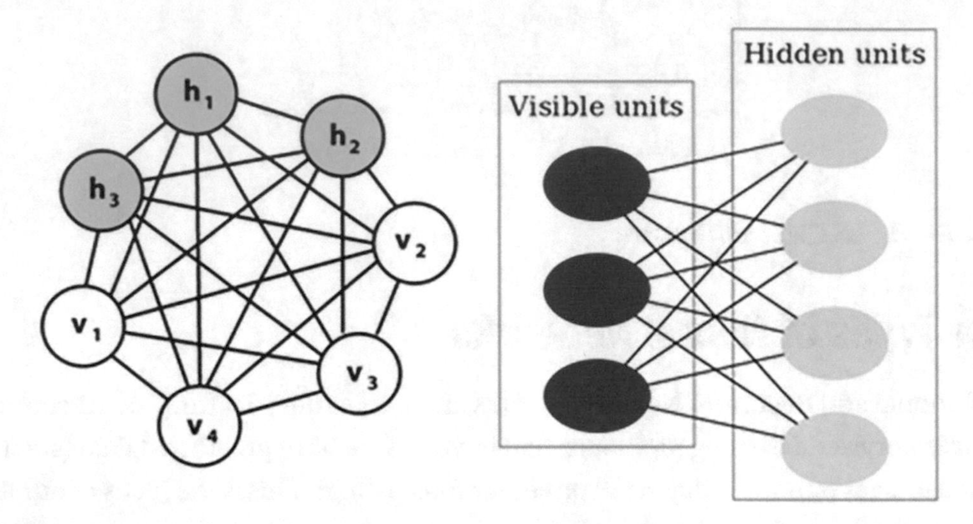

Figure 5-11. *Boltzmann machine vs. restricted Boltzmann machine*

Deep Belief Networks (DBNs)

If we stack Restricted Boltzmann Machines then fine-tune using **gradient descent** and **back-propagation**, then we have a Deep Belief Network. Essentially stacked layers of RBMs, the top two layers have undirected connections while lower layers are directed.

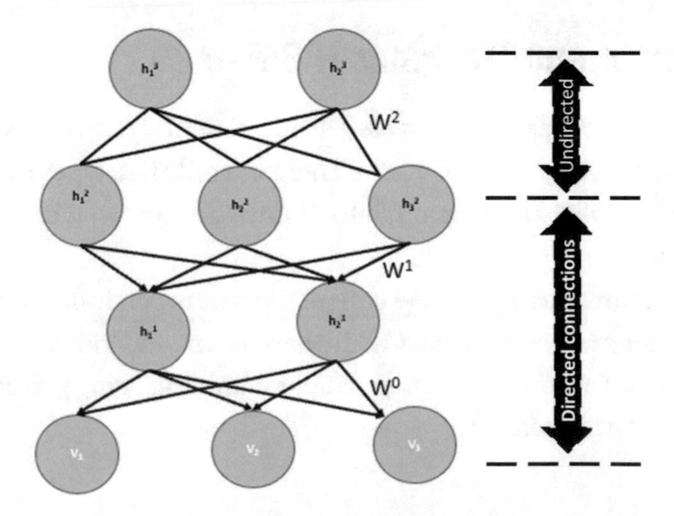

Figure 5-12. *Deep Belief Network (DBN)*

Deep Boltzmann Machines (DBMs)

A Deep Boltzmann Machine (DBM) is an **undirected** probabilistic graphical model with multiple layers of hidden random variables.

Like Deep Belief Networks, DBMs can **learn complex and abstract internal representations** of the input in tasks such as object or speech recognition. However in contrast to DBNs, training and inference processes are done in both directions, bottom-up and top-down, which allows DBMs to better interpret input data/features.

The learning process in Deep Boltzmann Machine can be slow, limiting their performance potential and functionality.

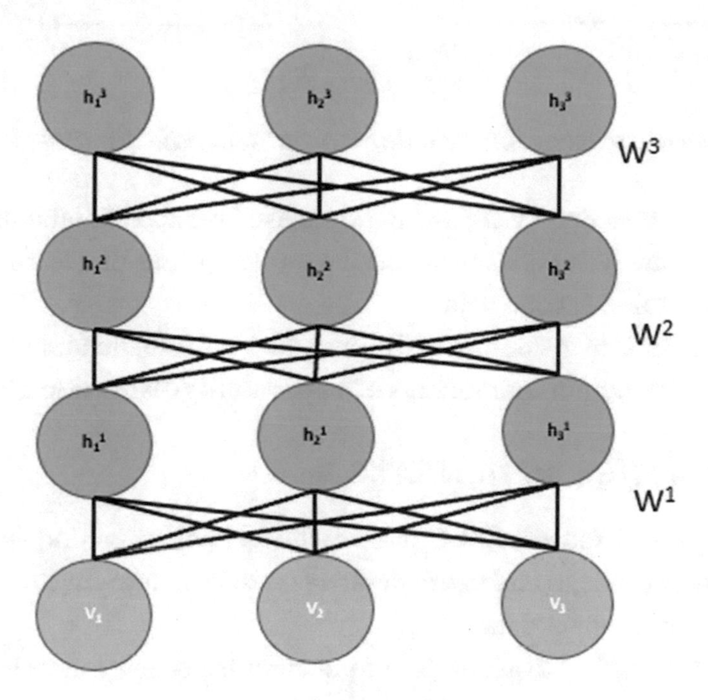

Figure 5-13. *Deep Boltzmann Machines (DBM) Example*

Autoencoders

An autoencoder neural network is a type of feedforward neural network. Specifically, it's an **unsupervised learning algorithm** that applies backpropagation and sets the target values to be equal to the inputs. Autoencoders are one component of OpenAI's exciting DALL-E Generative AI model which we will take a look at in a hands-on lab in Chapter 8.

Autoencoders compress the input into a lower-dimensional code then reconstruct the output from this representation. Architecturally they consist of three components: an encoder, code, and a decoder. The encoder compresses the input and produces the code while the decoder then reconstructs the input from the code as shown in Figure 5-13.

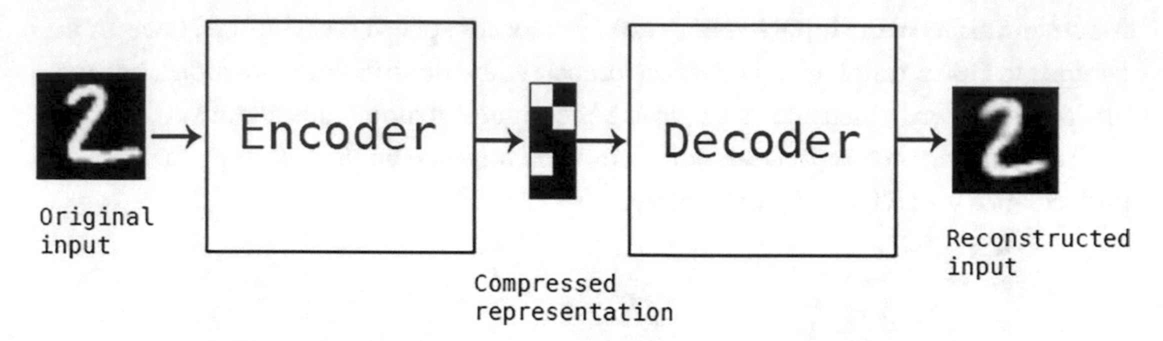

Figure 5-14. *Reconstructed image using an Autoencoder (Source: keras.io)*

Variational autoencoders (VAE) are an improvised version of autoencoders. Like autoencoders they are unsupervised artificial neural networks that learn how to efficiently compress and encode data

Where they differ is in the output of the encoder - an autoencoder (AE) outputs a vector whereas a VAE outputs parameters of a **probability distribution** for every input.

Generative Adversarial Networks

Generative Adversarial Networks, or GANs, are the technology behind Deepfakes. They create synthetic images and use generative modeling, auto-discovering/learning regularities/patterns in input data.

A GAN model is applied to generate/output completely new examples that are "credible" in comparison with the original dataset.

Although they employ unsupervised deep learning, building a Generative Adversarial Network is partly "framed" as a supervised deep learning problem. The underlying generator and discriminator submodels are trained together, with the generator generating a batch of samples which are then passed with a real dataset to the discriminator for classification. Because of the reliance on a real dataset in the discriminator model, the training process is benchmarked on a "supervised loss." The two models (generative and discriminator) are trained together in a zero-sum, adversarial game.

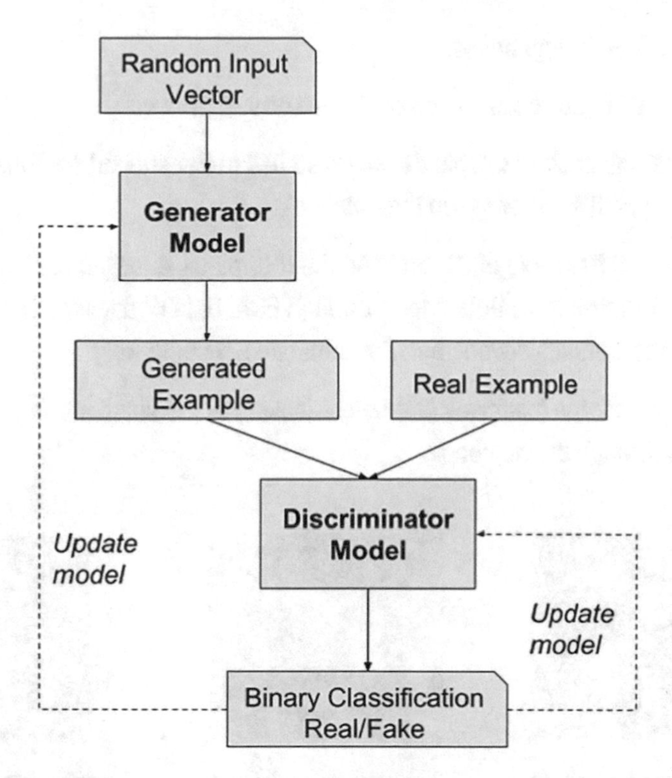

Figure 5-15. *Generative Adversarial Network – Modeling Process*

A Simple Deep Learning Solution – MNIST: Hands-on Practice

A FIRST LOOK AT THE DEEP LEARNING MODELING PROCESS

The MNIST (Modified National Institute of Standards and Technology) handwritten digits dataset is a standard dataset used in computer vision and deep learning.

Consisting of 60,000 small square 28×28 pixel grayscale images of handwritten single digits between 0 and 9, it's great as a basis for learning and practicing how to develop, evaluate, and use convolutional deep learning neural networks for image classification.

As such, the goal of this exercise is to implement a Convolutional Neural Network in Python (with Google Colab) to correctly identify (classify) the handwritten numbers.

1. Clone the GitHub repo below:

   `https://github.com/bw-cetech/apress-5.3.git`

2. Import the notebook to Google Colab. Colab has better support for TensorFlow (no install!) so it's easier to run the notebook

3. Take a look at the Table of Contents on the RHS panel, showing each part of the modeling process from Data Import through EDA, Data Wrangling, Building and Running the Neural Network, and Performance Benchmarking

4. Try to complete the exercises first before looking at the solutions as you progress through the notebook

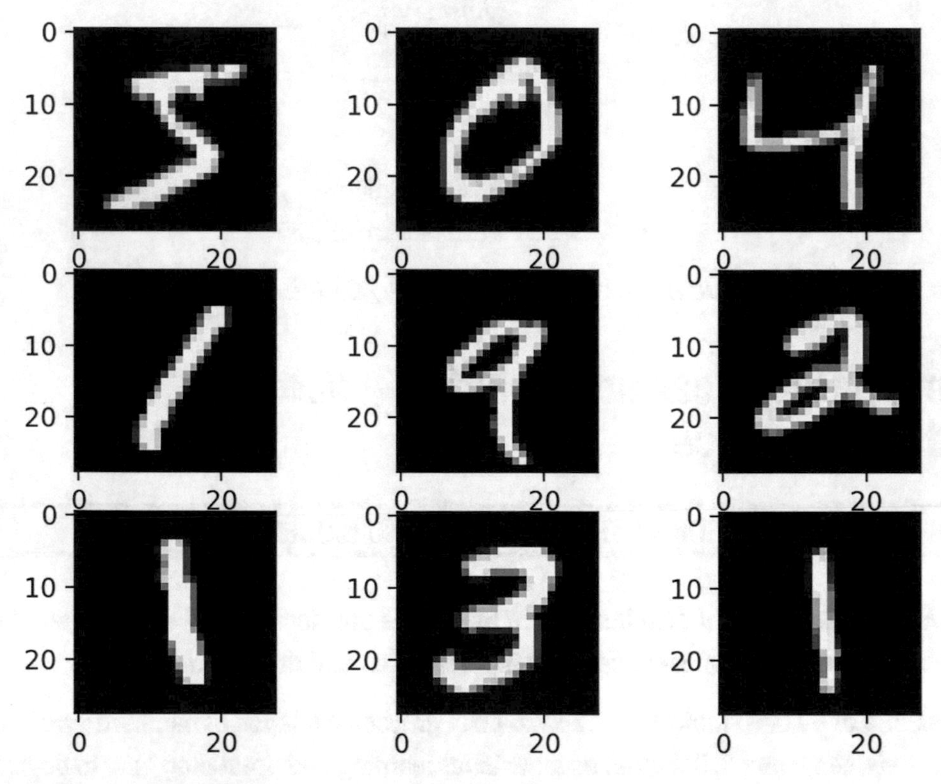

***Figure 5-16.** MNIST dataset*

Autoencoders in Keras: Hands-on Practice

UNSUPERVISED DEEP LEARNING

Our follow-up lab in this section explores Unsupervised Deep Learning using the MNIST Fashion dataset. The idea is to leverage the encoder, code and decoder architecture of an autoencoder to effectively "re-construct" specific images for a given class (here articles of clothing but the code is easily adapted to other image sets):

1. Clone the GitHub repo below:

   `https://github.com/bw-cetech/apress-5.3b.git`

2. Import and run the notebook in Colab
3. Run through the notebook and built-in exercises

   a. Perform EDA

   b. Normalize image data

   c. Add noise to the images for a more generalizable model

   d. Build and configure the model

   e. Reducing image noise and reconstruct images using the trained autoencoder

4. Compare runtime using Colab GPU runtime vs. a normal (CPU) runtime

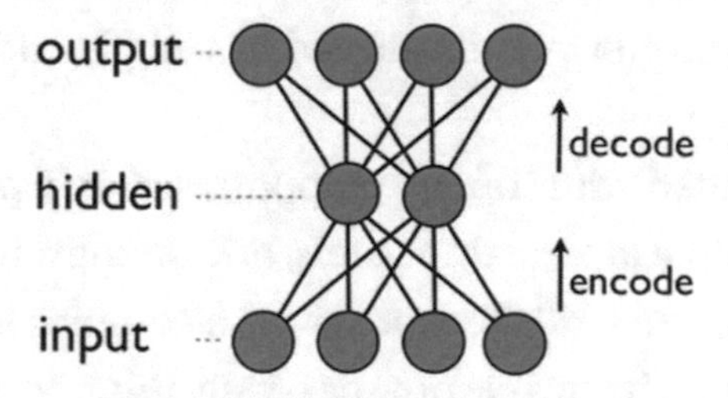

Figure 5-17. *Autoencoder: Encoder-Decoder model*

Deep Learning Tools

Before proceeding to how to implement an artificial neural network in Python, let's take a look at some of the main libraries and tools required to support Deep Learning.

Tools for Deep Learning

Any cursory glance at the latest TIOBE Programming Community index shows continuing Python popularity and as of June 2022 it is still the number 1 software programming language.

As the "go-to" programming language for Machine and Deep Learning, investment in key integration with all the main Deep Learning libraries and APIs is central to the development pathway as is strong functionality and features for data mining, particularly with the main data wrangling library, Pandas.

Given programming AI in Python is the theme of this book and we have already covered several hands-on Python labs, we assume a certain amount of knowledge of the language already. The rest of this section is concerned with the open-sourced tools we need to build neural networks with convolutional layers or recurrent loops and simplify the deep learning process.

TensorFlow

The favorite tool of many industry professionals and researchers, TensorFlow is an end-to-end deep learning framework developed by Google and released as open source in 2015.

Though well-documented with training support, scalable production and deployment options, multiple abstraction levels, and support for different platforms (such as Ubuntu, Android, and iOS), TensorFlow is a complex low-level programming language, coding of which on its own is not massively intuitive to Data Scientists.

A focus in the last 5-10 years on TensorFlow development means that accelerated improvements in research, flexibility, and speed have led to uptake by many of the major brands and corporations recognized today, including Airbnb, Coca Cola, GE, and Twitter.

No longer just the low-level language, Google improved integration with Python runtime in its last major (TensorFlow 2.0) release in 2019. Today's TensorFlow ecosystem encompasses Python, JavaScript (TensorFlow.js), and mobile (TensorFlow Lite) development as well as TensorFlow Extended (TFX) for ML Production Pipelines/ deployment, TensorBoard for model and results visualisation and TensorFlow Hub for pretrained models.

There are multiple ways and IDEs in which to install and import TensorFlow, but in this book we will use what we consider is by far the easiest – Google Colab. Because of its linkage to its creator, enhanced user workflow and TensorFlow support on Colab means we can just simply import TensorFlow as a library without installing the code. Colab also comes with GPU support, important for accelerating, scaling, and productionizing Deep Learning models.

Keras

Developed by MIT, Keras was open-sourced in 2015 to address the challenges around ease of use coding deep learning model training and inference in less user-friendly "back-end' languages, such as the syntax used in TensorFlow, Theano, and CNTK,[4] the latter two of which are now deprecated.

A high-level, modular and extensible API written in Python focused on fast experimentation, Keras is the high-level API of TensorFlow 2 and the most used Deep Learning framework among top-5 winning teams on Kaggle.

It's "drag-and-drop" style of programming maintains back-end linkage to TensorFlow and accelerates the design, fit, evaluation, and use of deep learning models to make predictions in a few lines of code.

The actual implementation of the Keras API on TensorFlow is codified as tf.keras with set up in Python is simple:

```
import numpy as np
import tensorflow as tf
from tf import keras
```

The core data structures of Keras are **layers** and **models** with the simplest type of model a **sequential model**, a linear stack of layers. We will discuss these layers and the implementation of a deep learning model in Keras more in the following section of this chapter.

[4] aka Microsoft Cognitive Toolkit

PyTorch

PyTorch is the last of the main Deep Learning "tools" we will cover in this book and is predominantly used for Deep Learning with Natural Language Processing.[5]

Developed by Facebook's AI research group and open-sourced on GitHub in 2017, PyTorch claims to be simple and easy to use and flexible, as well as possessing efficient memory usage with dynamic computational graphs. PyTorch is faster and has excellent debugging support, however claims to be easier to use should be taken with a pinch of salt- it is a low-level language. PyTorch integration with a simpler DL framework like Keras does not yet exist although PyTorch Lightning as a lightweight wrapper providing a high-level interface to PyTorch may come into its own as an alternative in the coming years.

In this author's opinion, implementation of PyTorch is easier using Jupyter Notebook on Windows. We will install PyTorch using the one-off shell command (directly in a cell in a Jupyter notebook):

```
%pip install torch
```

After commenting out the above Python code line, the installation can then be verified by running the sample PyTorch code:

```
import torch
x = torch.rand(5, 3)
print(x)
```

where, if successful, a random tensor output should be displayed on the screen

Other Important Deep Learning Tools

Of course, TensorFlow (with Keras) and PyTorch are not the only Deep Learning frameworks in use today. An abundance of tools exist, many of which in their own right may produce better results or performance when focused on a specific use case.

Eclipse Deeplearning4J, also known as **DL4J**, for example, is a set of projects intended to support all the needs of a Java Virtual Machine (JVM) based deep learning application.

[5] We will cover more on NLP in Chapter 10

In terms of Python libraries, while both **Theano** (developed by Université de Montréal) and **Microsoft Cognitive Toolkit (CNTK)**[6] have seen their day and are now deprecated, **Caffe**, developed by the University of California, Berkeley remains in use, particularly for Image classification/segmentation. Caffe is strong on NVIDIA GPU integration, although somewhat difficult to debug.

MXNet, from the incredibly large Apache toolset ecosystem, has multilanguage support for Deep Learning and high-level API support.

Apache Spark

While MXNet is more of a niche Deep Learning tool, Apache Spark is a popular unified analytics engine for big data processing, with built-in modules for streaming, SQL, machine and deep learning, and graph processing.

Because of its inherent "scalability" around Big Data processing and its built-in interface for programming entire clusters with implicit data parallelism (so a great fit for distributed training on large datasets) and fault tolerance, Data Scientists have jumped on Apache Spark as an accelerator for productionizing Deep Learning.

Many APIs exist to simplify the implementation and usage of Apache Spark in Python, including PySpark (defacto Python API for Apache Spark) and SparkTorch (for running PyTorch models on Apache Spark).

Probably the most important of the "other" tools important for productionizing Deep Learning, Apache Spark is the focus of a hands-on lab in the next chapter of this book.

Frameworks for Deep Learning and Implementation

Now that we have established the main Python integration tools for Deep Learning, we can proceed to the main focus of this chapter – how to implement a Deep Learning model.

[6] Azure Machine Learning and Azure Cognitive Services are now main Microsoft's main AI tools. We will discuss these more in Chapter 6: The Employer's Dream AutoML, AutoAI, and the rise of NoLo UIs

Tensors

Both TensorFlow and PyTorch utilize **tensors**, so it's helpful to understand this important data structure before proceeding to how they are constructed and used in a deep learning model.

In effect, a **tensor is a** multidimensional array with a uniform data type. They are similar to numpy arrays but are immutable, much like a standard Python "tuple." Tensors can also be stored in GPU memory as opposed to standard CPU memory, optimizing usage for deep learning.

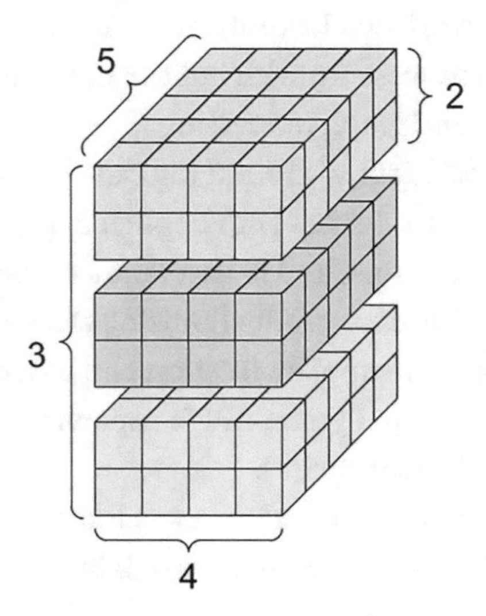

Figure 5-18. *rank-4 tensor with shape: [3, 2, 4, 5]*

Key TensorFlow Concepts

TensorFlow 2.0 uses **eager execution** by default to evaluate operations immediately, and therefore calculates the values of tensors as they occur in the code. Similar to the concept of imperative programming, eager execution involves step-by-step operations executed by Python then results returned back to Python (immediately).

The other type of execution method in TensorFlow is **graph execution.** Here tensor computations are executed as a dataflow graph **tf.Graph** through **tf.operations** graph nodes.

As this is a book on Python for productionizing AI applications, we focus on eager execution in TensorFlow rather than graph execution. Best practice in general is to develop and debug in eager mode, then decorate/productionize with TensorFlow's **tf.function** which converts regular Python code to a callable Tensorflow graph function generally offering better performance.

TensorFlow also makes use of various **execution pipelines** for performing multiple/concurrent tasks in the deep learning process:

- **tf.data** - API enables reuse of complex input pipelines to, for example, aggregate data from multiple files, apply stochastic "noise" to images and batch random images for training
- **TFX Pipelines** - extends this to end-to-end MLOps (input ➤ train ➤ deploy ➤ track)

The Deep Learning Modeling Lifecycle

Moving on to initial implementation, as we go through the data import, vectorize with numpy, model building, and model running processes, we know that modeling will fail unless the data is transformed into a tensor before using TensorFlow eager execution to experiment, visualize immediate results and debug.

In order to traceback coding (and troubleshooting) to the underlying process, it is useful to think of the Deep Learning modeling lifecycle as having five distinct phases:

- Define the model
- Compiling the model
- Fitting the model
- Model evaluation
- Predictions/Inference

These phases can be further subdivided into subprocesses broadly relatable to Keras/TensorFlow operations and defined via a single line of Python code.

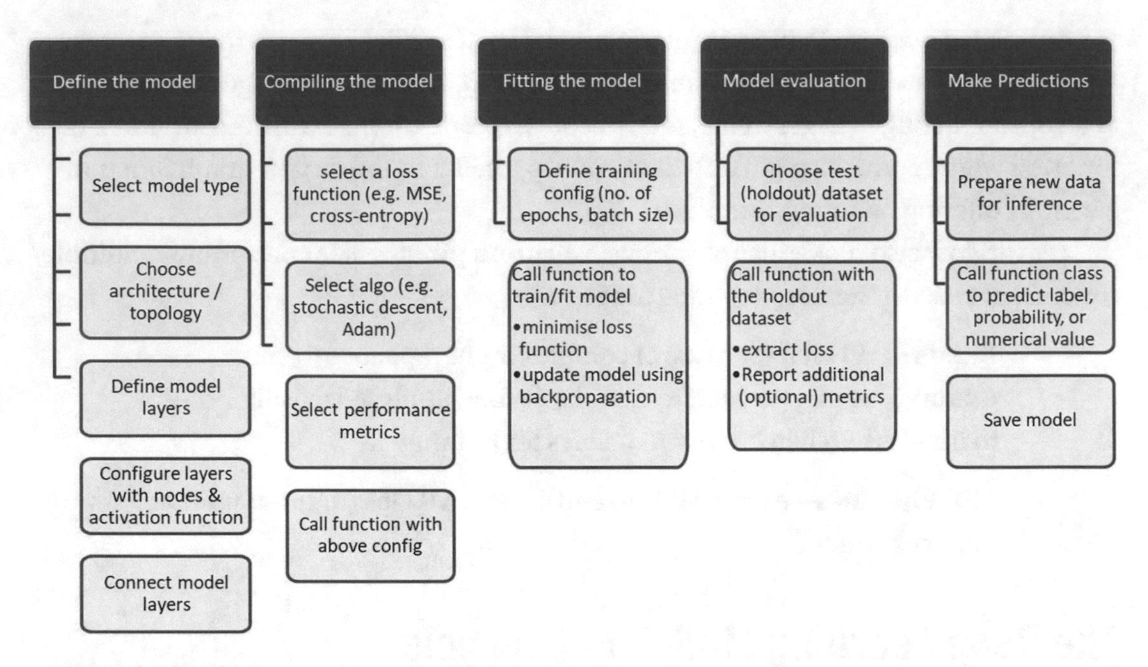

Figure 5-19. *Deep Learning modeling lifecycle (Source: machinelearningmastery)*

Example code mirroring the above DL phases:

```
# define the model
model = ...
# compile the model
opt = SGD(learning_rate=0.01, momentum=0.9)
model.compile(optimizer=opt, loss='binary_crossentropy')
# fit the model
model.fit(X, y, epochs=100, batch_size=32, verbose=0)
# evaluate the model
loss = model.evaluate(X, y, verbose=0)
# make a prediction
yhat = model.predict(X)
```

Sequential and Functional Model APIs

In the process and Python code above, in order to build a model we first need to define it. In the Keras API (tf.keras) there are in fact two main ways to do this: the sequential model and functional model.

Sequential Model API

The simplest approach, the Sequential model works on neural network layers where each layer has exactly one input tensor and one output tensor.

To implement this in Python, as shown in the code below, we add layers to the model one by one in a linear manner using model.add().

Example of a model defined with the sequential API

```
from tensorflow.keras import Sequential
from tensorflow.keras.layers import Dense
# define the model
model = Sequential()
model.add(Dense(10, input_shape=(8,)))
model.add(Dense(1))
```

Functional Model API

The Keras functional API is more complex but more flexible than the sequential API. Rather than linearly adding layers, here we use a model object specifying the input and output layers and explicitly connect ALL outputs of one layer to the inputs of another layer (each connection is specified)

Example of a model defined with the functional API

```
from tensorflow.keras import Model
from tensorflow.keras import Input
from tensorflow.keras.layers import Dense
# define the layers
x_in = Input(shape=(8,))
x = Dense(10)(x_in)
x_out = Dense(1)(x)
# define the model
model = Model(inputs=x_in, outputs=x_out)
```

Implementing a CNN

The above sections cover the generic deep learning implementation process but how does the process go when we are specifically building, for example, a Convolutional Neural Network for Image Classification? Machine and Deep Learning modeling is highly iterative and building a performant Convolution Neural Network is about as difficult as it gets.

As we will see shortly in the hand-on lab following this section, adopting a step-by-step process while budgeting for a degree of iteration around EDA and Data Wrangling is key to getting (a) the model to run and (b) performance to meet acceptable criteria (e.g., fbeta > 0.9).

After importing the data and performing basic Exploratory Data Analysis, there are a number of requirements unique to the data wrangling and model set up for a CNN. These are described below with the technical subprocesses covered in the next section in this chapter.

1. **Preprocessing and partitioning**
   - Image to data matrix/tensor conversion
   - Partitioning of the dataset into training and test sets
   - Conversion of the tensor data format to a float and rescaling
2. **Build network**
   - Initialize filters for the convolutional base (convolutional and **pooling**[7] layers)
   - Define activation functions to compute gradients/enable backpropagation
   - Update weights for the dense layers
3. **Compile network**
   - Combine forward and backward operations and build NN

The remaining steps are for the most part in line with the generic process shown in the Deep Learning lifecycle above, from training the network, through evaluation on the test set (checking such metrics as loss, accuracy, confusion matrix, and classification report). Often training and evaluation is repeated until we hit our "target" performance, with **dropout** often added to address overfitting.

[7] More on pooling in the next section

Typical Keras CNN implementation:

```
model = models.Sequential()
model.add(layers.Conv2D(32, (3, 3), activation='relu', input_shape=(32,
32, 3)))
model.add(layers.MaxPooling2D((2, 2)))
model.add(layers.Conv2D(64, (3, 3), activation='relu'))
model.add(layers.MaxPooling2D((2, 2)))
model.add(layers.Conv2D(64, (3, 3), activation='relu'))

model.add(layers.Flatten())
model.add(layers.Dense(64, activation='relu'))
model.add(layers.Dense(10))
model.compile(optimizer='adam',
          loss=tf.keras.losses.SparseCategoricalCrossentropy(from_
          logits=True), metrics=['accuracy'])

history = model.fit(train_images, train_labels, epochs=10, validation_
data=(test_images,test_labels))
```

Implementing an RNN

There are actually three types of Recurrent Neural Network "layers" we can leverage in Keras:

keras.layers.SimpleRNN

This is the "vanilla" version – essentially a fully connected RNN where the output from previous timestep is to be fed to next timestep.

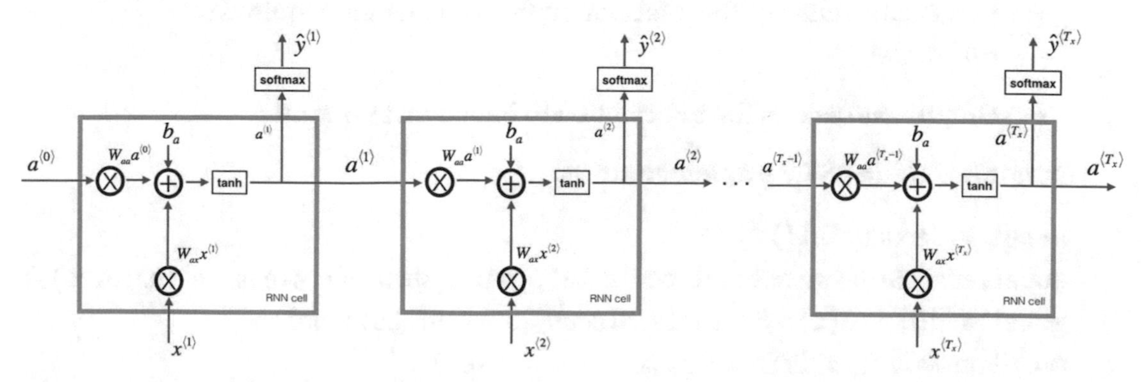

Figure 5-20. *Recurrent Neural Network – SimpleRNN*

keras.layers.GRU

Gated recurrent units (GRUs) are gating mechanisms in recurrent neural networks intended to solve the **vanishing gradient** problem in standard RNNs. Via an update gate and a reset gate, a GRU "vets" information before passing to an output, thus avoiding diminishing gradients and inflexible (unchanging) weights in a simpleRNN. We will discuss vanishing gradients in more detail in the next section 6 "Tuning a DL Model" in this chapter

Example Keras RNN implementation of GRU and SimpleRNN layers:

```
model = keras.Sequential()
model.add(layers.Embedding(input_dim=1000, output_dim=64))

# The output of GRU will be a 3D tensor of shape (batch_size,
timesteps, 256)
model.add(layers.GRU(256, return_sequences=True))

# The output of SimpleRNN will be a 2D tensor of shape (batch_size, 128)
model.add(layers.SimpleRNN(128))

model.add(layers.Dense(10))
model.summary()
```

keras.layers.LSTM

As we know from Section 3 in this chapter, LSTMs generally help achieve better forecasts with "longer-term effects." This "memory" in the network is achieved through the persistence of hidden state through three gates:

- Input gate adds information to the cell state
- Forget gate removes the information that is no longer required by the model
- Output gate selects the information to be shown as output

Example Vanilla LSTM implementation:

```
model = Sequential()
model.add(LSTM(50, activation='relu', input_shape=(n_steps, n_features)))
model.add(Dense(1)) # single hidden layer of LSTM units
model.compile(optimizer='adam', loss='mse')
```

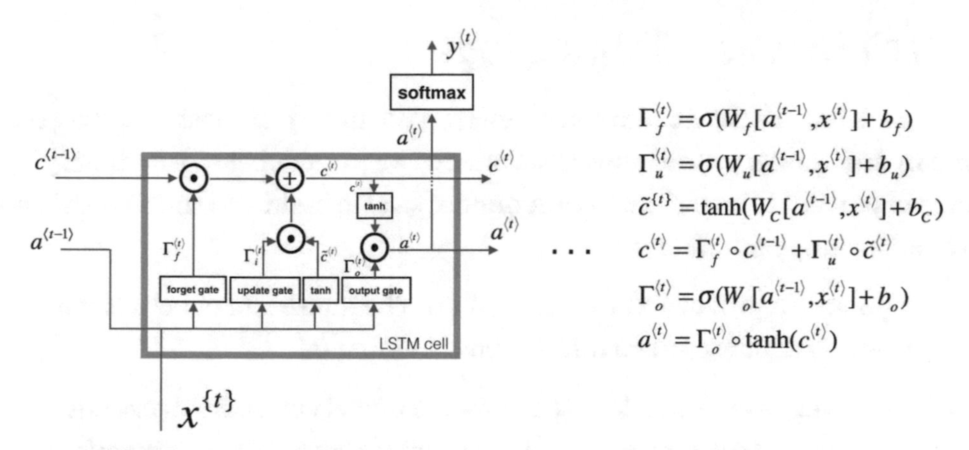

Figure 5-21. *LSTM Architecture*

LSTM Implementation for Time Series

We will be building an LSTM in our second lab in this section shortly but before we do so, the wrangling process for data used in time series forecasting with LSTMs (or any RNNs) should be addressed.

In time series forecasting, previous time steps actually become "features" for the deep learning process, with the target being the current (training) or next (inference) time step. If we are going to forecast forward (for example) 60 time steps, then each row of data presented to the LSTM needs to include the data from previous time steps going back to T-60. This is shown in the following table, where the column labeled "Time Step" is effectively our target variable and the other columns are our Feature set.

Time Step	Feature 1	Feature 2	.	.	.	Feature 60
T-60	T-120	T-119	T-118			T-61
T-59						
T-58						
.						
.						
T-0	T-60	.	.			T-1

Neural Networks – Terminology

In the next section, we will go into some detail on the more technical modeling concepts involved in fine-tuning neural networks. For now, we present below a summary reference of the main "levers" we have at our disposal to build and run a model and achieve satisfactory performance:

- **Epoch** — a pass over the entire dataset. The number of epochs is the number of times you go through your training set
- **Learning rate** — a scalar used to train a model via gradient descent. Learning Rate determines how fast weights change for each iteration
- **Activation function** — the output of a node/neuron, given a set of inputs, which is then used as input for the next node. Examples include ReLU, tanh, sigmoid, and linear
- **Regularization** — a technique (and hyperparameter) which changes the learning algorithm to prevent overfitting. Regularization Rate is also used to specify the rate of regularization
- **Batch size** — a small, randomly selected subset of the data run in a single iteration. 1 epoch (a pass thru entire training set) = batch size * no. of iterations/steps
- **Hidden layer** — a layer between input and output layers, where neurons take weighted inputs and use the activation function to produce an output
- **Loss function** (or a cost function) - by how much does the prediction given by the output layer deviate from actual (ground truth)? The objective of a neural network is to minimize this

Computing the Output of a Multilayer Neural Network

A final word in this section is reserved for some guidance around the computational process within a neural network. This is described below for calculating the output of a multilayer neural network $\hat{y}$ (y_hat).

For an input X with m features:

- m features have m weights (w1, w2, ..., wm)
- we can take the dot product of the input and the set of weights then add bias:

z = w1x1 + w2x2 + ... wnxn + bias

- z is then fed into an activation function to get hidden output h: (h1, h2, ..., hn)
- hidden layer has n neurons
- n neurons have n weights (w1, w2, ..., wn)

Which gives us a final output y_hat = h1w1+ h2w2 +...+ hnwn

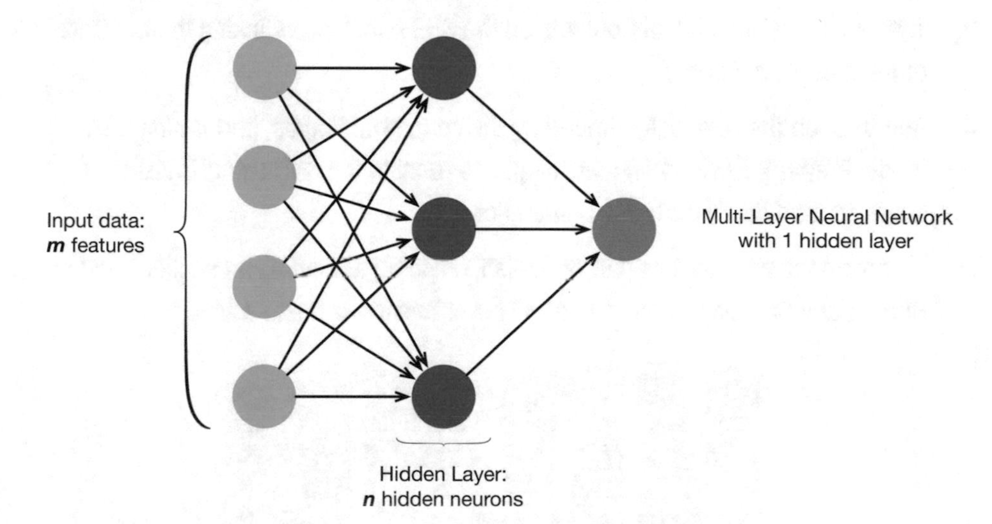

Figure 5-22. *Multilayer Neural Network – how do we get our output from the input data?*

Convolutional Neural Networks with Keras and TensorFlow: Hands-on Practice

GERMAN TRAFFIC LIGHT IMAGE CLASSIFICATION

In this lab, we will build and train a Convolutional Neural Network in Keras and TensorFlow to solve a predictive (multiclass classification of complex images) problem:

1. Clone the GitHub repo below:

   `https://github.com/bw-cetech/apress-5.5.git`

2. Import the notebook to Google Colab. Colab has better support for TensorFlow (no install!) so it's easier to run the notebook
3. Take a look at the Table of Contents on the RHS panel, hyperlinking to each part of the modeling process
4. Run through the notebook – importing the code from Kaggle, performing EDA, model Prep/image wrangling, building, and running the CNN using Transfer Learning, and finally performing inference
5. Complete the exercises as part of the lab, stopping at markdown section "2nd Run – Early Stopping Criteria" (we will cover this in the next section)

Figure 5-23. *Image Classification – German Traffic Signs*

Recurrent Neural Networks – Time Series Forecasting: Hands-on Practice

STOCK PRICE FORECASTING WITH UNIVARIATE RNNS

The goal of this lab is to build and train an LSTM for stock price (univariate) forecasting, that is, where the forecasted stock price depends only on previous time steps:[8]

1. Clone the GitHub repo below:

   `https://github.com/bw-cetech/apress-5.5b.git`

2. Import the notebook to Google Colab
3. Run through the notebook – uploading the data into a (temporary) Colab folder as described and importing, performing EDA and Data Wrangling in order to transform the stock price data into the format needed to forecast with RNNs
4. Complete the exercises as part of the lab
5. Exercise - retrain the model with a different batch size and compare the (root mean square) error
6. Exercise – retrain the model with a different activation function (tanh) and compare the test result (tanh generally regulates better the output of a recurrent neural network than ReLU)
7. Stretch Exercise – adapt the TATA univariate forecast to carry out a live stock price forecast by completing the exercise commented out under "Live Stock Price Scenario". Run thru the steps to import the latest stock price price for a leading Tech Stock (e.g. Apple or Tesla) from Jan-21 to D-1 and performing a forecast 30 time steps into the future

[8] Multivariate would involve the target variable (stock price) being dependent on multiple other variables, such as macroeconomic factors, weather and previous time steps

Tuning a DL Model

Our final section in this chapter is reserved for the multitude of "performance-enhancing" levers we have at our disposal in order to build a credible deep learning model.

Activation Functions

As discussed in previous sections, neurons produce an output signal from weighted input signals using an activation function. So in effect, an activation function is a simple mapping to the output of the neuron where weighted inputs in the neural network are summed and passed through the activation (or transfer) function.

The term "activation" is related to the threshold at which the neuron is activated and the corresponding strength of the output signal. The Heaviside step function used in the Simple Perceptron is a simple step activation function where if the summed input exceeds 1 then the neuron would output a value of 1.

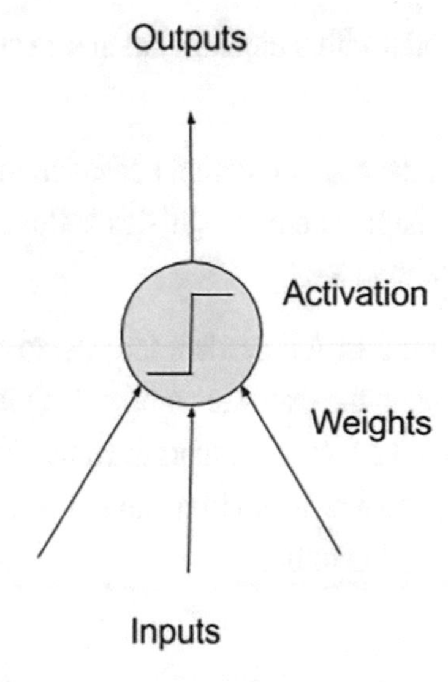

Figure 5-24. *Activation/Transfer Function architecture (showing Heaviside step function)*

The four main nonlinear activation functions which we actually use in practice provide much richer functionality:

(Logistic) sigmoid function

The sigmoid activation function has a characteristic "S" shaped curve and is used for binary classification problems. Real value inputs are "squashed"/mapped into a range from 0 to 1 representing the probability of a specific binary output.

Hyperbolic Tangent Function (tanh)

Tanh (pronounced "tanch") is also an "s" shaped curve, but this time rescales the logistic sigmoid function. The outputs range from -1 to 1. Because the gradient is stronger for tanh than sigmoid, tanh is generally preferred over sigmoid as an activation function.

Rectified Linear Unit (ReLU)

Both tanh and sigmoid are prone to "saturation" or vanishing gradients. For this reason, the most commonly used activation function in deep learning models is the simpler Rectified linear unit (ReLU). ReLU speeds up the convergence of stochastic gradient descent (covered in the next section), by virtue of its simpler mathematical operations, returning 0 for negative values, and x for positive values.

In the case of ReLU activation leading to "dead neurons" (where neurons become inactive and output 0 for any input), a variation, **Leaky ReLU** is sometimes used. Leaky ReLU has a small slope (set as a hyperparameter[9]) for negative values instead of a flat slope.

It should be pointed out that leaky ReLU does not always outperform ReLU - if a model using ReLU activation is performant, stick with it.

Softmax

The softmax activation function (or normalized exponential function) is a special case for multiclass classification problems where there is a discrete (non-binary) output, such as in the case of most image classification problems (e.g., is the image a person, a building, a vehicle, a road, etc.).

Softmax turns logits (the final layer in neural network) into probabilities that sum to 1.

[9] Predetermined before training, as opposed to learnt during training

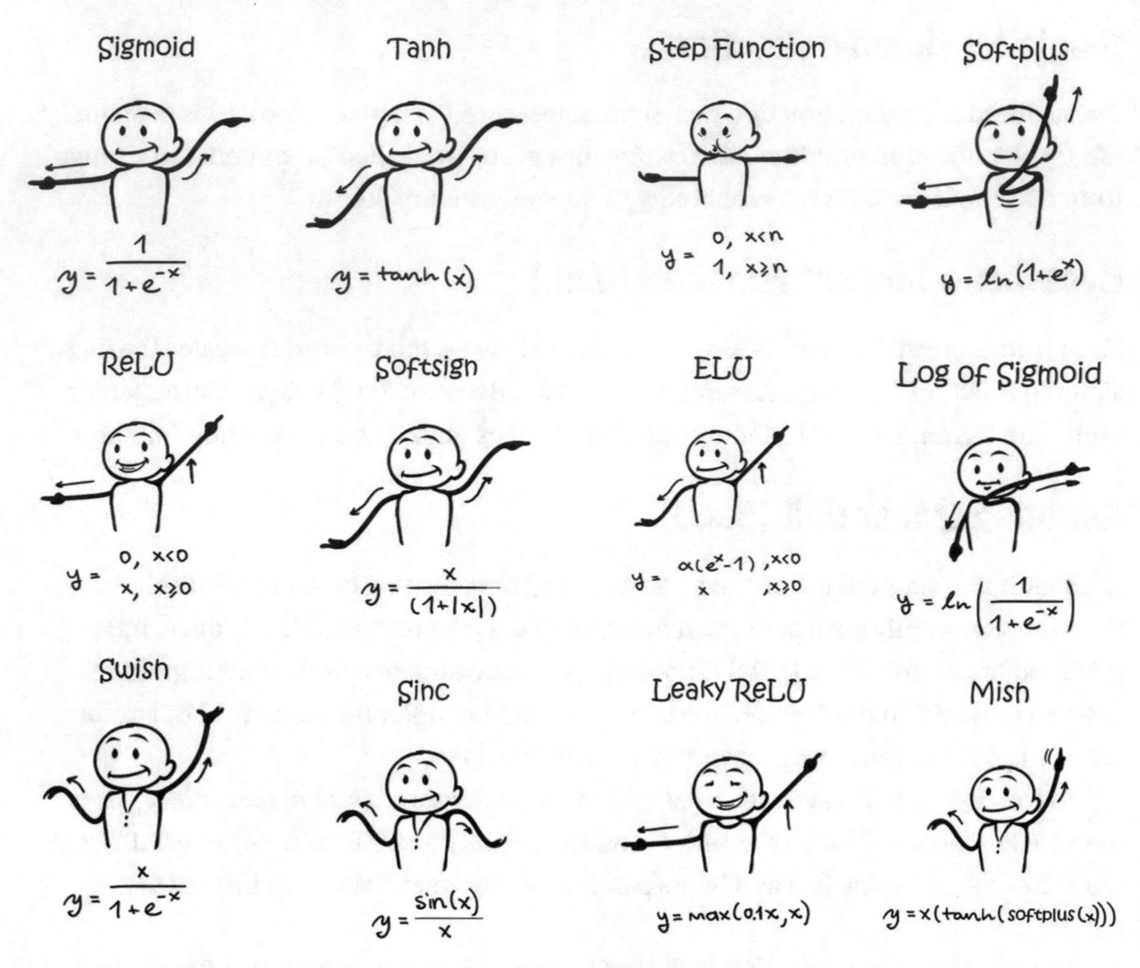

***Figure 5-25.** Activation functions – shapes and formulae*

Gradient Descent and Backpropagation

Gradient descent is the main process by which neural networks are trained. It is essentially a method to minimize **loss** (or **cost function**) by updating weights in the neural network.[10]

For most big data problems, **stochastic gradient descent (SGD)** is used – rather than using the entire dataset for each iteration, a sample of the data is used.

[10] See Loss Functions

A **forward pass**[11] is then carried out where the input data gets transformed through the network, activating neurons through hidden layers to ultimately produce an output value. This output is then compared to the expected (actual) output and an error is calculated.

Backpropagation

Backpropagation takes place when the above error is propagated back through the network, one layer at a time, and the weights are updated according to the amount that they contributed to the error.

The process repeatedly adjusts weights so as to minimize the difference between actual output and desired output and is repeated for all rows/examples in the training data (an **epoch**). A neural network is typically trained over multiple epochs, with 100 generally being an upper limit.

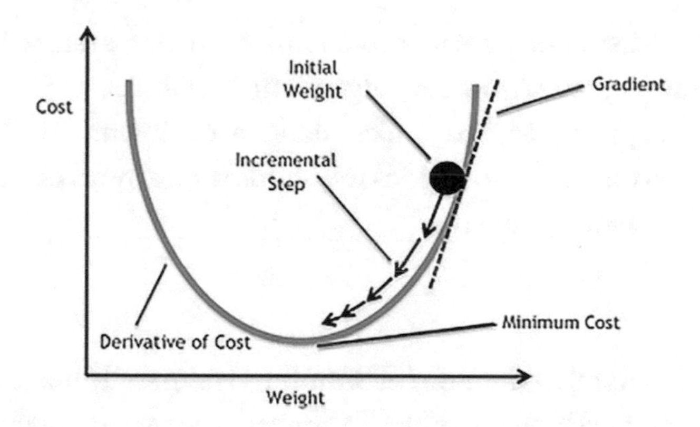

Figure 5-26. *The Gradient Descent process used to minimize loss in a neural network*

Other Optimization Algorithms

In practice, Stochastic Gradient Descent can get stuck at local minima, or take too long to converge, and so more sophisticated variants are used to achieve better model performance

[11] A forward pass is also used in the inference process, after the network is trained, in order to make predictions on new data

SGD with Momentum

As we saw in the previous section, Stochastic Gradient Descent addresses a large memory disadvantage of Gradient Descent (loads the entire datasets to compute loss derivative). SGD with Momentum is an enhancement of the vanilla SGD algorithm, helping accelerate gradient vectors in the right directions in order to speed up convergence.

AdaGrad, Adadelta, and RMSProp

SGD involves manually tuning the learning rate of the network which can be slow and cumbersome – the below three algorithms more efficiently update network weights.

AdaGrad or Adaptive Gradient Algorithm is another stochastic optimization method that attempts to address this by adapting the learning rate to the network parameters (weights).

Adadelta on the other hand uses an exponentially moving average learning rate to avoid very small learning rates/slow convergence in AdaGrad.

Root mean square prop (RMSProp), like Adadelta, deals with AdaGrad's radically diminishing learning rates but also divides the learning rate by an exponentially decaying average of squared gradients.

Adam

Adam (derived from **adaptive moment estimation**) is generally the choice of optimization algorithm in Deep Learning. To converge faster (use of this algorithm can be the difference in getting quality results in minutes, days, or months), Adam uses both Momentum to accelerate the stochastic gradient descent process and Adaptive Learning Rates to reduce the learning rate in the training phase.

Adam combines the best properties of the AdaGrad and RMSProp algorithms to handle sparse gradients on noisy problems.

Loss Functions

Before we move on to look at best practice in improving Deep Learning performance, let's take a closer look at the concept of "loss" or the cost function in a neural network.

Minimizing a loss function in Deep Learning equates to minimizing the training error or lowering the cost of the neural network/weight calibration process.

The three most common loss are **binary cross entropy** for binary classification, **sparse categorical cross entropy** for multiclass classification and **mse (mean squared error)** for regression, but as with all things in deep learning there are several variants which under certain conditions can provide better results:

Table 5-1. *Loss function comparison*

Loss Function	Type	Description	Pros	Cons
Squared Error (L2) Loss / Mean square error (MSE)	Regression	square of the difference between the actual and the predicted values	penalizes large errors by squaring them	Not robust to outliers
Absolute Error Loss / Mean Absolute Error (MAE)	Regression	distance between the predicted and the actual values	more robust to outliers as compared to MSE	Penalization of large errors may be insufficient
Huber Loss	Regression	Combined MSE and MAE - quadratic for smaller errors, linear otherwise	more robust to outliers than MSE	Slower convergence
Binary Cross-Entropy	Binary Classification	Uses Log-Loss and sigmoid function	Ideally suited for binary classification models	Sigmoid can saturate and kill gradients
Hinge Loss	Binary Classification	primarily used with Support Vector Machine (SVM)	penalizes wrong predictions as well as low confidence right predictions	Limited to SVM models
Multi-class Cross Entropy Loss / Categorical Cross entropy	Multi-class Classification	generalization of the Binary Cross Entropy loss	Works well with one hot encoded target variables	Each sample should have multiple classes or is labelled with soft probabilities

(*continued*)

Table 5-1. *(continued)*

Loss Function	Type	Description	Pros	Cons
Sparse categorical cross entropy	Multi-class Classification	Sparse target variable classification version of Categorical Cross entropy	Reduces memory and computational overhead through use of single integers for classes	Only works for classes which are mutually exclusive
Kullback Leibler Divergence Loss	Multi-class Cross Entropy Loss	measure of how a probability distribution differs from another distribution	Based on probabilities	more commonly used to approximate complex functions than in multi-class classification

Improving DL Performance

The above internal workings of a deep learning model help to understand what is happening "under the hood" and the options described to improve results are somewhat experimental, tied to the specific dataset and business or organizational problem with which we are presented.

Where do we start though if we are embarking on improving neural network performance? We should take a best practice approach, almost always commencing (and ending) with a review of the underlying data:[12]

[12] Source: `machinelearningmastery.com`

1. Review Data
   - Invent More Data
   - Rescale Your Data
   - Transform Your Data
   - Feature Selection
2. Include more sophisticated algorithms
3. Tune hyperparameters
4. Try Algorithm Ensembles
5. Get More Data

Besides these core principles for achieving higher model performance, periodic refactoring should be built into model maintenance to ensure code is as efficient as possible and (python) libraries and functions are not deprecated.

Deep Learning Best Practice – Hyperparameters

The above high-level steps are informative, but on exhaustion of a data review process, what specific hyperparameters do we have at our disposal to tune and calibrate the model output? Below we group these together into two categories: network and process tuning.[13]

Network Tuning

Deeper Network/More Layers/More Neurons

Adding more hidden layers/more neurons per layer means we add more parameters to the model, and therefore allow the model to fit more complex functions

[13] a good idea is to reference again in parallel TensorFlow Playground (Hands-on Practise) to see how these "levers" are configured

Activation Function

As described above, generally **with CNNs,** ReLU is used to address vanishing gradients with Sigmoid (binary) or Softmax (multiclass classification) in the outer layer. For RNNs, tanh is used to overcome the vanishing gradient problem, as its second derivative can sustain for a long range before going to zero.

Neural Network Ensembles

In much the same way that a random forests is an enhancement of a decision tree model to prevent overfitting in machine learning, we can use ensemble methods in deep learning -different neural networks with different model configurations to prevent the same issue. However, this does come at the computational expense of training and maintaining multiple models. A better approach is to use dropout as described in the section below.

Batch Normalization

The deep neural networks training process is also sensitive to the initial random weights and configuration of the learning algorithm. Normalization of the layers' inputs (**batch normalization)** can help to make artificial neural networks faster and more stable by recentering and rescaling of the data.

Pooling

Multiple convolutional layers in a CNN can be very effective, learning both low-level features (e.g., lines) and higher-order features, like shapes or specific objects in the outer layers.

However, these "feature maps" are tied to the EXACT position of features in the input. This "inflexibility" can result in different feature maps for minor image changes such as cropping, rotation, shifting, and a resultant loss in model generalization to new data.

Pooling (implemented using pooling layers) is used to down sample convolutional layers in a CNN and avoid overfitting to exact/precise image features. This lower-resolution version of an input signal which retains only important structural elements is similar to the effect of "pruning a decision tree" in machine learning.

Image Augmentation

Image data augmentation is used to artificially expand the size of a training dataset for Convolutional Neural Networks. The idea is that training on more data means a more skillful model. Using the **ImageDataGenerator class in Keras to generate batches of augmented images, we can achieve this** in a number of ways:

- Random rotation of the image
- Shifting the objects within the images
- Shear (distortion along an axis)
- Flipping the image (e.g., upside down)

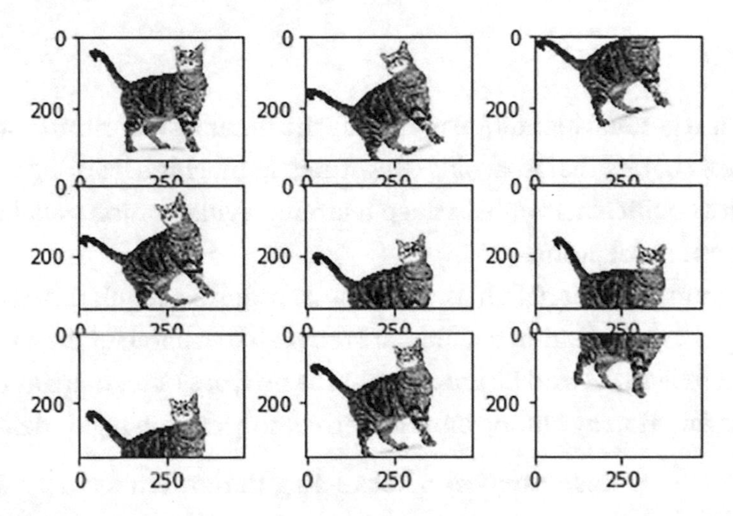

Figure 5-27. *Image Augmentation in Keras to augment an image datasets with rotated, offset, or distorted copies of the original image*

Process Tuning

Number of Epochs and Batch Size

Increasing the number of epochs (a complete sampling of the entire training set) will generally produce better performance, but only up to a point. Evidence of validation (test) accuracy decreasing even when training accuracy is increasing should put an upper limit on the number of epochs, otherwise we are essentially overfitting our model. Rather than trial and error, we can control setting the number of epochs at too high a level by implementing early stopping criteria to avoid overfitting on the training set.

Our batch size – the size of the input data sample used in a forward pass is another lever we have at our disposal. Using too large a batch size can have a negative effect on generalization, that is, the test accuracy of the network is worse as we have reduced the stochasticity (randomness) of gradient descent during the training process. A rule of thumb for largish datasets (e.g., > 10,000 images) is to use the default value of 32 first, then increase to 64, 128, and 256 if the model is underfit or the training time is onerous.

Learning Rate

Learning rate is a hyperparameter used to impact the speed of the gradient descent process. Setting this too high can mean the algorithm will bounce back and forth without reaching a local minimum, while too low and convergence can take some time.

Regularization

Regularization is a general technique to modify the learning algorithm such that the model generalizes better, that is, avoids overfitting. In machine learning regularization penalizes feature coefficients while in deep learning regularization penalizes the weight matrices of the nodes/neurons.

A regularization coefficient (a hyperparameter) controls regularization – we get underfitting if this regularization coefficient is so high that some of the weight matrices are nearly equal to zero. L1 and L2 are the most basic types of regularization, updating the general cost function by adding another term known as the regularization term:

Cost function = Loss + Regularization term

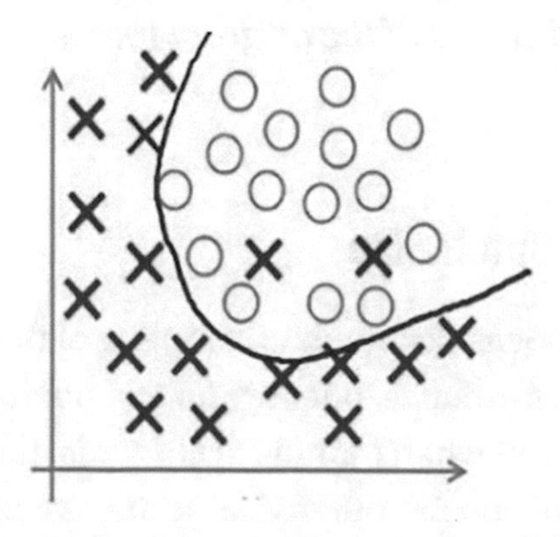

Figure 5-28. *Appropriate model fitting*

Dropout

Deep learning neural networks are prone to overfitting when datasets are relatively small. Dropout is a computationally cheap and effective regularization method to reduce overfitting and improve generalization error. In effect, dropout simulates in a single model a large number of different network architectures by randomly dropping out nodes (neurons) during training.[14]

A good starting point for dropout is to set this @ 20% and increase to, for example, 50% if the model impact is minimal. Too high will result in under-fitting.

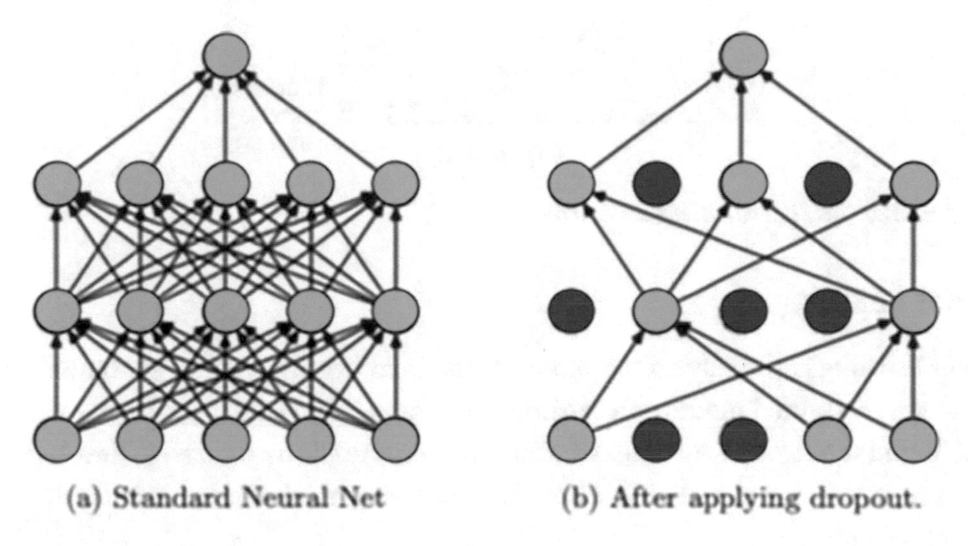

Figure 5-29. *Illustration of Dropout (`medium.com`)*

Early Stopping

We should stop training at the point when performance on a validation dataset starts to degrade, for example, by:

- No change in metric over a given number of epochs
- An absolute change in a metric
- A decrease in performance observed over a given number of epochs
- Average change in metric over a given number of epochs

[14] Nodes may be input variables in the data sample or activations from a previous layer

This is implemented in Keras using the **EarlyStopping** function. The below example monitors and seeks to minimize validation loss across epochs:

EarlyStopping(monitor='val_loss', mode='min')

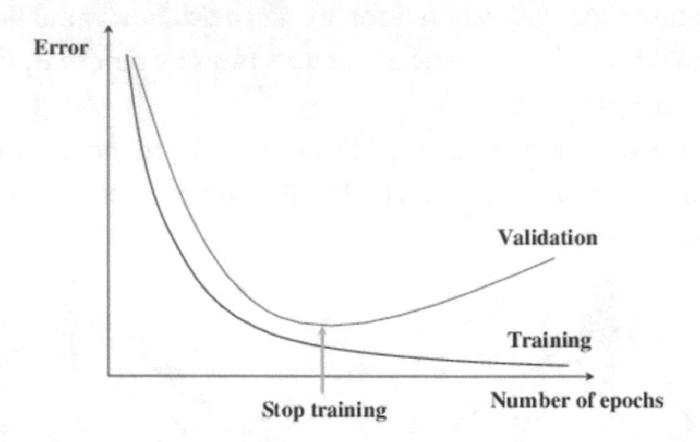

Figure 5-30. *EarlyStopping in Keras*

Transfer Learning

Transfer learning is actually an "accelerator" for deep learning model - essentially we are taking a pretrained model as a starting point for another (different, but related) model. Popular in Deep Learning to reduce the computation required to develop neural network models from scratch, there are a number of research models often used for transfer learning:

- Oxford VGG Model
- Google Inception Model
- Microsoft ResNet Model
  - See also `https://github.com/BVLC/caffe/wiki/Model-Zoo`

The VGG model is used in the Transfer Learning exercise in our Hands-On lab above (Convolutional Neural Networks with Keras & TensorFlow).

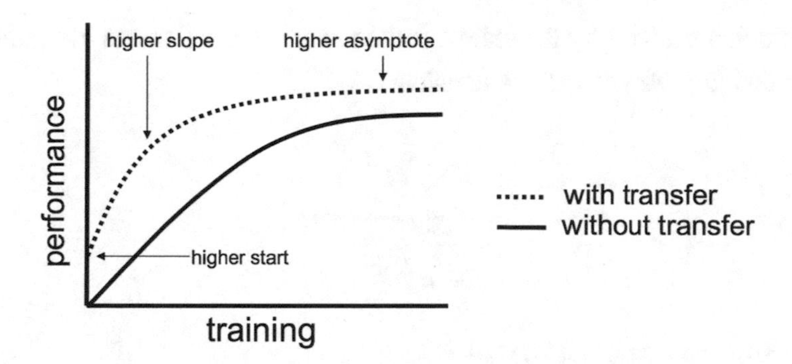

Wrap-up

The exhaustive neural network and process tuning that we have just walked through is clearly complex. The best practice techniques described are often rather vague and the range of options available/settings to configure can be difficult to pin down, especially if results are required relatively quickly. But quick wins are always possible and perseverance in achieving better results is usually rewarded provided a structured approach is adopted and performance iteratively monitored.

While we may have reached the end of this chapter, our next chapter will take a look at how we can couple best practice in deep learning with training (and testing) automation to help accelerate the process of fine-tuning both machine and deep learning models. Before we go there though, we complete this chapter with two deep learning model tuning labs.

Softmax: Hands-on Practice

ACTIVATING LOGITS WITH SOFTMAX

Simple hands-on lab to see how the Softmax activation function takes the last neural network layer and turns output into probabilities that sum to one:

1. Clone the GitHub repo below:

   `https://github.com/bw-cetech/apress-5.6.git`

2. Run through the Python code in Jupyter Notebook to see how softmax is computed.

3. As a stretch exercise try to create a Python function which stores the softmax activation formula and call the function

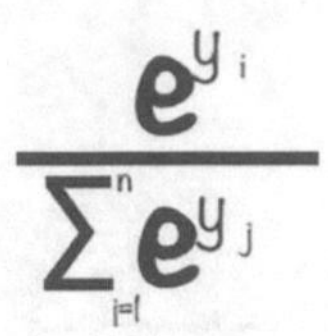

Figure 5-31. Softmax activation function

Early Stopping: Hands-on Practice

AVOID OVERFITTING IN DEEP LEARNING

In this lab, we will continue with our German traffic light image classification dataset and look at the impact of early stopping criteria on model performance:

1. Continue with the notebook from earlier (Convolutional Neural Networks with Keras and TensorFlow: Hands-on Practice)
2. Run through the remainder of the notebook from “2nd Run –Early Stopping Criteria”
3. Complete the exercises as part of the lab

CHAPTER 6

AutoML, AutoAI, and the Rise of NoLo UIs

In what is still so far a relatively short space of time, growth in machine and deep learning implementations in organizations across the world has been extraordinary. However, this hasn't always translated into commercial success, with disappointingly low adoption rates in the retail sector (11.5% in the UK[1]) and just over 50% of prototypes ending up in production across all industries.[2] Historically many solutions have been operationally siloed with PhD-level statisticians left to explain code-heavy technical models.

Coupled with a reliance on dummy or synthetic datasets for training, an absence of, or worse, broken interfaces with training and testing done in a Python notebook (Jupyter, Colab, etc.) and the "technical debt" associated with these poorly designed applications starts to become a burden to the organization.

Roll forward to today and we are starting to see a rise in the rollout of "AutoML"[3] and "AutoAI" tools far better equipped for an enterprise-wide deployment – from fully automated data import, through interface orchestration, machine/deep learning, and deployment. Moreover, these tools are increasingly more usable, and, importantly, understandable to multiple stakeholders across departments by virtue of built-in user-friendly "NoLo" GUIs and embedded data traceability and auditability.

---

[1] https://assets.publishing.service.gov.uk/government/uploads/system/uploads/attachment_data/file/1045381/AI_Activity_in_UK_Businesses_Report__Capital_Economics_and_DCMS__January_2022__Web_accessible_.pdf. For Gartner stat, see www.gartner.co.uk/en/newsroom/press-releases/2020-10-19-gartner-identifies-the-top-strategic-technology-trends-for-2021

[2] www.gartner.co.uk/en/newsroom/press-releases/2020-10-19-gartner-identifies-the-top-strategic-technology-trends-for-2021

[3] The global AutoML market is projected to grow @ 43% CAGR and exceed $5b by 2027 (Source: businesswire)

B. Walsh, *Productionizing AI*, https://doi.org/10.1007/978-1-4842-8817-7_6

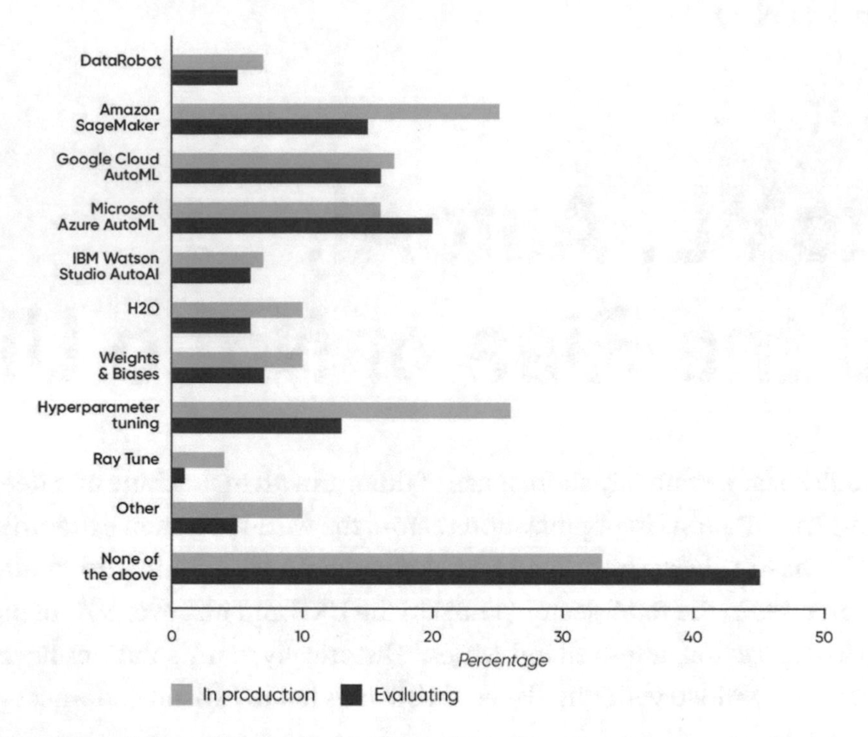

Figure 6-1. *Use of AutoML tools (Source: O'Reilly AI Adoption in the Enterprise 2022)*

The step change in use, and collaboration across AutoML/AI tools and so-called NoLo, or no/low code applications,[4] is coming at a time companies shift from "rule-based" robotic process automation (RPA) to enhanced cognitive robotic process automation (CRPA) with AI-infused "context."

Perhaps this is nowhere more apparent than in the evolution of chatbots where intelligent virtual agents (IVAs) or "conversational" chatbots have supplanted legacy, rule-based chatbots, but the trend is also evident in the proliferation of augmenting tools such as Microsoft PowerAutomate on top of PowerBI, NLP/text analytics on top of optical character recognition (OCR) and specific sectoral enhancements such as (X-Ray) diagnostics (AI) on top of patient screening (RPA) in the healthcare industry.

[4] Or "LCNC" Low-Code/No-Code platforms. Although growth is expected to be at a slower c. 23% CAGR than AutoML, the market size for low-code development platforms is bigger and expected to reach $35b by 2030 (Source: Grand View Research, Inc)

The drive for this business transformation is in essence a pursuit of an "Employer's Dream" – the democratization of tools which are universally popular, highly visual, and collaborative yet are able to perform often complex predictions on future outcomes for the organization. As such the aims of this chapter are to start readers on a path to meet this goal, developing a "production mindset" to deliver "enterprise-wide, full-integration, full-stack" AI applications and solutions.

After first revisiting the end-to-end machine learning process and an introduction to Bayesian Optimization as the basis for AutoML, we will take a hands-on look at Python model automation libraries before focusing on the growing ecosystem of automated AI tools. By no means exhaustive, our labs in this chapter will focus on a number of the leading best-of-breed NoLo tools for AutoML and AutoAI including IBM Cloud Pak for Data, Azure Machine Learning, and Google Teachable Machines.

Machine Learning: Process Recap

We start by revisiting Data Pipeline Orchestration and the end-to-end machine learning process. As the basis of most built-in AutoML functionality today – almost all AutoAI tools are anchored to the series of steps below – a grounding here will provide a reference for the rest of the chapter – from important Bayesian optimization and inference to the automated Python modeling libraries and AutoML/AutoAI tools covered in our hands-on labs.

In practice, not everything is in scope of AutoML, but as shown in Figure 6-2 we can group these steps into premodeling and postmodeling processes. Starting with raw data import, AutoML/AutoAI relies on slick Data Pipeline Orchestration through these pre- and postmodeling steps all the way to hyperparameter tuning and final algo selection. Today's leading products for AutoAI even include automation of the "retraining" process where "data drift" is monitored and a new training process triggered for statistically significant deviations in the underlying dataset.[5]

[5] See also Chapter 9 AI Project Lifecycle - Data Drift

Figure 6-2. *Automated processes in Machine Learning*

As we shall see later,[6] typically the above automations are boiled down to four core processes[7] for building and evaluating candidate model pipelines:

- Data preprocessing
- Automated model preselection
- Automated feature engineering
- Hyperparameter optimization

Global Search Algorithms

While many of the processes listed above can be, and are, automated in AutoML and AutoAI, once data is imported, the model training process actually "optimizes" the set of seemingly infinite feature and feature weight combinations[8] using a **global search algorithm.**

Random sampling is one method, as is grid search, whereby samples are drawn more evenly from the **search space** of parameters/features. In both cases, the goal is to minimize a cost or **objective function** – typically a difference/delta "score" between actual predictions/forecast and modeled output.

[6] See AutoAI in IBM Cloud Pak for Data section below

[7] See e.g. `https://dataplatform.cloud.ibm.com/docs/content/wsj/analyze-data/autoai-overview.html`

[8] Or variable search space. See also `https://machinelearningmastery.com/what-is-bayesian-optimization`

Bayesian Optimization and Inference

Both random sampling and grid search have their limitations, however – neither makes use of previous sample results to direct/improve sampling in the next iteration. A better/more sophisticated approach receiving a lot of attention today[9] is to use Bayesian Optimization where the "tuning algorithm" optimizes parameter selection in each iteration according to the previous iteration "score," that is, Bayesian Optimization "adaptively" samples data probabilistically more likely to be "optimal" while Bayesian Inference "infers" the outcome/score.

Under the hood, Bayesian Optimization seeks to find a **surrogate function** that optimizes data, features, algorithm, and hyperparameters in order to approximate the objective function.[10] Typically a surrogate function summarizes the conditional probability of an objective function using supervised regression techniques such as random forest or a Gaussian process regression. For the latter, a kernel function is required to control the shape of the objective function – by default a Radial Basis Function (RBF) is used but performance with different datasets can dictate use of different kernel functions.

Bayesian Optimization works best when used on problems with less than 20 dimensions/features so for BIG datasets, a dimensionality reduction technique such as Principal Component Analysis should be undertaken first.

Bayesian Inference: Hands-on Practice

SEARCH OPTIMIZATION PERFORMANCE BENCHMARKING

The goal of this lab is to compare machine learning model performance when using (a) random sampling, (b) grid search, and (c) Bayesian optimization

1. Clone the GitHub repo below:

    ```
    https://github.com/bw-cetech/apress-6.2.git
    ```

[9] Out of scope of this book, but also gaining in application, is the computationally intensive use of Genetic Algorithms

[10] Objective functions are nonconvex, nonlinear, noisy, and computationally expensive, hence the need to approximate with a surrogate function

2. Walk-through the notebook which imports the dataset (also provided in repo above), sets up a data wrangling pipeline then uses the three different techniques above to search for the most optimal model parameters for predicting a heart signal
3. Exercise – try to plot the mean test score (AUC) for each of the three techniques
4. Exercise (stretch) – import a larger IoT or retail dataset, update the wrangling pipeline and modeling hypothesis and view the mean score to see how Bayesian inference outperforms the other techniques

Python-Based Libraries for Automation

Bayesian Optimization is the primary technique used in AutoML and AutoAI to probabilistically search through multidimensional space of parameters available to the underlying (surrogate) model.

While clearly neither as usable nor accessible to nonprogrammers as the NoLo code tools we will cover later, there are a wealth of Python libraries available for AutoAI that we will now run through first.

PyCaret

While experience with Python is required, PyCaret is marketed as low-code machine learning due to its accelerated approach to machine learning model training. The USP is in democratization of machine learning and as the example of training a dataset for anomaly detection in Figure 6-3 shows, a minimal amount of end-to-end coding is required.

```
1 # load dataset
2 import pandas as pd
3 data = pd.read_csv('data.csv')
4
5 # init setup
6 from pycaret.anomaly import *
7 s = setup(data, normalize = True)
8
9 # train isolation forest model
10 iforest = create_model('iforest')
11
12 # assign anomaly labels on training data
13 iforest_results = assign_model(iforest)
14
15 # assign anomaly labels on new data
16 new_data = pd.read_csv('new_data.csv')
17 predictions = predict_model(iforest, data = new_data)
18
19 # save iforest pipeline
20 save_model(iforest, 'iforest_pipeline')
```

Figure 6-3. *PyCaret for anomaly detection*

We will look at PyCaret as part of an extensive hands-on lab deploying an insurance premium calculator to Azure in our final chapter.

auto-sklearn

Auto-sklearn automates the Data Science libraries from scikit-learn in order to determine effective machine learning pipelines for supervised classification and regression datasets

Machine Learning for the "Enterprise," prioritizing Data Team efficiency and productivity, is more easily accessible to non-Data Scientists, but like PyCaret, still involves Python coding. It comes with built-in preprocessing and data cleaning, feature selection/engineering, algo selection, hyperparameters optimization, benchmarking/performance metrics, and postprocessing.

A variation of auto-sklearn is **Hyperopt-Sklearn** which uses **Hyperopt**[11] to describe a search space over possible configurations of sklearn preprocessing and classification modules.

Auto-WEKA

Auto-WEKA is in fact a Java application for algorithm selection and hyperparameter optimizations, which is built on the University of Waikato, New Zealand's WEKA[12] Machine Learning. pyautoweka is the Python wrapper.

In contrast to auto-sklearn, Auto-WEKA simultaneously selects a learning algorithm and configures hyperparameters with the goal of helping nonexpert users more effectively identify ML algorithms and hyperparameter settings appropriate to applications, as well as improve performance.

TPOT

TPOT (Tree-based Pipeline Optimization Tool) uses a tree-based structure/genetic programming to optimize machine learning pipelines and is designed to train on large datasets over several hours. The TPOT API supports both supervised classification and regression, carrying out, as shown in Figure 6-4, data wrangling and PCA before iterative training, testing, and recursive feature elimination to arrive at a pipeline with the highest score.

[11] Distributed Asynchronous Hyper-parameter Optimization – an open-source Python library for Bayesian optimization. See `https://hyperopt.github.io/hyperopt/`

[12] Waikato Environment for Knowledge Analysis

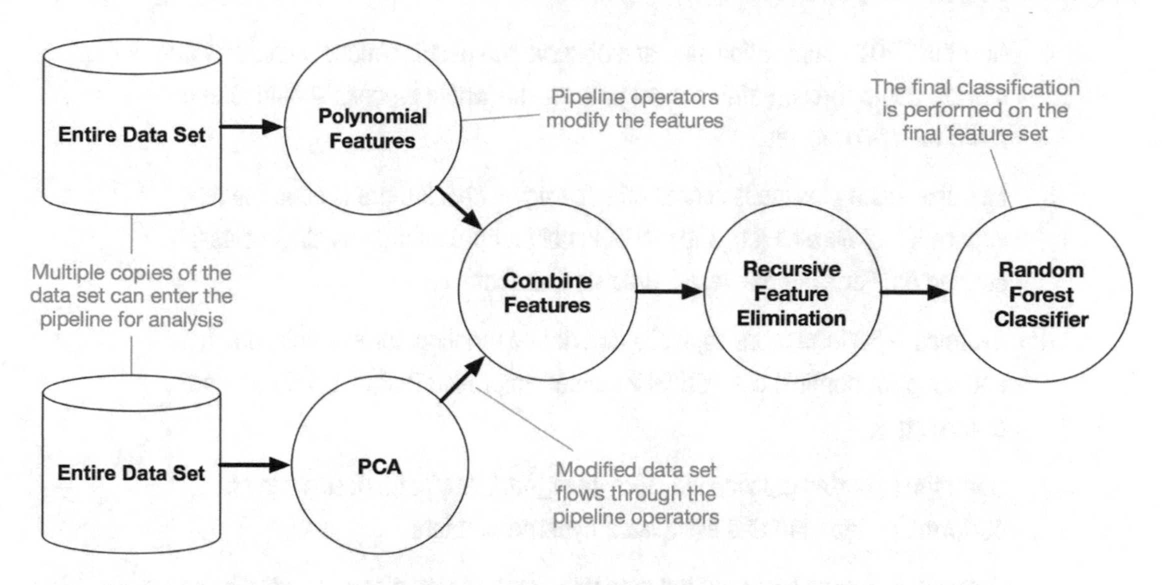

Figure 6-4. *TPOT operation (Source: towardsdatascience)*

The **generations** parameter used in the pipeline_optimizer variable is the number of iterations to run the pipeline optimization process – although a decent pipeline (and ML model) can be achieved relatively quickly, more generations are required to ensure the most performant pipeline is achieved, particularly for large datasets.

Python Automation with TPOT: Hands-on Practice

GENETIC PROGRAMMING AUTOML

The goal of this lab is to use TPOT optimization to find the best-performing pipeline and algorithm on a synthetic classification problem. The code sample also shows how to automate in Python a direct connection, unzip, and read of a web dataset:

1. Clone the GitHub repo below:

   `https://github.com/bw-cetech/apress-6.3.git`

2. Walk-through the notebook step by step to read in the dataset directly from the UCI website, unzip the files, and import the smaller csv dataset
3. Carry out basic EDA, data wrangling, partition the data and configure KFolds cross-validation (STEPS 2, 3, and 4)

4. Run the TPOT optimization step and observe the pipeline/model scores as they start to come through after a few minutes. The whole process should take no more than 30 minutes

5. Exercise – the pipeline is scored on accuracy – check there is good model separation of classes (i.e., the model is not just predicting one single class) by scoring on. For example, recall, precision, or fbeta

6. Exercise – perform more sophisticated data wrangling, for example, one hot encoding for nominal categorical variables, improved feature selection, and/or scaling

7. Open the exported pipeline file "tpot_best_model.py" and observe the best-performing algorithm and associated hyperparameters

8. Exercise (Stretch): swap out the data with the larger banking dataset. Speed up runtime by executing instead on Colab using a GPU accelerator and compare pipeline performance over 10 generations

AutoAI Tools and Platforms

We now arrive at our main section in this chapter - a look at several key "NoLo" tools for AutoAI.[13] The intention is to give the reader a flavor of how these tools are winning over business leaders with their simplicity and highly visual, translatable, and collaborative USPs and are increasingly adopted by AI Engineers and Data Scientists to fit/achieve "enterprise-wide" business goals.

The section is bookended with several hands-on labs covering the operation of these tools applied to specific use cases in AI.

IBM Cloud Pak for Data

Dressed as a data and AI platform with a "data fabric" architecture, IBM Cloud Pak for Data essentially brings together a number of legacy tools (including Watson Studio, Decision Optimization, and Watson Assistant) into one platform solution, adding AutoAI on top.

[13] We cover five main AutoAI/AutoML "platforms." Out of scope of this book but certainly worth a look are tools such as c3, DataRobot, Peltarion, Ludwig, and KNIME in what is fast becoming an increasingly fragmented ecosystem.

The product's "Data Fabric" support means the product supports multiple APIs into structured and unstructured data sources spread across multicloud environments, whether IBM Cloud, AWS, Azure, or GCP. This underpins of the USP of the product - IBM claim that Data Fabric architecture means data access is 8* faster, while reduced ETL requests amount to a 25-65% productivity boost. There are data governance benefits as well - $27m saving in costs by virtue of the products inbuilt smart data negating the need for manual cataloging.

AutoAI is the graphical tool, previously built into Watson Studio, which automates the AI process - analyzing data, discovering data transformations, algorithms, and parameter settings that work best for a specific predictive modeling problem.[14] As shown below, these automations essentially fit the four core AI automation processes described earlier:

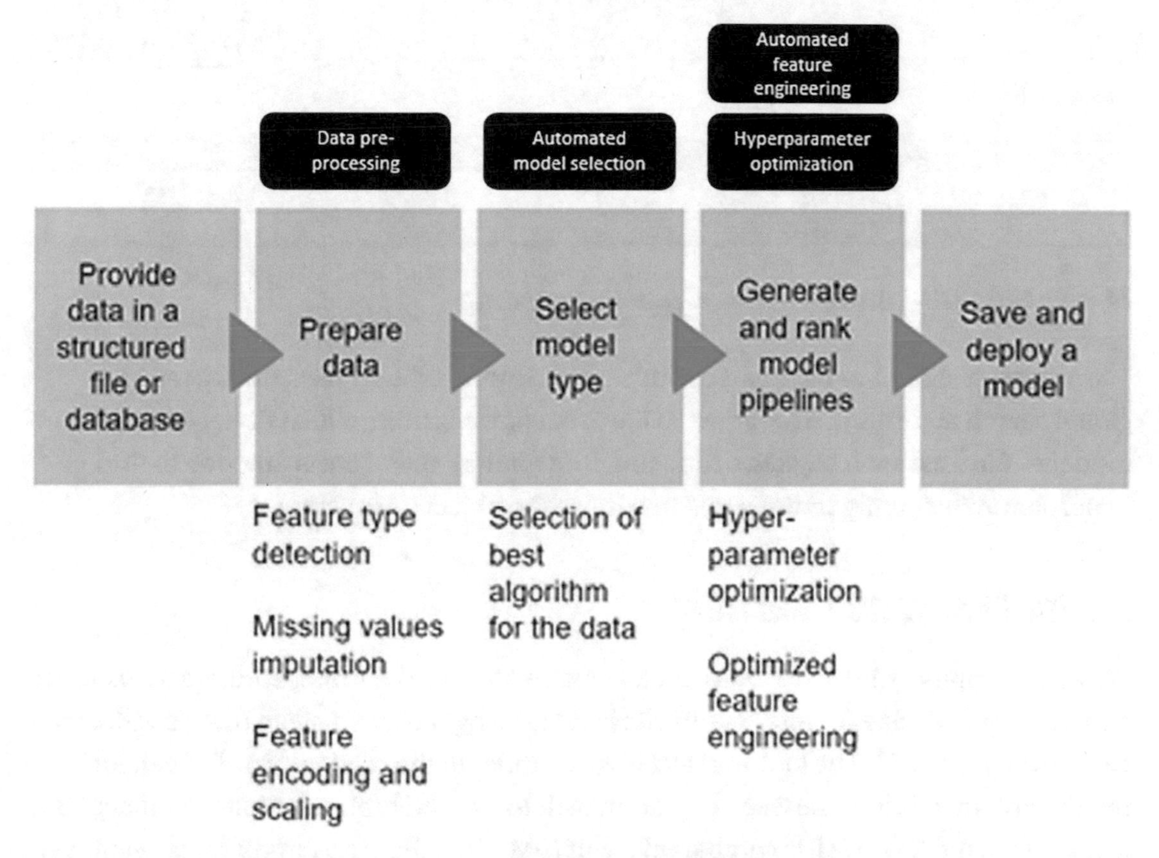

Figure 6-5. *IBM Cloud Pak for Data: AutoAI automation*

[14] As well as meshing with IBM "ModelOps" best practice for monitoring model/data drift and re-training

AutoAI displays the results of the underlying automations as model candidate pipelines ranked on a leaderboard for the end user to select.

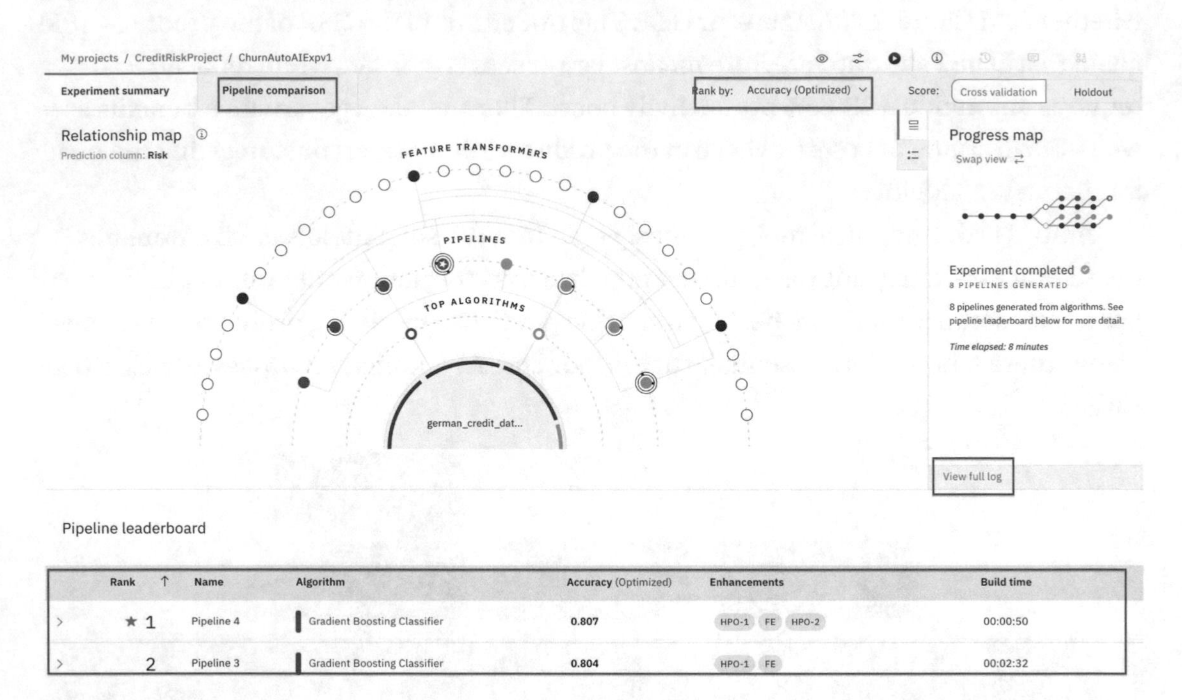

***Figure 6-6.** IBM AutoAI – model pipeline ranking*

Although similar to Bayesian optimization, AutoAI actually uses **RBfOpt** as its global search algorithm. In contrast to Bayesian optimization, which fits a Gaussian model to the unknown objective function, RBfOpt fits a radial basis function to find hyperparameter configurations that maximize the objective function.[15]

Azure Machine Learning

Back in Chapters 1 and 4, we took a look at Azure Machine Learning Studio. Microsoft are deprecating this "classic" interface in 2024 and moving the tool instead to Azure Machine Learning (AzureML). The look and feel is very similar to the original "Studio" version but the enhanced cloud service integration with Azure and built-in automated machine learning means Azure ML is comparable with IBM Cloud Pak in terms of functionality.

[15] See also `https://dataplatform.cloud.ibm.com/docs/content/wsj/analyze-data/autoai-details.html`

Microsoft promote AzureML in terms of an enterprise-grade service for building business-critical machine learning models at scale. The product comes with MLOps model governance and control and claims to 70% fewer model training steps and (somewhat contradictory as a no-code tool), 90% fewer lines of code.

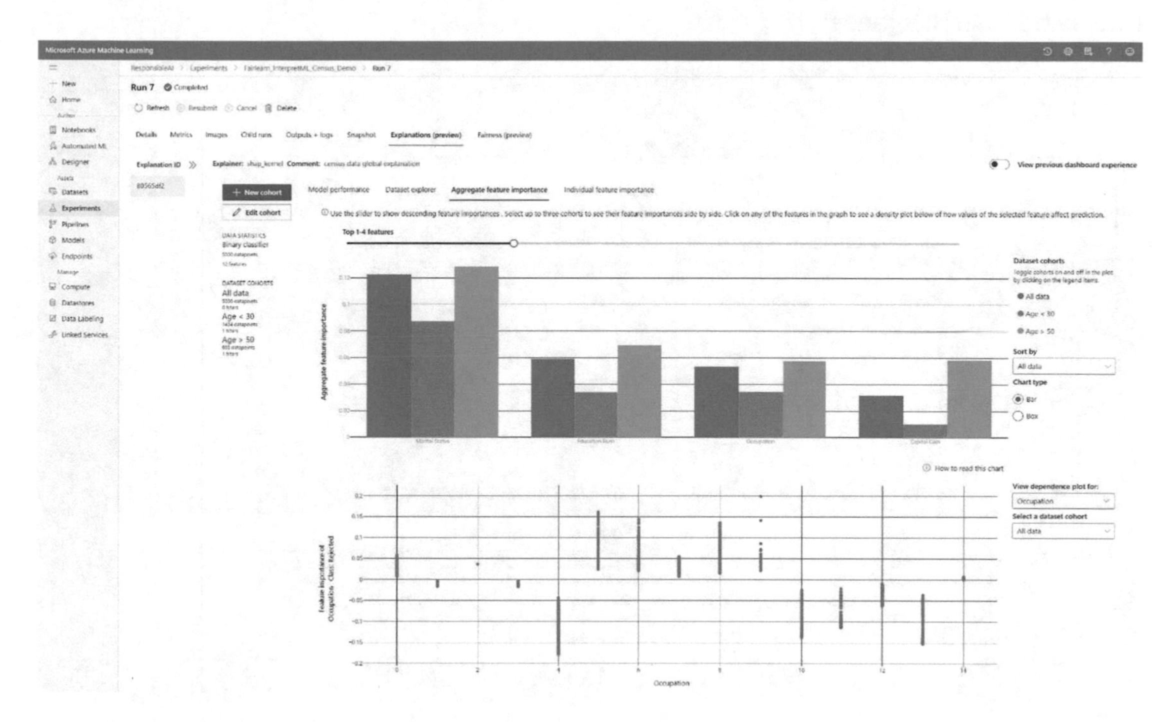

Figure 6-7. Azure Machine Learning UI

Google Cloud Vertex AI

Vertex AI is now Google's main platform for API-driven AI automation. It is part **AutoML** for training models on image, tabular, text, and video datasets without writing code and part **AI Platform** for running custom training code. Google Teachable Machine, which we will look at in a hands-on lab below, is part of Google's AutoML platform.

The entire ecosystem of MLOps tools within the Vertex AI platform is too broad to mention here[16] but of note is **Vertex AI Pipelines** – a serverless service that runs both TensorFlow Extended, which we will cover shortly, Kubeflow pipelines. In keeping with the theme of unified data and machine learning, Vertex AI also comes with multiple integrations to BigQuery[17].

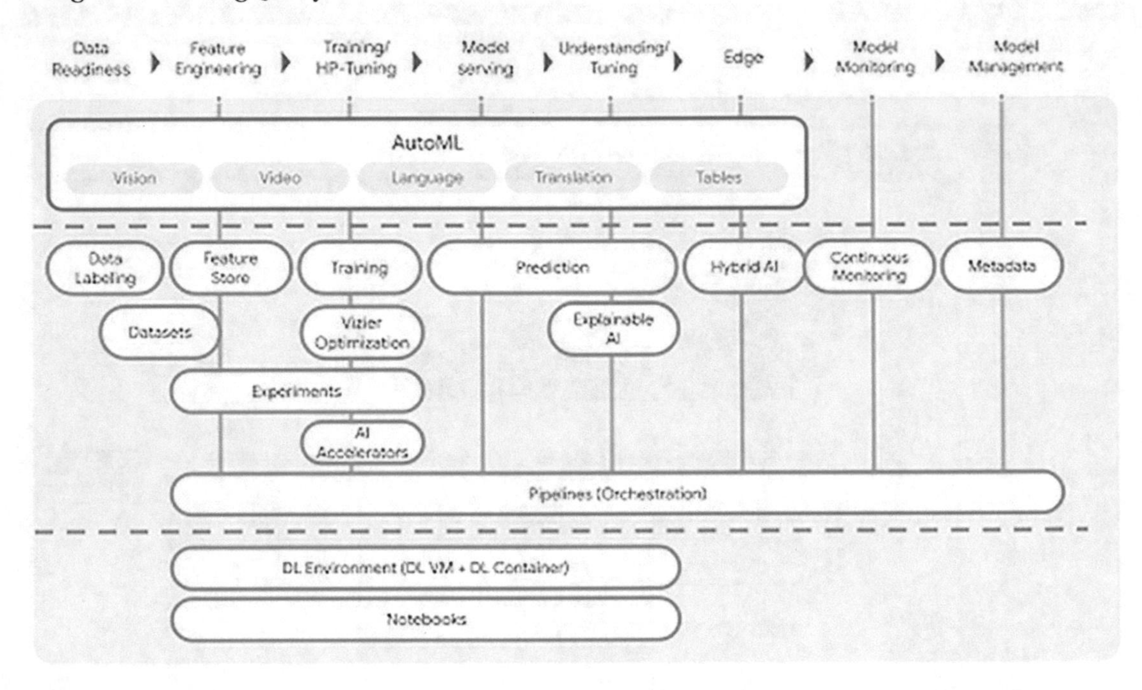

Google Cloud Composer

As the creator of TensorFlow, it's no surprise the scope of Google's AI automation extends beyond Vertex AI. Google Cloud Composer is a managed version of Apache Airflow for orchestrating data pipelines. The workflow and GCP architecture consists of **Dataprep** for import and wrangling, **Cloud Dataflow** for transforming data, **BigQuery ML** for model training, and **Cloud Composer** for pipelining/orchestrating the data.

[16] See `https://cloud.google.com/vertex-ai`

[17] See `https://cloud.google.com/blog/products/ai-machine-learning/five-integrations-between-vertex-ai-and-bigquery` for more information

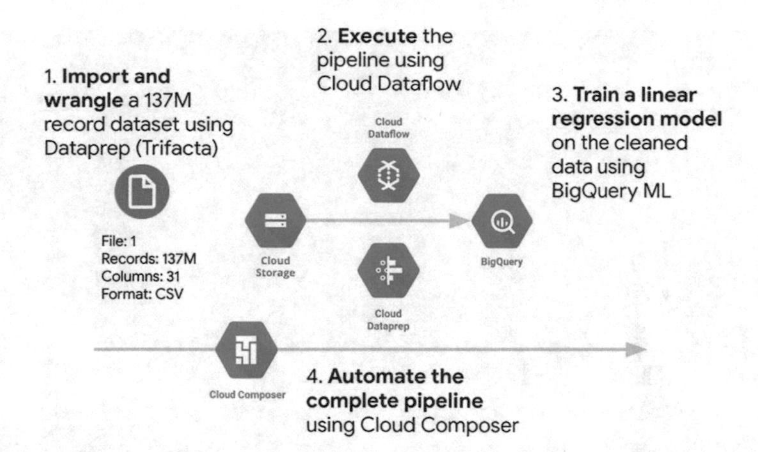

Figure 6-8. *Using Google Cloud Dataprep API to trigger automated wrangling jobs with Cloud Composer*[18]

AWS SageMaker Autopilot

Like Google Vertex AI, the Amazon SageMaker stack consists of multiple tools for AutoML,[19] the main ones being **Amazon SageMaker Studio**, **Amazon SageMaker Autopilot**, and **SageMaker Data Wrangler**.

Amazon SageMaker Studio comes with recommended models which can be used as building blocks for customization while Autopilot simplifies the ML model build process after Data Wrangler performs the necessary wrangling tasks such as automatically populating missing data, displaying column stats, encoding nonnumeric columns, and extracting date and time fields.

[18] See `https://medium.com/google-cloud/automation-of-data-wrangling-and-machine-learning-on-google-cloud-7de6a80fde91`

[19] `https://aws.amazon.com/machine-learning/automl/`

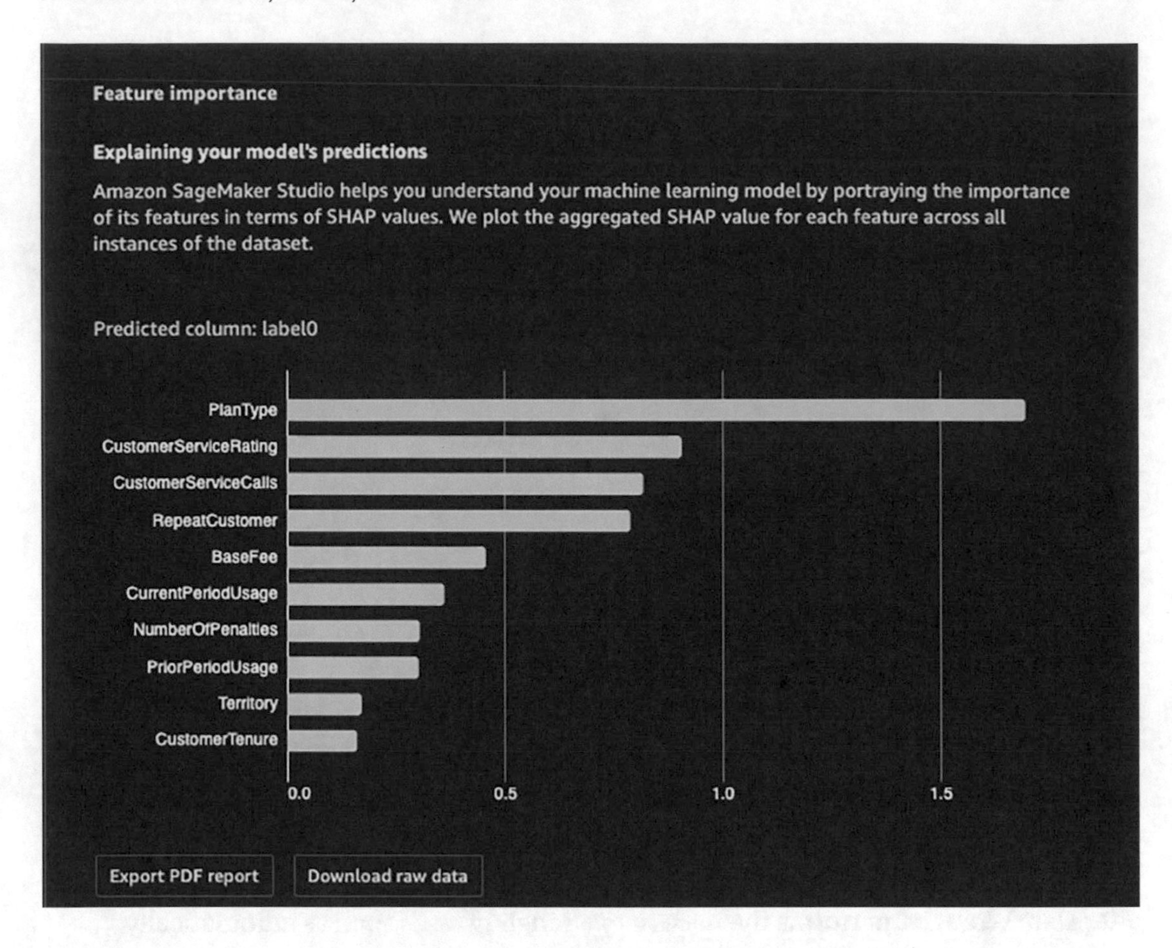

TensorFlow Extended (TFX)

Our final AutoML tool mentioned in this section goes to TFX. Although developed by Google, TensorFlow Extended (TFX) is an open source tool and considered here in its own right. Built for scalable, high-performance ML Production Pipelines/deployment, TFX extends TensorFlow execution pipelines and the tf.data API to end-to-end MLOps.

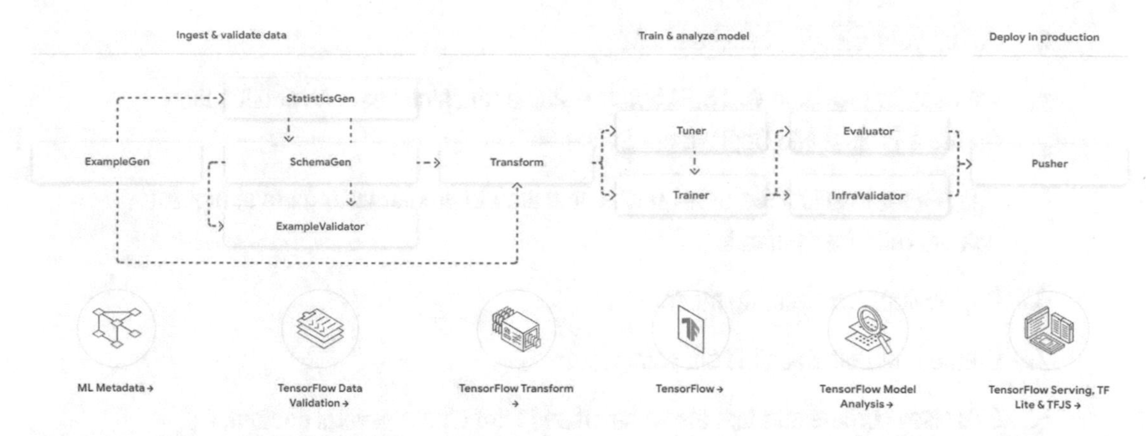

Figure 6-9. *TFX Pipeline operation*

Several big brands use TFX, including Spotify (for personalized recommendations) and twitter (for ranking tweets) and although TFX is not strictly a no-code tool for AI automation, we will take a look at how this works in one of our hands-on labs below.

Wrap-up

The above coverage of the main AutoAI tools being used today completes this chapter. While many strive to incorporate NoLo user interfaces to ensure stakeholder engagement extends beyond data team silos, the need for many of these applications to support model customization with Python remains. In the next chapter, we will turn our attention to developing AI applications - specifically taking back-end models beyond simple scripts to front-ended "full-stack" solutions.

AutoAI with IBM Cloud Pak for Data: Hands-on Practice

AUTOAI FOR PREDICTING CREDIT LOAN DEFAULTS

We now proceed to have a look at our AutoAI tools – first up is IBM Cloud Pak where we will run an AutoAI experiment for predicting customers likely to default on loans:

1. Download lab assets from `https://ibm.github.io/ddc-2021-development-to-production/setup/`

2. Sign up/log in to IBM Cloud Pak at the link below

   `https://dataplatform.cloud.ibm.com`

3. Set up your CloudPak environment
4. We need to provision WATSON Studio – add a Machine Learning service then Create a Project and Deployment Space
5. Run AutoAI to select features/choose best algo to predict customers at highest risk of credit loan defaults
6. Deploy your best scoring model
7. Create and test an online endpoint
8. Exercise – create and test also a batch endpoint where several customer records are entered as a batch and predictions returned for all of them
9. Exercise (stretch) – try to integrate your deployed model with a sample (Flask[20]) application
    a. copy .env file
    b. add API key and endpoint
    c. install and initialize virtual environment, install dependencies
    d. run app locally on your machine
    e. test app
10. **NB make sure to stop your environment runtime after** use by following the instructions "Stop the Environment" here:

    `https://ibm.github.io/ddc-2021-development-to-production/ml-model-deployment/batch-model-deployment/`

[20] Flask is covered in more detail in Chapter 7

Healthcare diagnostics with Google Teachable Machines: Hands-on Practice

NO-CODE ML – XRAY IMAGE CLASSIFICATION

This lab takes a look at how Google Teachable Machines operates by loading in xray images from Kaggle and training a predictive model to detect health issues from the scans (in this case pneumonia).

See also `https://towardsdatascience.com/build-a-machine-learning-app-in-less-than-an-hour-300d97f0b620` for reference:

1. Download training images from Kaggle at the link below

   `www.kaggle.com/datasets/paultimothymooney/chest-xray-pneumonia`

   Note this is a large 2GB dataset so may take several minutes to download.

2. Unzip the images in your local drive – this should complete in < 5 minutes.
3. Go to the url `https://teachablemachine.withgoogle.com/train/image` to train images on Google Teachable Machines:
4. Upload normal xray scans from the unzipped training folder to Class 1 on Teachable Machines, upload the pneumonia cases to Class 2. Note that the xrays with pneumonia have areas of "abnormal opacification" – the xray is more "opaque"/less transparent
5. Select "Train Model," changing Batch Size from 16 to 128 to speed up the training process
6. Test the model on completion using images from the test set in the unzipped image folder
7. Export the model, choosing TensorFlow and Keras. We will need this in a later lab[21]

[21] The lab is expanded on in Chapter 9 where we will build with Streamlit and deploy on Heroku a full stack application integrated with our model trained here on Google Teachable Machines

TFX and Vertex AI Pipelines: Hands-on Practice

AUTODL WITH TFX, KERAS, AND VERTEX AI

This lab is based on the TensorFlow tutorial `www.tensorflow.org/tfx/tutorials/tfx/gcp/vertex_pipelines_simple` – the goal is to use Google Cloud Vertex Pipelines to automate a TFX pipeline for training a deep learning model:

1. Activate your free trial on GCP by pressing the blue ACTIVATE button in the top right if you have not already done so. You will need to enter credit card details to activate the $300 of free GCP credit over three months – take care to monitor usage in your Google Cloud Portal[22] although Google claim they won't auto-charge when credit limits are exhausted.

2. After activating your free trial, go back to the Create a Vertex AI dashboard `https://console.cloud.google.com/vertex-ai` and create a project. Make note of your project ID.

3. Create a Cloud Storage bucket in the region closest to your location by following the four steps here: `https://cloud.google.com/storage/docs/creating-buckets`

   Make note of your bucket name and region for step 6b below

4. Enable Vertex AI and Cloud Storage APIs by confirming your project, then enabling at the link below:

   `https://console.cloud.google.com/flows/enableapi?apiid=aiplatform.googleapis.com,storage-component.googleapis.com`

5. Download the notebook from the GitHub repo below and run it in Colab:

   `https://github.com/bw-cetech/apress-6.4-tfx-vertex-ai.git`

[22] Go to Google Cloud Console `https://console.cloud.google.com/` and select Billing from the Navigation (hamburger) Menu in the top left corner. Remaining credits are shown in the lower right corner of the billing dashboard.

6. Run through the notebook steps making sure to restart the runtime after installing the dependencies at the start

   a. Log into your google account from the notebook

   b. Set up your variables (project, region, and bucket name)

   c. Prepare the example data from the Palmer Penguins sample dataset

   d. Create the TFX Pipeline

   e. Run the pipeline on Vertex AI Pipelines

7. The TFX pipeline at the end is orchestrated using Vertex Pipelines and Kubeflow V2 dag runner. Make sure to click on the link shown in the last cells output to see the pipeline job progress in Vertex AI on GCP, or visit the Google Cloud Console: `https://console.cloud.google.com/` to see the API requests.

NB make sure to delete resources on GCP after finishing the lab, namely your pipeline run, Colab notebook, Cloud Storage bucket, and project

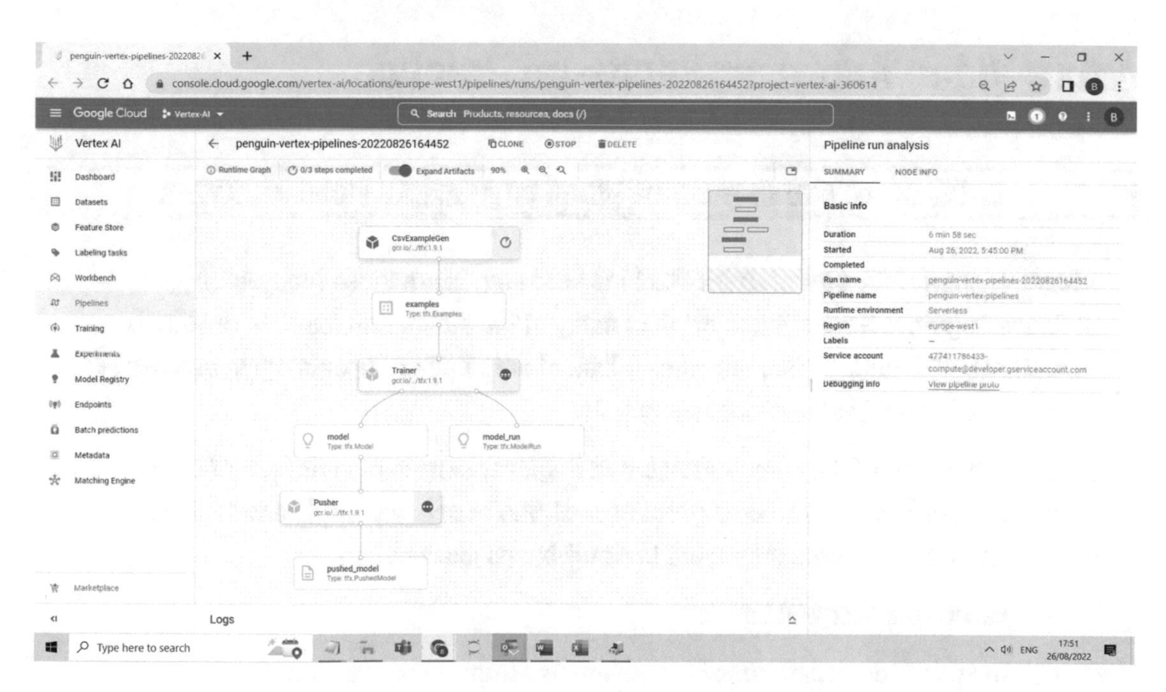

Figure 6-10. *Vertex AI (TFX) Pipeline progress on GCP*

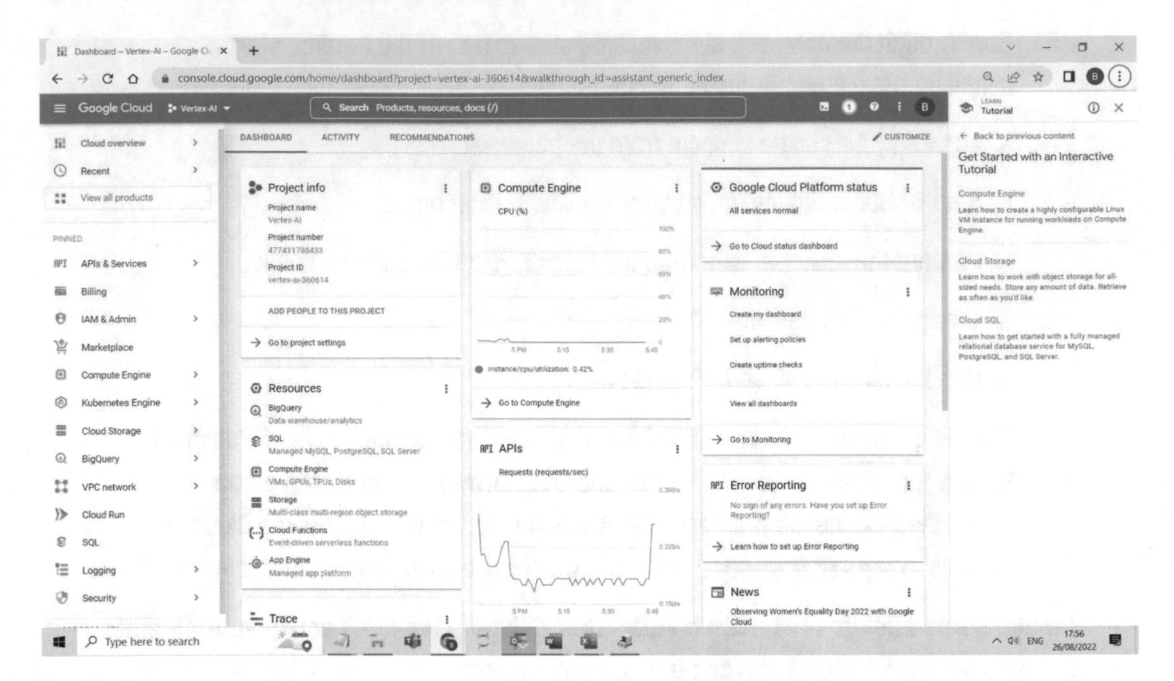

Figure 6-11. *Google Cloud Console Vertex/TFX Pipeline API calls*

Azure Video Analyzer: Hands-on Practice

CATEGORIZE AND CATALOGUE VIDEO WITH AZURE VIDEO ANALYZER

Although not strictly part of Azure Machine Learning, Azure Video Analyzer is part of Azure Cognitive Services and exhibits many of the automated features central to AutoML/AutoDL/AutoAI. This lab takes a look at automating the cataloging process of video metadata and video sections/snippets.

NB this lab is part of the Microsoft Azure AI Engineer certification `https://docs.microsoft.com/en-us/learn/certifications/azure-ai-engineer/` – highly recommended for readers interested in cloud-based certification:

1. Launch the Microsoft Lab below

   `https://docs.microsoft.com/en-us/learn/modules/analyze-video/5-exercise-video-indexer`

2. Login and Launch Virtual Machine
3. Clone the GitHub repo
4. Upload Video to Video Analyzer at `www.videoindexer.ai` (on your local machine, not the VM) – you will need to login using an Azure account (sign up here if not already done: `https://azure.microsoft.com/en-gb/free/`)
5. NB the video upload may take some minutes (5-10) as the video is indexed
6. Review Video Insights, selecting Transcript on the RHS of the screen and observing a moving transcript with speakers, topics discussed, named entities, and keywords as the video plays
7. Exercise – get your API key from `https://api-portal.videoindexer.ai` and use the Video Analyzer REST API via Visual Studio. The JSON response from the REST service should contain details of the Responsible AI video previously indexed
8. Exercise – run the REST API again, this time to get finer-grained insights. Check your solution against the PowerShell script at the GitHub location below:

   `https://github.com/bw-cetech/apress-6.4.git`

NB make sure to close Visual Studio and logout of the Virtual Machine on completion of the lab to avoid costs incurred on Azure

CHAPTER 7

AI Full Stack: Application Development

IDC estimate AI currently is a $341 billion market, and if global management consultants are to be believed,[1] could contribute $13 trillion (McKinsey) or $15.7 trillion (PwC) to the global economy by 2030.

Clearly, the market is booming but how do companies overcome historical "technical debt" associated with the accumulation of, and overreliance on standalone, highly technical scripted models and poorly managed Data Science teams?

The "Apex" skillset today is in converting an "Enterprise" AI Vision into a deployed reality. With the global pandemic having pushed AI to the top of the corporate agenda, empowering business resilience and relevance, the emphasis is now on how to refactor and embed demos and prototypes as full-stack AI solution across the Enterprise.

IDC predicts (Figure 7-1) AI service client demand for technical expertise to develop, implement, and manage AI applications is expected to grow in the next five years at a CAGR of 18.4%. Starting from key business/organizational needs for AI, the goal of this chapter is to identify the correct ML or DL solution and technologies to develop and deliver these "Full-Stack AI" solutions while developing some of these as boilerplate solutions in our hands-on labs.

[1] McKinsey famously predicted in 1980 that the cellphone market would reach 900,000 subscribers by 2000, less than 1% of the actual figure of 109m. `www.equities.com/news/a-look-at-mckinsey-company-s-biggest-mistakes`

B. Walsh, *Productionizing AI*, https://doi.org/10.1007/978-1-4842-8817-7_7

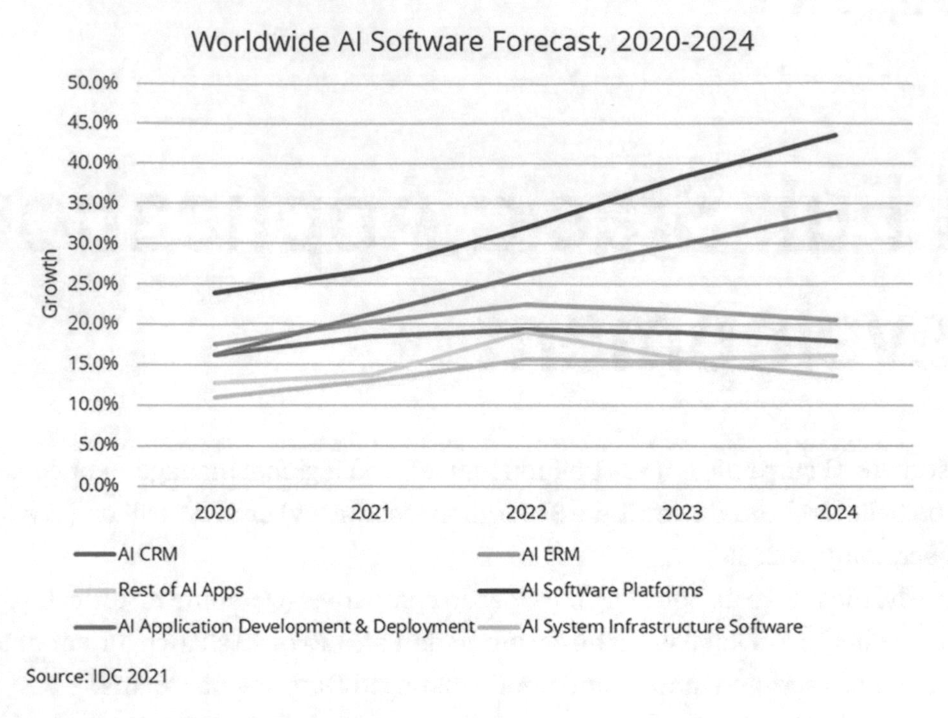

Figure 7-1. *IDC AI Growth Forecast to 2024*

Introduction to AI Application Development

In this first section, we take a look at the current drivers for AI application development, some basics in moving from script to solution as well as accelerators for Big Data handling: parallel processing, clusters, and GPUs.

Developing an AI Solution

Agile Software Development is the number 1 job on itjobswatch.[2] Stemming from a frenzied demand for AI solutions, this presents a huge opportunity for Full-Stack Data Scientists or AI Engineers.

[2] `www.itjobswatch.co.uk/jobs/uk/agile.do`. AI, Cyber and Cloud and Software Engineering/Development are also Indeed's top 5 "In-Demand Tech Skills for Technology Careers." See `www.indeed.com/career-advice/career-development/in-demand-tech-skills`

In 2022, most AI apps fit into one of three categories: Machine Learning, Computer Vision, or NLP but due to their technical complexity, machine and deep learning implementations have historically been operationally siloed with PhD-level statisticians left to explain code-heavy technical models.

Quick-fix or mock-up dummy/synthetic datasets have been relied on rather than implementing a Data Fabric approach to connect to (or create) sophisticated prelabeled datasets. Training and testing has often been limited to a Python notebook (Jupyter, Colab, etc.) and inference (often without an API) done as an afterthought. The end result is significant technical debt - certainly not an “app” as we know it.

The evolution is toward an “Enterprise AIaaS” solution, perhaps coupled with TinyAI: new algorithms to shrink existing deep learning models without losing their capabilities.[3]

AI Apps – Up and Running

How do we go about developing an AI app? From a process perspective, we recommend following an agile approach, start small, but don’t leave technical debt behind, the end goal should be Enterprise AI, even if that is a longer 2-5-year journey. The below seven-step plan is a good framework for success:

1. Pick an interesting topic/problem
2. Develop the low fidelity solution - find a quick solution and keep it open source - if possible! Use google to help (seriously!), framing the problem with **precise** keywords
3. Improve your simple solution - look for similar problem formulation and example Python code on Kaggle or on blogs/websites of one of the many great thought leaders in AI/DL or ML[4]
4. Share your solution and solicit feedback - make use of same specific forums to ask questions and for support/debugging

[3] Not mutually exclusive in our opinion as Enterprise AI is hard and takes time, so easier, incremental developments are still vital for commercial buy-in

[4] See recommended reference list

5. Repeat steps 1-4 for different problems – turn ambiguous/vague business/organizational objectives into concrete problems solvable with ML/DL

6. Complete a Kaggle competition and collaborate

7. Aim for a high-value solution – does it solve the organizational problem, is it measurable?

Successful deployment of an AI solution requires a huge amount of supporting infrastructure and an understanding of the target/To-Be architecture. We discuss two aspects critical to scaling a solution below – running machine and deep learning code as web services behind APIs and distributed computing.

APIs and Endpoints

Application Programming Interfaces (or APIs) provide a standardized way of communication between two software applications.

An API opens up certain **user-defined URL endpoints**, which are then used to send or receive data requests. REST (representational state transfer) APIs are one of the most popular specialized web service APIs which make use of URIs, HTTP protocol, and JSON data formats, but there are others, with Google's gRPC high-performance Remote Procedure Call (RPC) framework and Facebook's GraphQL suitable alternatives.[5] Both are now open-sourced.

There are clear benefits in exposing AI/ML models as an API, from UI/UX or providing a user-friendly analytics/model interface and workflow to endpoint stability, ability to include data validation[6] checks and security handling, separation of Data Science and IT functions, usability across a wider organization as well as multi-app reusability.

Besides internal productivity, there is additional external value creation from API-supported AI solutions also exposing Data Science models to a wider customer base. API endpoints/responses can be tested using cURL and/or Postman – both simplifying the steps in building an API and troubleshooting/debugging connection issues. Figure 7-2 illustrates this.

[5] There are multiple libraries for implementing GraphQL in Python including ariadne, graphene, and strawberry

[6] Data querying via GraphQL restricts the API response to providing only that information that is required, cf. the analogy of ordering from a menu rather than eating from a buffet. See also e.g. `https://medium.com/@kittypawar.8/alternatives-for-rest-api-b7a6911aa0cc`

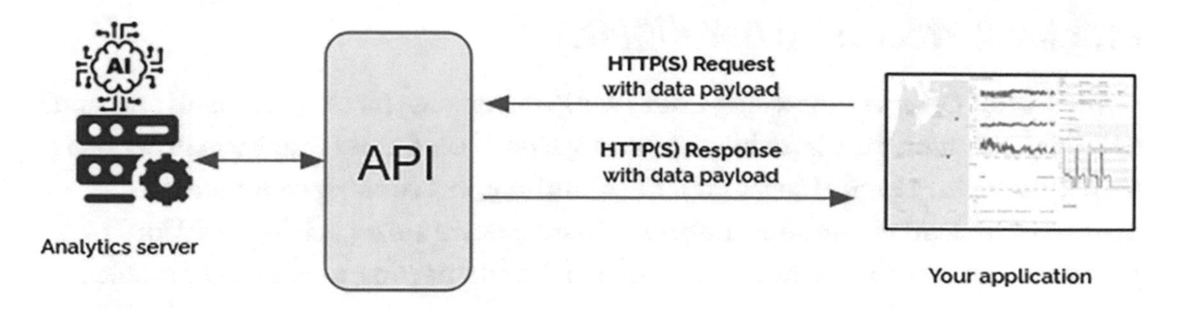

Figure 7-2. *Typical API architecture (source: ubiops)*

Distributed Processing and Clusters

As Deep (and sometimes Machine) Learning requires matrix computations on large training sets, distributed computing makes it possible to perform these computations in parallel, thus saving on time. Most production-grade solutions today require this big data processing capability to support the underlying AI application.

Clusters

Clusters are a group of nodes (computers) in a high-performance computing (HPC) system (Figure 7-3) which process distributed workloads using parallel processing.

The most famous example is Apache Hadoop - a framework that allows for the distributed processing of large datasets across clusters. See Chapter 3 for more on Hadoop.

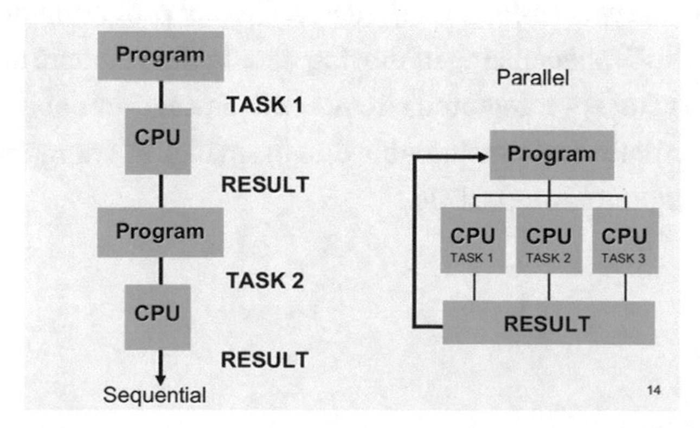

Figure 7-3. *Parallelization in computing - use of multiple CPUs to split ML/DL processes into concurrent tasks*

Graphical Processing Units (GPUs)

While CPUs (Central Processing Units) work well enough for simple models and small datasets, execution time is prohibitive for larger datasets. **GPUs (Graphical Processing Units) have more logical cores than CPUs and support running concurrent processes instead of one after the other. As such there** are especially suited for problems that can be expressed as data-parallel computations, such as an AI problem that has many hundreds of thousands of parameters.

GPUs are more famously utilized in modern multiplayer games but their evolution to handle applications that involve arithmetic on large amounts of data has seen all CSPs offering pay-per-use high-performance GPUs on their cloud platform.

Under the hood, GPUs are generally dependent on CUDA (Compute Unified Device Architecture) parallel programming developed by NVIDIA.

TensorFlow Processing Units (TPUs)

TPUS (TensorFlow Processing Units) are Google's hardware accelerators specialized in deep learning tasks.

Typically used by researchers and developers, they are more expensive than GPUs but highly optimized for large batches and CNNs. On the whole they are faster and more energy efficient, but for mid-large datasets GPUs can still be most performant.

Parallel processing/computing is a fast-moving area of research and development, with all the main high-profile media AI projects (self-driving cars, robotics, for example) reliant on the latest technology. Two of interest are likely to be competitive in the near future: DPUs (data processing units) – a new class of programmable processor or "system on a chip" (SoC) specializing in moving data in data centers and FGPAs (field programmable gate arrays) – integrated circuits with a programmable hardware fabric. FGPAs are also potentially more sustainable due to savings in energy consumption compared with (energy-guzzling) GPUs.

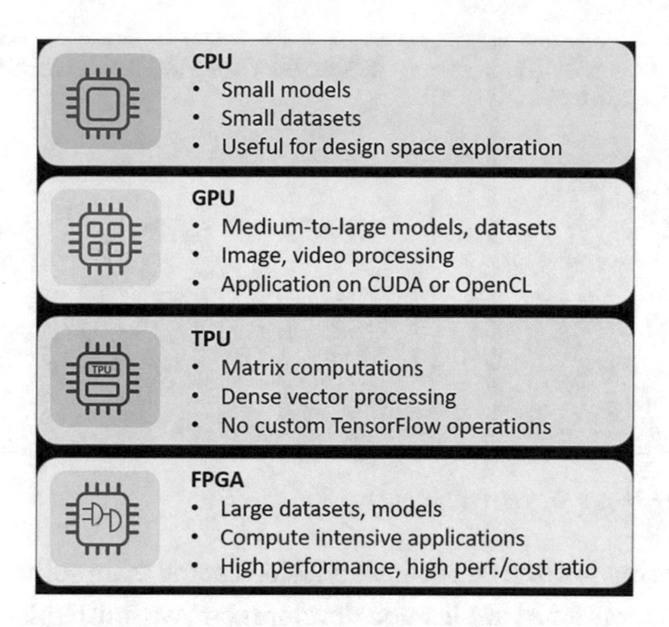

Figure 7-4. *Comparison of latest parallelization technology (Source: inaccel)*

Sharding

Sharding is the process of breaking up large tables into smaller chunks called shards that are spread across multiple servers.[7] A shard is essentially a horizontal data partition that contains a subset of the total dataset, and hence is responsible for serving a portion of the overall workload.

Sharding in Deep Learning is capable of saving over 60% in memory and training models twice as large in PyTorch.

[7] Similar to the partitioning process served through Resilient Distributed Datasets (RDDs) in Apache Spark. See Chapter 5

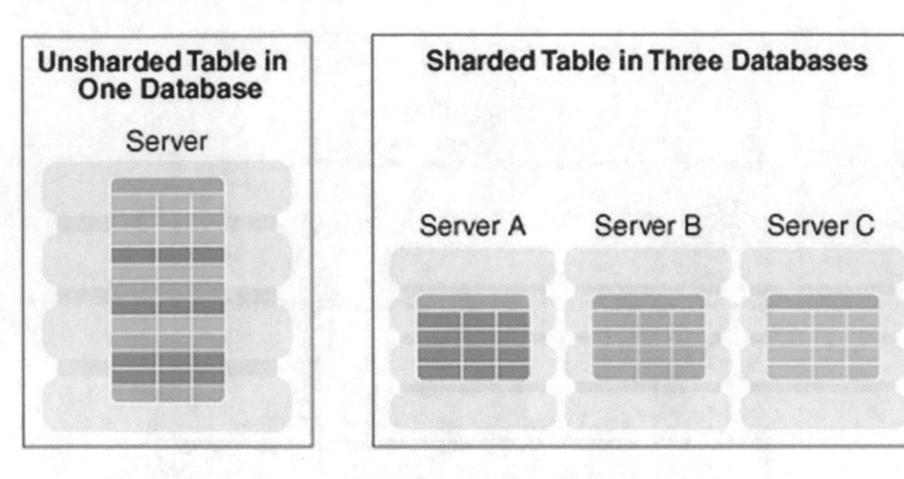

Figure 7-5. *Sharding (Source: Oracle)*

Before proceeding to our next section and a look at the main software vendors and their ecosystem of tools for AI application development, we finish this section with a look at creating isolated (virtual) environments for application development and some hands-on practice around scripting and working with APIs in Python before wrapping up with the use of GPUs in Colab.

Virtual Environments

Virtual Environments offer isolated Python package installations for specific apps allowing apps to coexist independently on the same system. Because these are self-contained Python (and Python library) installations where an application does not share dependencies with any other application, they are particularly useful for developing AI solutions.

A virtual environment in Python is created using **virtualenv** (or **venv** from Python 3.3+) which includes a Python binary and essential tools for package management including pip, for installing Python libraries. The following hands-on lab describes how to set up a virtual environment from terminal.

Running Python from Terminal: Hands-on Practice

GETTING AWAY FROM NOTEBOOKS – PYTHON SCRIPTING (.PY)

Many AI applications utilize raw Python scripts (.py files) instead of Jupyter notebooks. Using standalone Python, the goal of this exercise is to get familiar with these scripts, create a virtual environment and run the script:

1. Install a stable release of Python from `www.python.org/downloads/windows/`
2. Add the path of the installed location on your laptop as a (system) environment variable – this will enable both Python and pip (library installer) commands to work on terminal
3. Create a "test" folder in a suitable location on your local directory. Clone the Python script to your local drive from GitHub below:

   `https://github.com/bw-cetech/apress-7.1.git`
4. On File/Windows Explorer go to the cloned local folder and type "cmd" in the pathname to open terminal. Create a virtual environment using the commands below (one after the other):

   ```
   python -m venv env
   env\Scripts\Activate
   ```
5. Your virtual environment should now be enabled. Run the simple game with the command
   `python python-guessing-game.py`

   when finished, close the virtual environment with "deactivate"
6. Stretch exercise – try to run the random walk exercise done in Chapter 5 (Neural Networks and Deep Learning) as a standalone Python script, after reactivating the same virtual environment above. NB for any ModuleNotFoundError: No module named errors (e.g., numpy), run the following command:

   ```
   pip install numpy
   ```

API Web Services and Endpoints: Hands-on Practice

MACHINE LEARNING FOR CYBERSECURITY DDOS ATTACK

The goal of this exercise is to deploy, then test/perform inference via an endpoint on a machine learning model for detecting cybersecurity (DDoS – Distributed Denial of Service) threats:

1. Starting from the experiment below trained on labeled network traffic events, we deploy a machine learning model from Azure Machine Learning Studio by first creating a predictive experiment

   `https://gallery.cortanaintelligence.com/Experiment/Cyber-DDoS-trained-model`

2. Next we set up an endpoint by deploying the Web Service
3. Finally consume the Web Service by invoking the API from excel. Test using the sample data at the GitHub link below. NB the first record is a "teardrop" denial of service (DoS) attack, the second record is benign network traffic

   `https://github.com/bw-cetech/apress-7.1b.git`

4. Exercise: walk through the model training process as well by following the link `https://gallery.azure.ai/Experiment/e7fb30de726e4e02b034233ec6c34ce4`. Note the training experiment originally used Azure Blob storage links to the datasets but as these are no longer supported, now uses instead the training network_intrusion_detection.csv and test sets network_intrusion_detection_test.csv at the same GitHub link above
5. Stretch: see if you can beat the model performance (AUC = 0.85) by changing the data (or the algorithms) used

AI Accelerators - GPUs: Hands-on Practise

PERFORMANCE TESTING COLAB GPUS

The final lab in this section compares runtime for a Big Data pipeline – specifically the time to download a zip file directly from Kaggle then unzip the file.

1. Obtain an API key from Kaggle (kaggle.json file) or use the same key used in labs from Chapter 4 and 5
2. Now download the python script "GPU_test.ipynb" at the GitHub link below and open in Colab

   `https://github.com/bw-cetech/apress-7.1c.git`
3. Drag and drop your kaggle.json file to the Colab default (content) folder
4. Run the Colab notebook to copy the json file to a root > .kaggle folder
5. Connect directly to a BIG dataset on Kaggle – the example given will download a 350 MB dataset containing 50,000 images. Do this initially with the standard (no hardware accelerator / CPU) runtime and time how long it takes
6. Unzip the images to a "review" folder – still using the standard runtime, timing how long it takes
7. After completion (or after interrupting the code if taking too long), repeat steps 5 and 6 above with Runtime type set to GPU. Compare the time taken for both steps with the standard runtime.

   NB Although the lab is focussed on the data import process, the same parallel process efficiencies extend to machine and deep learning, providing substantial runtime savings when training and deploying models
8. Exercise – run the additional python notebook "Autoencoders.ipynb" at the GitHub link above first using a local runtime, then with a GPU. Verify that the model training time is substantially quicker using a GPU.

Software and Tools for AI Development

Building on our first section, we now take a look at practical AI application/software development, comparing the main cloud vendors and their tools and services available to support AI projects and infrastructure, develop industry-specific use cases and solutions in order to ultimately deploy a successful application.

AI Needs Data and Cloud

The importance of cloud for AI application integration, and data to feed the application cannot be underestimated. While Cloud is a key enabler of AI, Cloud only works for AI if the Data Strategy is underpinned by rich, BIG data sources and/or training data.

Successful Data Projects today are now (Enterprise) AI-Infused, requiring an end-to-end Cloud Infrastructure and in particular, Storage and Compute as the main Cloud components for Big Data processing.

While Enterprise Machine Learning projects can run with low overheads on both, Deep Learning projects cannot.

As we will see below, all major cloud service providers provide a catalog of AI services and tools which greatly simplify the process of building applications. But care must be taken around hidden costs.[8] The big 3 of AWS, Azure, and GCP have made recent attempts to make pricing models less opaque, but anyone wishing to do experimentation without a supporting corporate budget is clearly disadvantaged.

[8] Despite the hidden costs, 2021 saw a 33% increase in global cloud spend driven by intense demand to support remote working and learning, ecommerce, content streaming, online gaming and collaboration

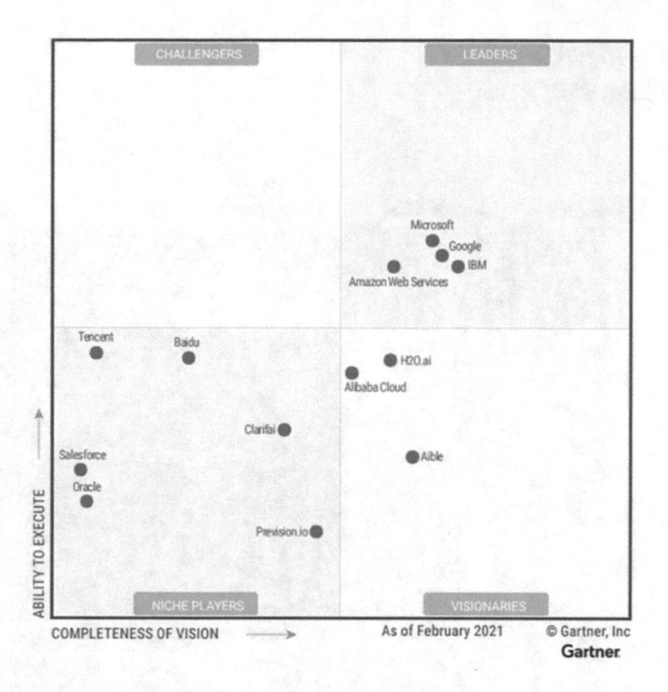

***Figure 7-6.** Gartner Magic Quadrant – leading CSPs*

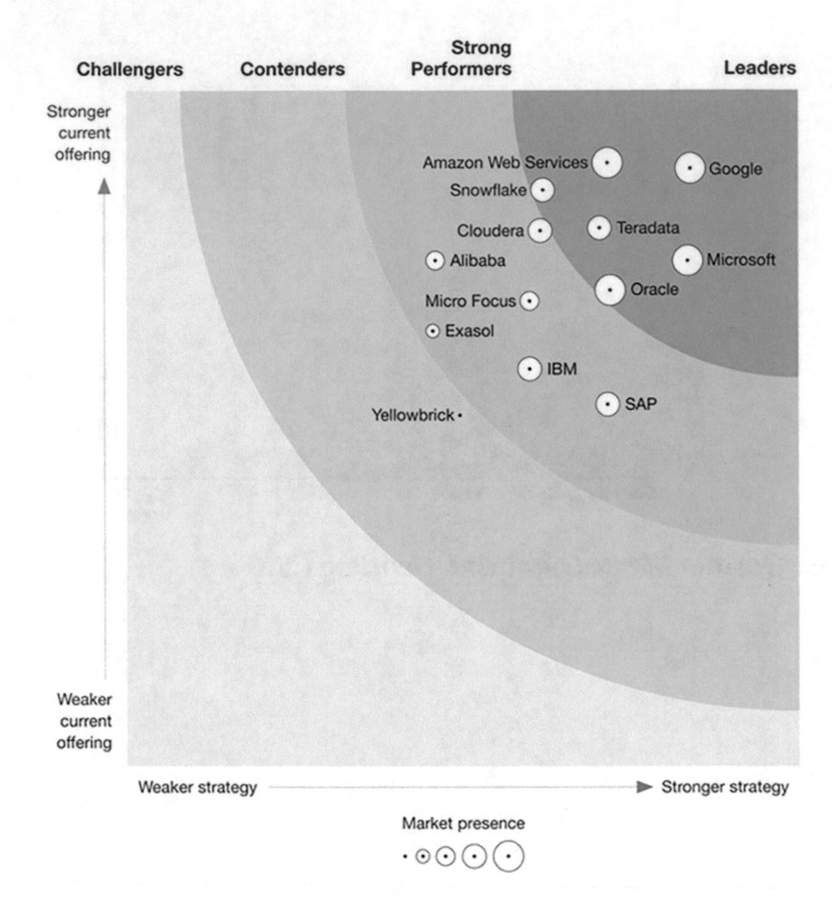

***Figure 7-7.** Forrester Wave – leading Cloud Data Warehouses*

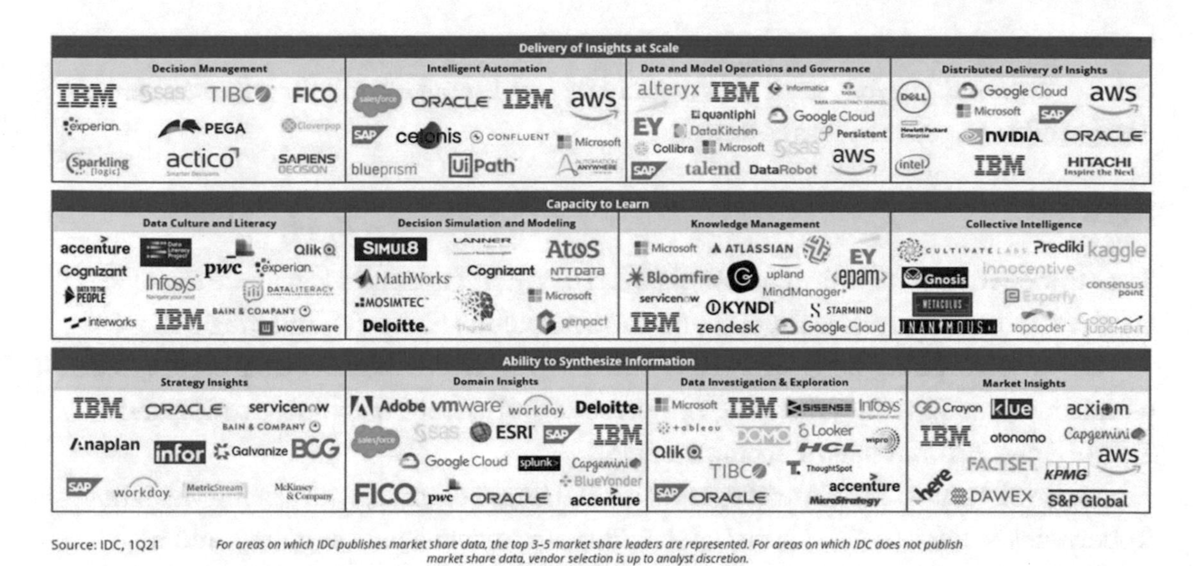

Figure 7-8. *IDC Market Glance – future of intelligence*

Cloud Platforms

There is general consensus on the leading cloud platforms, but we will now drill down further into the main services and resources offered by each, starting with Amazon Web Services.

AWS

AWS claims to offer the broadest and deepest set of machine learning services and supporting cloud infrastructure. SageMaker is the main tool for Data Science due to its scalability in machine learning while for NLP, AWS also offers Amazon Polly (text-to-speech) and Lex (Chatbots).

Evident from the list of customers using AWS, such as Siemens, FICO, Formula1, pwc, and Netflix, is a strong set of industry use cases for AI including Document Processing, Fraud Detection, and Forecasting.

As the world's biggest cloud service provider, many AI applications rely on AWS's breadth of supporting infrastructure, from Simple Storage Service (S3) cloud storage to Elastic Compute (EC2) instances, Elastic MapReduce (EMR) – managed clusters for running Apache Spark, Redshift – AWS's cloud data warehouse, Lambda – serverless compute for processing events and Kinesis for real-time streaming.

However despite appearances, AWS's "free tier" isn't completely free. AWS products can be *"explored" for free*, *some* products are free and some are *free for a limited period.* Almost all have capacity/usage limits.

Azure

The strength of Microsoft's Azure AI platform lies in API access to key Azure cloud services and an impressive list of clients from Airbus, NHS, Nestle, and the BBC.

Using Azure SDKs, simple API calls from Jupyter Notebook and Visual Studio Code enable integration with underlying Python code and sklearn machine learning and TensorFlow/PyTorch deep learning models.

Azure SDKs also enable access to Azure Machine Learning, with scaling via Azure Kubernetes Service (AKS), Azure Databricks (with Apache Spark support), and high-quality vision, speech, language, and decision-making models in Azure Cognitive Services including Anomaly Detector, Content Moderator, LUIS and QnA Maker (knowledge-based chatbots/conversational AI), speech to text to speech and Computer (prebuilt models) and Custom Vision (build your own models).

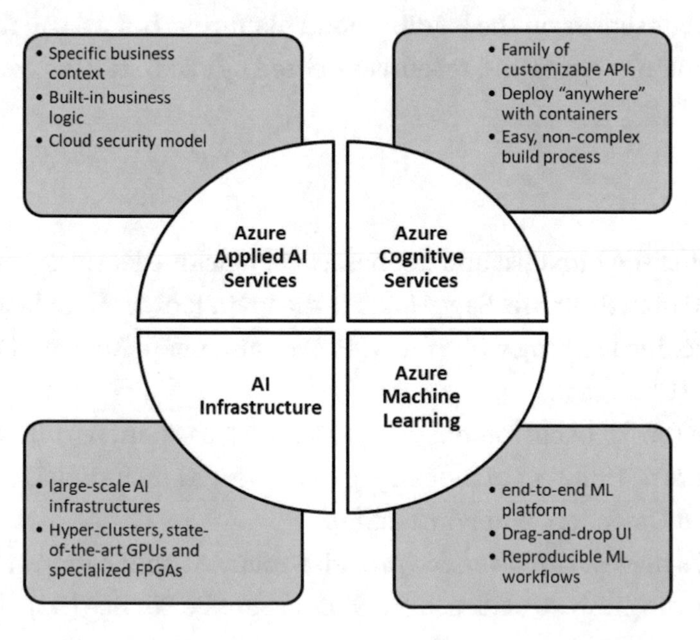

Figure 7-9. *Azure AI*

GCP

Although it's now open source TensorFlow is Google's child, is still developed by Google researchers and is probably still the USP for developing on Google Cloud Platform (GCP) for AI projects.

When coupled with BigQuery, Google's serverless Data Warehouse, Vertex AI, managed machine learning, and Colab for running Jupyter notebooks Google has a wealth of strong products for Data Scientists and AI Engineers.

Previously under the Google AI Platform umbrella, Vertex AI is the preferred "managed service" for taking sandbox machine learning models into production and includes Cloud AutoML high-quality low-code models with state-of-the-art transfer learning. It also supports Deep Learning containers and sharing of code from the centralized repos and ML pipelines on Google AI Hub.

The same ML pipelines on AI Hub can be deployed to highly scalable and portable Kubeflow pipelines based on Docker.

The Google Natural Language API provides app integration with Google NLP models while Dialogflow is used to integrate chatbots and IVAs into mobile and web apps.

IBM Cloud

While IBM may no longer be in the top tier of Big Tech companies, and IBM Cloud lies outside the "Big 3" cloud offerings, there is no doubt that the company continues to lead in terms of innovation, often rolling out new developments and services before mainstream catches on.

IBM's Watson platform has pioneered Commercial AI since winning the "Jeopardy!" quiz show in 2011 and continues to drive IBM's current suite of AI services provisioned on IBM Cloud. The main tools are shown below:

- WATSON Studio – Data Science platform
- WATSON Assistant – Chatbots/IVAs
- WATSON Discovery – Enterprise AI search
- WATSON Knowledge Studio – domain-specific NLP curation

More recently, IBM Cloud Pak® for Data (CPDaaS) has become IBM's go-to business data and AI platform-in-a-box, "infusing" applications with AI while automating (AutoAI) and governing data and the AI lifecycle.

Cloud Pak is a one-stop shop for collecting, organizing, and analyzing data assets for undertaking machine and deep learning. Composed of integrated microservices benefitting from running on a multinode Red Hat® OpenShift® cluster, Cloud Pak provides open and extensible REST APIs, support for hybrid cloud and on-prem resources and elastic resource management with "minimal" downtime.

Heroku

One of the first cloud platforms, Heroku is now owned by Salesforce. Lesser known but one of our favorites in terms of simplicity and elegance, apps can be deployed, managed, and scaled on cloud much like the other leading platforms.

In our opinion, it is the most cost-effective cloud for quick application deployment. The Free Plan doesn't force users/developers to spend credits on enabling tools for machine/deep learning, provided you limit monthly usage. Azure VMs and AWS SageMaker, for example, are not free, but you can deploy an ML/DL app on Heroku for free.

Scaling models is also intuitive - with a simple "pay as you go" option for app monthly uptime. The hosting models are done through Heroku **dynos** - building blocks that underpin/power Heroku apps. Essentially these are containers, but each dyno type comes with a specific number of cluster workers (free/hobby dyno 1 cluster, standard 2 clusters, medium-performance 5 clusters, high-performance 28 clusters).

Heroku brings to a close the main cloud platforms available to build an AI solution. The remainder of this section looks at front-ending these solutions with Python-based user interfaces.[9]

Python-Based UIs

We address three main Python frameworks for creating web apps below. We will take a look at another, Streamlit, in Chapter 9 which uses a simple API, supports interactive widgets defined as Python variables, and deploys pretty quickly.[10]

---

[9] For Big Data Engineering/parallel processing, specifically Apache Spark, that supports AI applications, see Chapter 5. We will also cover another parallel computing platform, Dask in Chapter 9. For front-ending solutions, see also use of react.js and VueJS in Chapter 2 - we will look at an end-to-end application deployment built with a React UI at the end of this chapter.

[10] Streamlit is used by a growing list of Fortune 50 companies including Tesla, IBM, and Uber

Flask

Flask is a micro web framework written in Python.

Simple, self-contained but scalable code, mostly suited for single-page applications, SQLAlchemy can be used for database connectivity and Flask comes with a wider range of database support (such as NoSQL) than Django (discussed below).

As we will see in the hands-on lab that follows this section, after installing Python for standalone scripting together with Visual Studio Code, the recommended workflow is to clone a flask app from GitHub, cd into the local copy of the app and create a virtual environment with

```
python -m venv env
env\Scripts\Activate
```

Flask is then installed within the virtual environment in the normal way, that is,

```
pip install flask
```

Dash

Dash converts Python scripts to production-grade business apps. Filling a perceived **Predictive Analytics gap** in Traditional BI/Tableau/PowerBI/Looker dashboards, Dash supports complex Python analytics /BI and is written on top of Flask, Plotly.js, and React.js.

Dash also provides a "point-and-click interface" to Machine Learning and Deep Learning models, greatly simplifying the process of front-ending AI apps with, for example, object detection and NLP user interfaces (e.g., chatbots).

The **pip install dash command will** also install a number of other tools:[11]

- dash_html_components
- dash_core_components
- dash_table
- plotly graphing library

The attraction of Dash is in the potential to quickly wrap a user interface Python code; it does this via a file structure of four files with html-like elements:

[11] Dash should be upgrades frequently with **pip install [package] --upgrade**

Layouts - describe how the dashboard will look

Components - Components make up a layout, comprising dash_core_components and dash_html_components

Callbacks - control the interactivity of a dash app

Bootstrap - prebuilt CSS Framework for creating interactive and mobile-ready applications

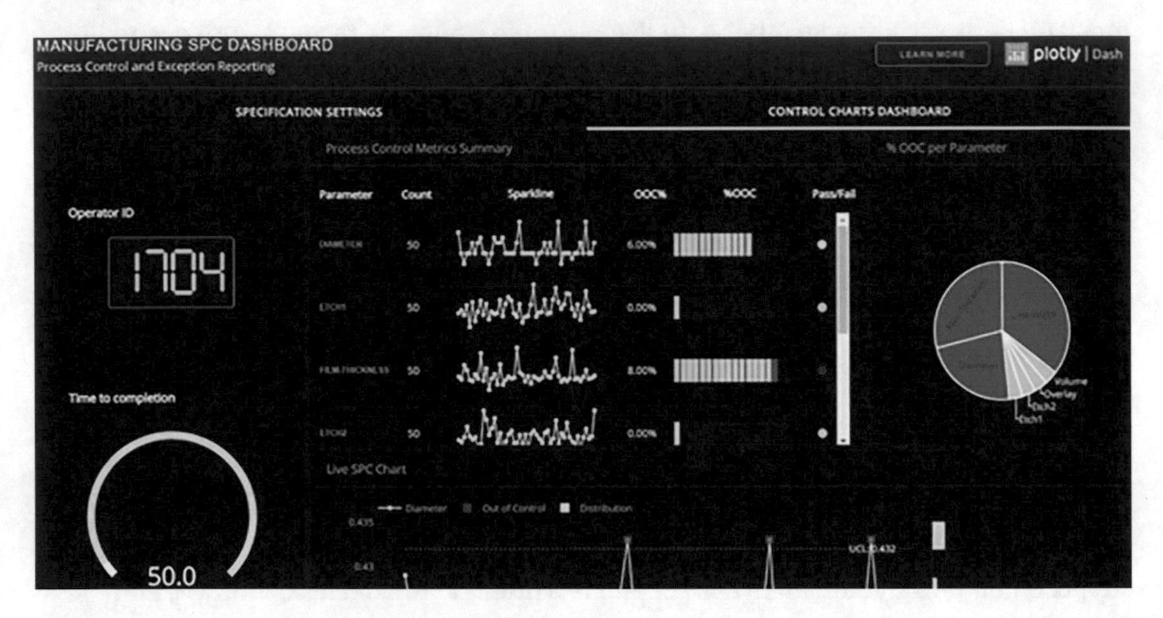

***Figure 7-10.** Dash UI for IoT streaming data alerts*

Django

Used by Facebook, Instagram, and Netflix, Django is designed to make it easier to build better Web apps more quickly and with less code. Like Flask, Django is a high-level Python Web framework for rapid development of web apps, but in contrast these apps tend to be full-scale and more powerful than simpler Flask limited-page apps.

Django is free and open source and has a "batteries-included" framework with most of the features preinstalled. It comes with automated tools to avoid repetitive tasks and a clean and "pragmatic" design which takes care of much of the hassle of Web development with a simplified three-step process for making model changes:

- Change models in **models.py**
- Run python **manage.py makemigrations** to create migrations for changes
- Run python **manage.py migrate** to apply those changes to the underlying database

Other AI Software Vendors

We take a final look in this section at some of the other AI software vendors that in their own way can support AI application building.

ONNX (Open Neural Network Exchange)

ONNX is an open-source model format and runtime for machine learning with a platform-agnostic design facilitating moving between different frameworks and hardware platforms. Its attraction in an age of Big Tech dominance/consolidation is its community model and interoperability.

C3

C3 is an AI and IoT software provider for building enterprise-scale AI applications. It offers out-the-box/prebuilt, industry-specific AI applications to optimize critical processes – C3 claims to run 4.8 million AI models and 1.5 billion predictions per day.

DataRobot

DataRobot's entire business model (and USP) is MLOps automation and "accelerating data to value." The product targets less technical users, for example, Business Analysts looking to build predictive analytics with no knowledge of Machine Learning.

We now move onto have a look at three of the tools covered above in our hands-on labs for this section.

Introduction to Dash: Hands-on Practice

DEPLOY AN IOT APP WITH DASH

Using sample Dash app boilerplates on GitHub, the goal of this exercise is to clone an IoT app's source code and run it locally. This lab includes a number of stretch exercises to edit source code, work with Dash callbacks and deploy a Dash app to Heroku to gain further experience with Dash:

1. We are going to clone the app shown at the link below:

   `https://dash-gallery.plotly.host/dash-manufacture-spc-dashboard/`

2. Go to `https://github.com/dkrizman/dash-manufacture-spc-dashboard`, click on green button to download a zip file containing the source code
3. Unzip the folder into a suitable location on your local drive
4. Copy the code from the app.py file to a notepad and save as app.ipynb
5. Open the new Jupyter notebook, change Debug = True in the last line to Debug = False and try to run it. NB you may need to install dash_daq first be running (then commenting out) %pip install dash_daq in the first code cell.
6. Click "Proceed to measurement" to see streaming metrics typical for a IoT sensor data. Press "stop" to stop streaming

   Exercise (Stretch): clone the Jupyter notebook Dash-Jupyter-getting_started.ipynb at the GitHub link `https://github.com/bw-cetech/apress-7.2.git` and run the code locally. Try to:

   a. Change the color shown in the scatter plot

   b. Increase the size of the plots

   c. Change the line plots to bar plots

7. Exercise (Stretch): clone Dash-InteractiveChart.ipynb from `https://github.com/bw-cetech/apress-7.2.git` and change the chart to a European chart
8. Exercise (Stretch): try to deploy the IoT app to Heroku

Flask: Hands-on Practice

DEPLOYING A FLASK DASHBOARD

In this lab, starting from a boilerplate on GitHub, we create a virtual environment, install some Python dependencies and create a Flask dashboard

1. Clone the source code at the link below to your local drive

   `https://github.com/app-generator/flask-black-dashboard.git`

2. Using a suitable Interactive Development Environment (IDE) such as IDLE or Visual Studio, open Powershell and create a virtual environment

3. Install dependencies listed in the requirements file by executing the command below:

   ```
   pip install -r requirements.txt
   ```

4. Run the app with the command (in command prompt/terminal)

   ```
   flask run --host=0.0.0.0 --port=5000
   ```

5. Exercise – try to add authentication to the app

6. Stretch: change the font and font size on LHS menu and replace the "Daily Sales" chart with a different dataset

Introduction to Django: Hands-on Practice

DJANGO APPLICATION DEVELOPMENT

Following on from the previous labs on Dash and Flask, the goal of this exercise is to familiarize ourselves with another Python front-end (web) framework: Django

- **Set up virtual environment**
- **Creating a simple polls app**
- **Create SQLite tables**

- **Playing with the Django API**
- **Admin oversight**

1. Set up a virtual environment
2. Install Django into the virtual environment (NB first check if you have this installed already with py -m django –version)
3. Start the app by cd'ing into the mysite folder then running

    ```
    python manage.py runserver
    ```

4. Create a polls app by following the steps at the tutorial link below, and for the remainder of this lab

    ```
    https://docs.djangoproject.com/en/3.2/intro/tutorial01/
    ```

5. Follow the steps in the tutorial to "Write your first view"

Exercise: try to complete "Writing your first Django app part 2" by following the instructions on the next page of the tutorial. Make sure to go through the steps below to understand how the application interfaces to the underlying datastore for the dashboard and how to manage the application development process

- Create tables in the "lightweight" SQLite database
- Create database models and activate
- Play with the API
- Create an admin user and explore functionality

ML Apps

We now turn our attention to the specific Machine and Deep Learning applications developed and deployed in organizations and businesses today.

As we have already covered the high-level overview of AI applications in our first chapter, we focus this section on how these applications are being built in organizations and the tools used. The specific "industry perspective" will be covered in the next chapter on AI Case Studies.

Developing Machine Learning Applications

Almost all organizations and businesses today are looking to build an AI strategy through implementable Machine Learning solutions. The drivers are varied, although the actual applications themselves tend to be concentrated around a relatively small subset of supervised and unsupervised machine learning problems. The uniqueness of the solution tends to reveal itself more in the tailored use of data and feature engineering undertaken to model the outcomes.

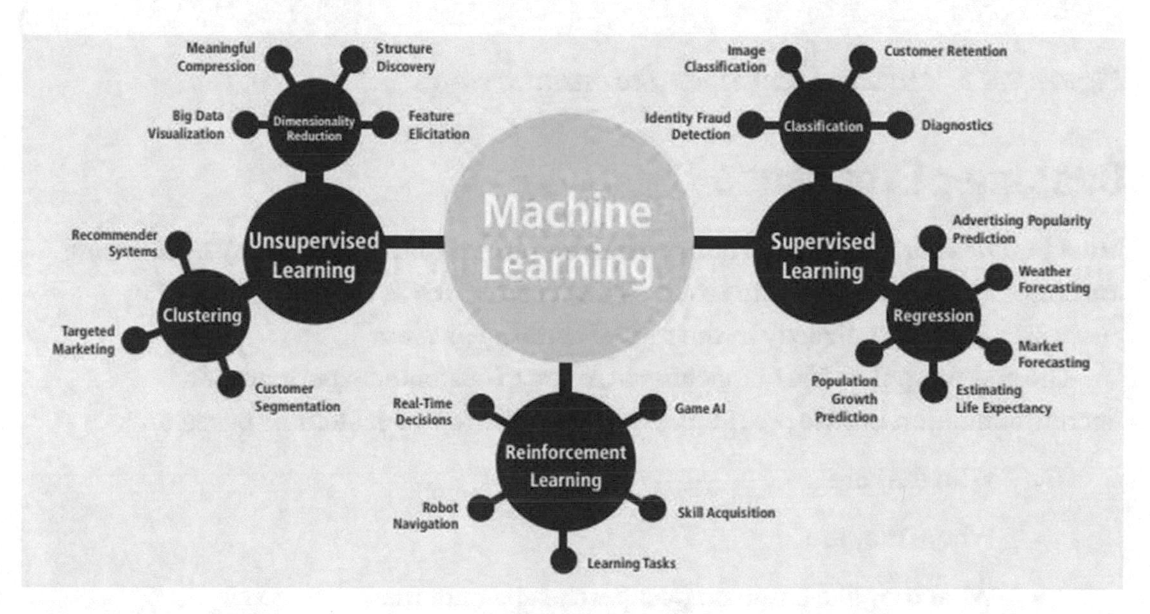

Figure 7-11. *Machine Learning Applications grouped by technique (Source: Forbes)*

By no means exhaustive, the table below describes what is driving businesses, what applications they are using, and the data enablers and sources that tend to be pulled on to get the job done. What is clear is that framing an organizational problem and documenting the delivery plan is key to successfully implementing Machine Learning – best practice is to embrace continual improvement over many iterative processes: from problem framing, through data collection and cleansing, EDA and data prep, feature engineering, model training, evaluation and benchmarking, inference and data drift.

Drivers

- Lower or Better Manage Costs
- Understand Revenue Opportunities
- Identify Market Trends
- Predict Customer Behavior
- Analyse Pricing Fluctuations
- Take the Pight Business Decisions

ML Applications

- Traffic Alerts / Maps
- Social Media Analytics
- Product Recommendations
- Virtual Personal Assistants
- Dynamic Pricing
- Google Translate
- Online Video Streaming
- Fraud Detection

Data Sources and Enablers

- Usage Log Files
- When (day/time) you Shop
- When you open/close/logon/logoff watch/pause/rewind/fast forward
- Customer Feedback Including:
  - Reviews
  - Ratings
  - Consumer Questionnaires
- Web Searches (about 3 million per day)
- Browsing and Scrolling Behaviour

Figure 7-12. *Machine Learning – the business case*

Customer Experience

Most businesses today are striving to achieve greater brand authority by tailoring and improving Customer Experience (CX) via predictive modeling and using customer feedback directly in the model training process

The classic application of machine learning to customer experience is a Recommendation Engine providing personalized offers to customers based on

- Who they are
- Where they are
- What they like based on past purchasing patterns
- Current conditions, such as the weather
- As well as outcomes from a retention model/their propensity to churn

Effective Customer Experience can provide bespoke predictions on the back of training a model across multiple transactional, demographical, and behavioral features such as

- **Transactional** – total customer revenue, minimum purchase, maximum purchase, time of last purchase
- **Demographic** – gender, age, address/city

- **Behavioral** – customer journey and browsing patterns including pages visited, time spent, device usage, likes, reviews, social media commentary, keywords used

Transactional and demographic data are long-established value levers, but in 2022, more sophisticated mining of customer journey and behavioral data, often unstructured, is what differentiates high-value retail organizations.

Fraud Detection and Cybersecurity

The $23.05b Fraud Detection and Prevention market is driving demand for Machine Learning and Fraud Analytics applications and shows no sign of let up. Fortune Business Insights forecast 27% CAGR to $142b by 2028 due to the increasing number of online transactions. And it's not just digital banking, insurance companies, and e-commerce increasing investment in secure online services, painful experiences of fraud across most private and public organizations have fast-tracked cybersecurity to the top of the corporate agenda.

The rationale is fairly obvious – leveraging the inherent strength of machine and deep learning algorithms to learn from historical fraud patterns and recognize them in future transactions.

The marketplace is somewhat congested with vendors such as FICO, AltexSoft, SAS, and DataVisor but the ability to leverage open source and Cloud fraud detection "boilerplates" using, for example, AWS SageMaker and IBM WATSON Studio is making cheaper, custom-developed solutions more accessible to AI Engineers and Developers.

Operations Management, Decision, and Business Support

Machine Learning use for Operations Management is largely focused on robotic process automation, cognitive or otherwise (RPA/CRPA), and planning and scheduling/rescheduling.

Decision Support systems built with a BI front-end (such as PowerBI, Tableau, Google Data Studio, or Looker) and predictive/prescriptive modeling (Python, CPLEX, etc.) often underpin usage in both Operations Management and SCM (Supply Chain Management).

Specifically in SCM end-to-end Machine Learning processes encompassing automated data collection through analysis, prediction and optimization are being adopted to deliver everything from automated dashboard engines to predictive maintenance, inventory management and materials planning, optimized purchasing and procurement, budgeting or customer/demand forecasting, and enhanced business process management (BPM) performance.

Risk Management, and Portfolio and Asset Optimization

Besides the risk management and forecasting examples mentioned in Chapter 1, there is considerable interest in the use of Machine Learning in Quantitative Finance, from constructing Investment Strategies to trading stocks.

The approach generally relies on forecasting to predict the movement of stocks and building bots (Robo-Advisors) that can look at the stock movement and directly recommend whether to buy/sell/hold. One of the more recent developments along these lines is the use of Reinforcement Learning in building optimal stock portfolios and comparing with portfolio theory-based approaches.

Other innovations are CVXPY, a Python-embedded modelling language for convex optimization problems which we will take a look at in the lab at the end of this section and DeepDow (Deep, rather than Machine Learning) - a Python package that focuses on neural networks to perform asset allocation in a single forward pass.

Developing a Recommendation Engine: Hands-on Practice

USING CURL TO TEST DEPLOYMENT TO A LOCAL ENDPOINT

We take a trained (content-based filtering[12]) Netflix movie / TV series recommendation engine model and deploy it as local endpoint and test the app with cURL:

1. Download GitHub source code from the link below `https://github.com/MAbdElRaouf/Content-based-Recommendation-Engine`

[12] **Content-based filtering** uses only the **existing interests of the user** as opposed to a **collaborative filtering model** which extends modelling to the entire user base and **seeks similarities between users and items** (movies here) simultaneously to provide recommendations. See also `https://developers.google.com/machine-learning/recommendation/collaborative/basics`

2. Set up a virtual environment
3. Install dependencies in the requirements.txt file
4. Run the app from the terminal
5. Add a JSON file to the local app folder with a movie title/TV series of interest

```
{
     "title" : "Narcos"
}
```

   Change the movie title to a movie/TV series

6. Test the app with cURL (installed by default with Windows systems):

```
curl -H "Content-Type: application/json" --data @test.json
http://127.0.0.1:5000/api/
```

   You should now see a list of recommended movies/TV series in a test.json file in your app folder based on the movie/TV series entered in step 5

7. Exercise: see what recommendations you get for "The Crown"
8. Stretch: Postman is another utility to explore and test APIs – set up a Postman account and send a request to the app above
9. Stretch: instead of content-based filtering, train and deploy a more sophisticated collaborative filtering model[13] which models similarity between users and movie / TV series purchases.

Portfolio Optimization Accelerator: Hands-on Practise

MAXIMISING PROFIT & MINIMISING RISK WITH CVXPY IN PYTHON

Although not strictly involving machine learning, prescriptive analytics problems requiring the use of complex optimisation techniques and solvers are often run in parallel (or as a post process) to ML/DL processes to maximise profit or minimise risk.

[13] See e.g. https://github.com/jaimeps/collaborative-filtering-netflix or https://pub.towardsai.net/recommendation-system-in-depth-tutorial-with-python-for-netflix-using-collaborative-filtering-533ff8a0e444

This lab takes a look at implementing Convex Programming with Python – in this case simplifying via a Linear Prigramming (LP) special case where the model constraints and objective function are linear.

1. Clone the below GitHub repo:

   `https://github.com/bw-cetech/apress-7.3.git`

2. Run through the python notebook steps in Colab:

   a. Import libraries including the CVXPY library

   b. Import the stock data from the monthly csv data file downloaded from GitHub

   c. Plot the data

   d. Compute expected risk and returns for each of the three stocks

   e. Optimise a $1000 across each of the three stocks to achieve a balanced portfolio

3. Exercise (stretch) – adapt the code to connect to live stock prices with Pandas DataReader and Yahoo Finance Python libraries, optimising / balancing £100k capital spend across three tech stocks

---

DL Apps

While certain machine learning applications are firmly established in many organizations, implementing Deep Learning applications has been far more an aspirational goal. That is starting to change, partly through the use of leading accelerator AutoAI tools we discussed in Chapter 6 and partly via experimentation and prototyping around the main use cases we discuss in this final section.

Developing Deep Learning Applications

Still relatively new in an "enterprise context," Deep Learning is starting to drive great advances in visual recognition, natural language processing, text analysis, and cybersecurity use cases.

Some of the drivers related to machine learning apply here too, but there is increased sophistication as well, such as in understanding customer journeys rather than just buying patterns. Some of these drivers are shown in the table in Figure 7-13, together with key business applications and benefits.

Drivers	DL Applications	Benefits
• Prospect to sales conversions • Employee efficiency • Quicker digitalization, archiving and image and video classification • Boost UX and CX • Faster processing of unstructured data sets • Enhanced cybersecurity	• Computer vision • Speech recognition and machine translation • Chatbots & IVAs • Disease detection and diagnosis • Bioinformatics & Drug Discovery • News Aggregation • Composing Music • Image Colouring • Robotics • Self Driving Cars	• Enhanced CX • Utilisation and realisation of potential of unstructured data • Automated feature engineering • Obsoletion of data labelling (both above in contrast to manual approaches in Machine Learning) • High performance / high-quality results • Waste and cost overhead elimination

Figure 7-13. *Deep Learning – the business case*

Forward planning on cloud storage and compute are vital in delivering a successful Deep Learning solution, and deep learning projects require a lot of iteration, a lot of time, and a lot of effort. But being disciplined, leveraging your resources to the max and monitoring progress along the way can help bring about success.

Like ML it starts by understanding the problem context and project lifecycle and a kaizen (continual improvement) approach is key. Before proceeding to our final section in this chapter, we outline below a DL-specific high-level framework for success (Source: neptune.ai):

- Define the data sources and collect
- Identify the high-level solution (CNN, RNN, GANs?)
- Establish a robust data pipeline for streaming/batch data
- Build a model based on ANNs
- Leverage Transfer Learning
- Training and inference
- Deploy to cloud: Heroku, IBM, AWS, GCP, Azure

Key Deep Learning Apps

Most of the below applications were covered in our introductions chapter so we wrap up this section with a brief summary of how major Deep Learning applications are generally being deployed and what tools are involved.

Computer Vision

No Computer Vision project today is developed without recourse to TensorFlow or OpenCV (open source computer vision tool). While TensorFlow has already been extensively discussed, OpenCV was originally developed by Intel and offers support for multiple programming languages and operating systems. OpenCV-Python is the Python library interface to the tool.

Most of the CSPs have APIs for Computer Vision, with IBM Watson Visual Recognition, API Google Cloud Vision, and Microsoft Computer Vision perhaps three of the best. Critical to these offerings is the huge storage requirement for Computer Vision coupled with low latency compute during critical training and inference processes, with ease of deployment (as containerized solutions) also increasingly important.

Forecasting

As described in Chapter 5, LSTMs have become highly accurate in forecast short- and mid-term horizons. These algorithms, combined (or compared) with fbprophet for longer-term forecasting with multiplicative seasonality where seasonality is added to the trend, provide a powerful multihorizon forecasting "arsenal."

The enhanced forecasting capability of neural networks means AI augmentation of traditional forecasting approaches is increasingly common – often LSTMs are compared with regressional techniques and exponential smoothing, ARMA/ARIMA/SARIMA/SARIMAX techniques, Monte Carlo, and VaR approaches.

IoT

More than 9 billion IoT devices currently exist online with some 50 billion to nearly 1 trillion devices expected online in the next decade.

This incredibly rich source of data is fueling the current trend toward the artificial Internet of things (AIoT) – the intersection of AI and IoT.

Effectively an embedding of predictive and cognitive capability in smart devices, this somewhat "Minority Report" – like dystopian "Smart Retail" use case is already globally prevalent. Cameras can be (and in many places are) equipped with computer vision while the bewildering granularity of data on our smart devices triggers real-time predictive (and attitudinal) analytics on consumer demographics and behavior as well as enablement of dynamic marketing/product placement.

There are plenty of other high-profile examples of IoT and AI working together including robotics, autonomous vehicles (self-driving/driverless cars), cashierless shopping (Amazon Go), drone traffic monitoring, and (telematics-equipped) fleet insurance.

Full-Stack Deep Learning: Hands-on Practice

DEPLOYING OUR FIRST AI APP WITH REACT, DASH, FLASK AND TENSORFLOW

The goal of this exercise is to deploy a deep learning application using a "hand-shake" between react.js and Flask:

1. Using a simple Early Stopping Criteria, train and export the VGG model from Chapter 5
2. Put the exported model above into a new local folder for the app called "dl-traffic-app"
3. Create a react.js front-end boilerplate by cd'ing into this folder from terminal and running the commands below (this will create a folder called react-frontend for the front-end source code we need):

   ```
   npm install -g create-react-app
   npx create-react-app react-frontend
   cd reactapp
      npm start
   ```

   NB install nodejs and npm from here if not already installed: `https://nodejs.org/en/download/`.
4. Create an (empty) Flask backend folder called "flask-backend," with two subfolders: static, templates, and a (blank) main.py file

5. Create a virtual environment

6. Install dependencies by creating a requirements.txt file with the following libraries:

```
numpy
flask #==1.1.2
dash
dash_bootstrap_components
matplotlib
tensorflow
opencv-python
```

7. To have the Flask backend serve the react.js front-end edit the blank main.py file with the contents of the file sample_flask.py here:

   https://github.com/bw-cetech/apress-7.4.git

8. A number of changes have to be made to the react front-end. Rather than detail these here, watch this video: www.youtube.com/watch?v=YW8VG_U-m48&feature=youtu.be

9. Fire up the Flask backend by cd'ing into the folder created in step 3 and running

```
python main.py
```

10. Taking the "additional-files" folder at the GitHub repo above https://github.com/bw-cetech/apress-7.4.git, replace those in the react-frontend and flask-backend folders. Add drag and drop capability to front end and model function call from flask. Additionally:

    a. Create a "python" folder under flask-backend and add the new file dlmodel.py (renamed from sample_dlmodel.py)

    b. Also update the path to the trained model in the Python script (dlmodel.py) to:

```
model = load_model("python/best-model-traffic-ESC.h5") # load model
```

11. Unpack the react-frontend folder with

```
npm run build
```

12. Test the app with some sample images from our GitHub repo

    https://github.com/bw-cetech/apress-7.4.git

13. Stretch: complete the steps in this lab using a VueJS front-end instead of react.js[14]

14. Stretch (HARD): separate out React and Flask in a Docker container

15. Stretch (HARD): Add a database to serve the app with training images

Figure 7-14. *Full-stack app to predict German traffic signs*

Wrap-up

This extensive lab bringing together back-end development (Flask and TensorFlow) with a front-end UI (Dash and React) hopefully gives a flavor of what to expect in building out "full-stack" AI application. We will walk-through several more of these labs in the next chapter when we cover solutions for specific industry case studies before a final look at end-to-end AI deployment in our penultimate chapter.

[14] for a simple VueJS front-end boilerplate, see hands-on lab in Chapter 2

CHAPTER 8

AI Case Studies

AI technologies are used today to improve customer service and offer personalized promotions to avert cyberattacks, to detect and prevent fraud, to automate management reports, to perform visual recognition, to mitigate insurance risks, and to sort and categorize documents and images.

Many of the technologies already have a major impact on everyday life, particularly the day-to-day "apps" and bespoke content services we engage with on our handheld devices.

This chapter takes a comprehensive (multisector, multifunctional) look at the main AI use uses in the last few years, in order to highlight what drives the needs and business requirements for AI in the workplace.

We will take a look at the key enablers for AI before embarking on a granular deep dive into specific vertical industry challenges in Telco, Retail, Banking and Financial Services, Oil and Gas, Energy and Utilities, Supply Chain, HR, and Healthcare. Ultimately the goal is to bring together the tools perspective in the last chapter with the business or organizational problem in this chapter in order to gain a deeper understanding of the Machine and Deep Learning applications implemented to address them.

The chapter includes some more advanced use cases involving multitool integration such as Social Media (Twitter API) Sentiment Analysis, Fraud Detection, and Supply Chain Optimization. The aim of these cases is to map the most suitable AI technologies and platforms for handling the underlying data and AI components: specifically data ingestion, storage, compute and modeling, and analytics.

As part of these enhanced use cases, we will look at how to set up a Twitter Developer account and implement the Twitter API to undertake Social Media Sentiment Analysis, how to leverage key cloud solution components such as AWS SageMaker, Lambda, S3 and QuickSight to train and deploy a Fraud Detection model and IBM Cloud Pak for Data, Watson Studio, Watson Assistant, AutoAI, and Decision Optimizer with CPLEX for Supply Chain Optimization.

B. Walsh, *Productionizing AI*, https://doi.org/10.1007/978-1-4842-8817-7_8

Industry Case Studies

Our first section in this chapter takes a high-level look at demand and enablers in the workplace for AI solutions, before a quick look at main use cases, framing the AI challenge and wrapping up with a look at To Be architectures for AI.

Business/Organizational Demand for AI

Proven scalability means that AI today is viewed by most private companies and public sector organizations as a transformative technology. But it can seem sometimes like the current hype is presenting AI as a solution to all our problems.

With this hype comes considerable fear, specifically growing concerns around job displacement/unemployment, inequality, and bias. The more recent trend toward "Enterprise AI" is attempting to address the imbalances, particularly with nontechnical staff.

From an innovation perspective, the focus is firmly on the "augmented intelligence" case – leveraging AI to solve many targeted/bespoke business and organizational problems. At the same time though, opening doors by fostering an internal AI learning culture that promotes experimentation and far better team collaboration via, for example, visual UIs are vital to ensuring a harmonious workforce.

AI Enablers

The main enablers of commercial AI solutions and use cases are today four-fold. From the **emergence of cloud-based services**, simplification and increased access to storage with limitless scaling and computing power (compute) enablement of complex calculation support is key.

The **large-scale use of sensors** means IoT (and mobile) devices now generate feature-rich, peta-scale data as a matter of course. This data is often integrated with **smart APIs,** meaning machine and deep learning is not always needed "from the ground up" – users can instead API into a pretrained endpoint solution.

And of course Digital Transformation itself and the driving force of "disrupt or be disrupted" enables Artificial Intelligence as a value-added profit center focused on the constant need to manage costs, improve productivity, and open up new revenue streams.

AI Solutions by Vertical Industry

In the sections that follow we will take a look at the AI use cases mentioned back in Chapter 1. Additionally, a more comprehensive segmentation of AI solutions is captured in Figure 8-1 for reference, many of which will be addressed further in the rest of this chapter.

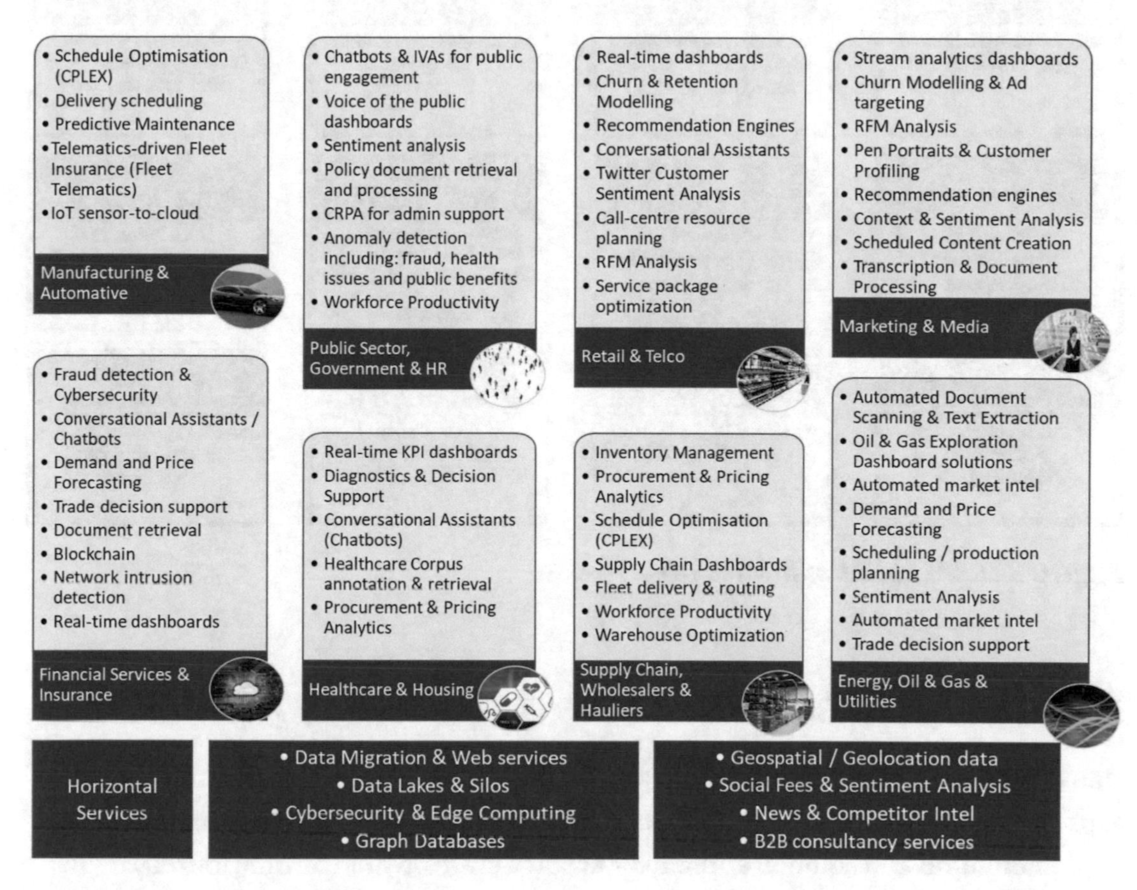

Figure 8-1. *AI use cases*

AI Use Cases – Solution Frameworks

None of the above use cases can be delivered without a comprehensive solution framework.

From defining the strategy and objectives through data strategy, security model, testing, implementation, and change management processing, it's critical to ask the right questions. Figure 8-2 shows some of the key ones we think are necessary in order to weigh the landscape and work toward a successful delivery.

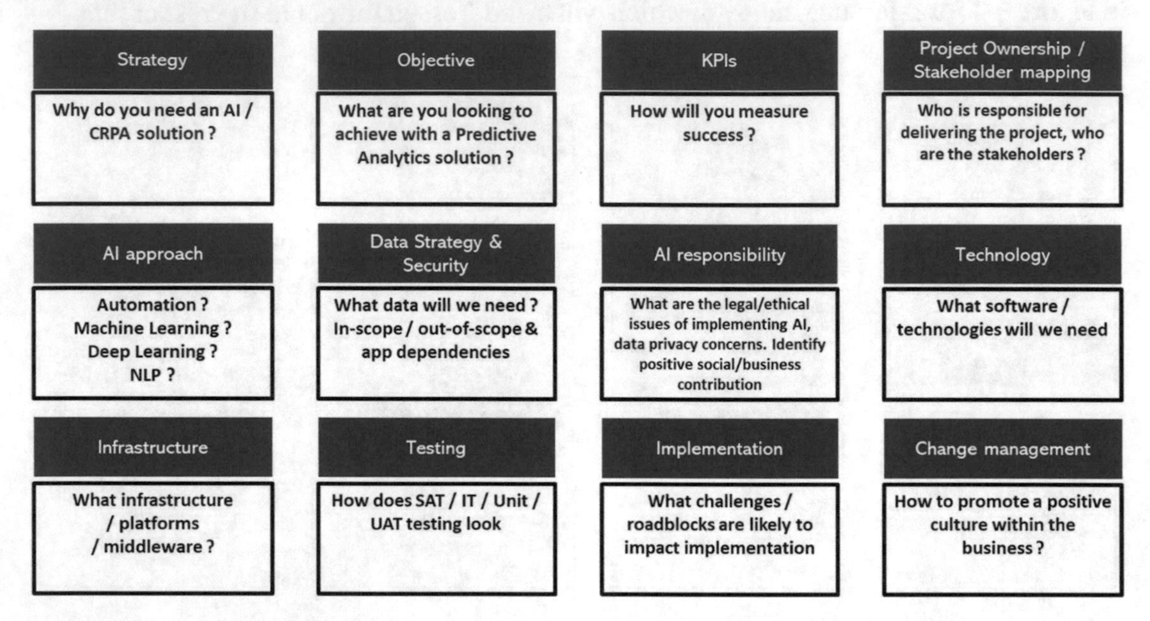

Figure 8-2. *AI solution framework*

Solution Architectures

Due diligence on the existing IT landscape is a necessary prerequisite to understand gaps in data, processes, technologies, and infrastructure. These gaps inform the implementation and solution approach – at some stage on this learning journey a "To Be" Solution Architecture will be required to articulate how the AI solution will interface to the supporting (hybrid) infrastructure, both on-prem and cloud-based.

We have already taken a look at some example architectures in our chapter on Data Ingestion (see Building a delivery pipeline) so we will wrap up this section with a quick look at one more – this time an Azure Cloud BI Architectures.

Example: Azure

Often AI projects are grown out of simpler descriptive analytics projects.

The simple Azure architecture in Figure 8-3 shows (in training phase I) an extract, load, and transform (ELT) pipeline executed by Azure Data Factory (ADF). ADF automates the workflow to move data from the on-prem SQL database to a cloud-based one – in this case Azure Synapse (SQL Data Warehouse).

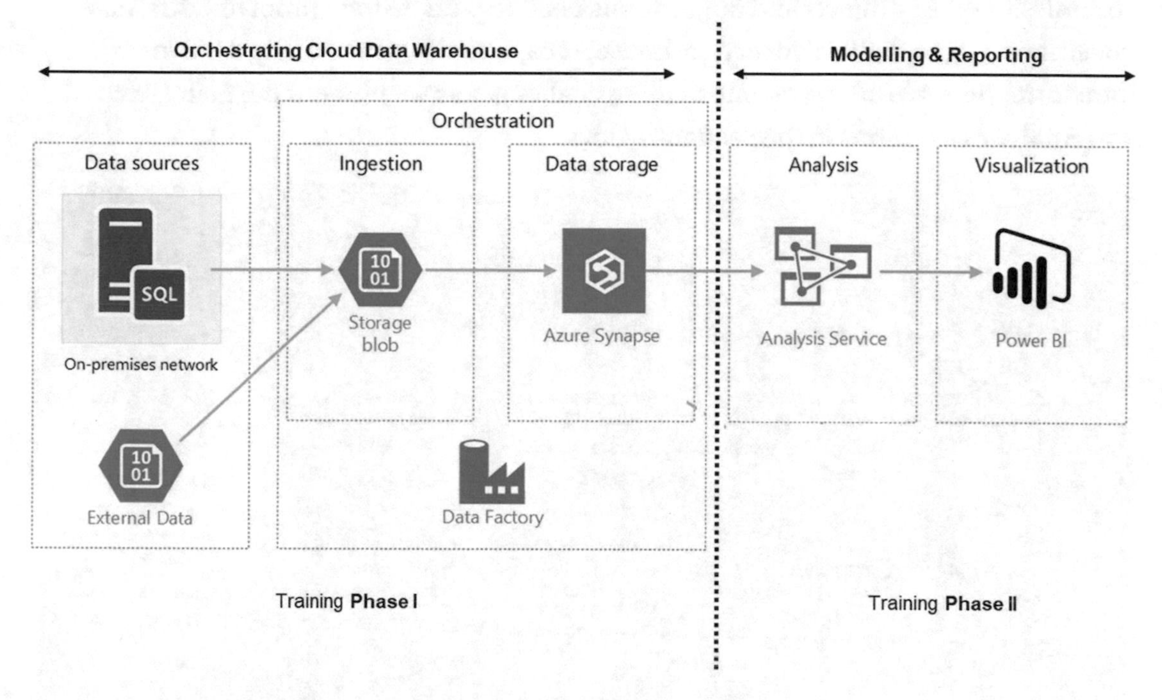

Figure 8-3. *Basic Data Analytics Azure architecture*

The data in Azure Synapse is then transformed for analysis using Azure Analysis Service, effectively creating a semantic model for Power BI to consume and for an end user to generate insights.

Telco Solutions

We now turn our attention to the leading AI use cases by specific industry, starting with the Telco sector.

Specific Challenges

Implementing AI solutions comes with specific challenges in the Telco sector, stretching across people, processes, and tools. As shown in Figure 8-4 from Boston Consulting, commissioned AI projects and applications need to demonstrate proof of value, data governance, agile digital delivery, in-house AI capabilities, and managed business transformation. We address some of these challenges as we pick out the main Telco AI and Analytics solutions in the sections below.

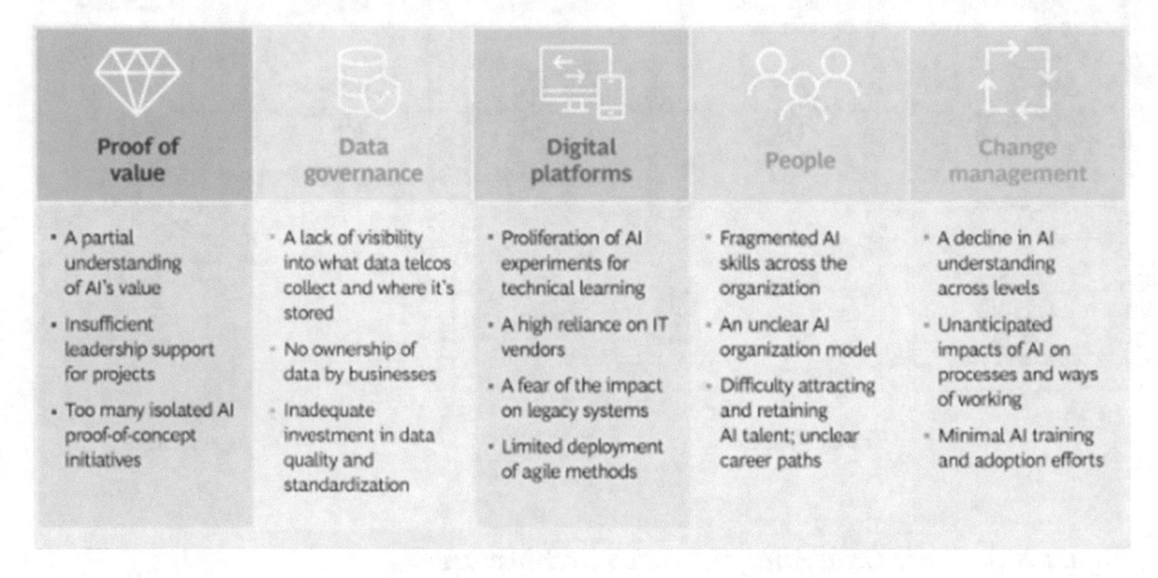

Figure 8-4. *BCG: Five AI Roadblocks in TELCO*

Solution Categories

The Telco sector is unique in its reliance on famously "feature-rich" datasets – information on customers often turns in hundreds of variables regarding demographics (age, gender, etc.), transactional (e.g., total spend per month), and attitudinal parameters (such as visited upgrade plans).

But AI solutions are not always focused on the end-customer. Broadly we see three main categories of AI solution in the Telco sector. These are grouped as shown in Figure 8-5 into Data and Governance and Oversight, Dashboards-driven Insights, and Predictive Analytics.

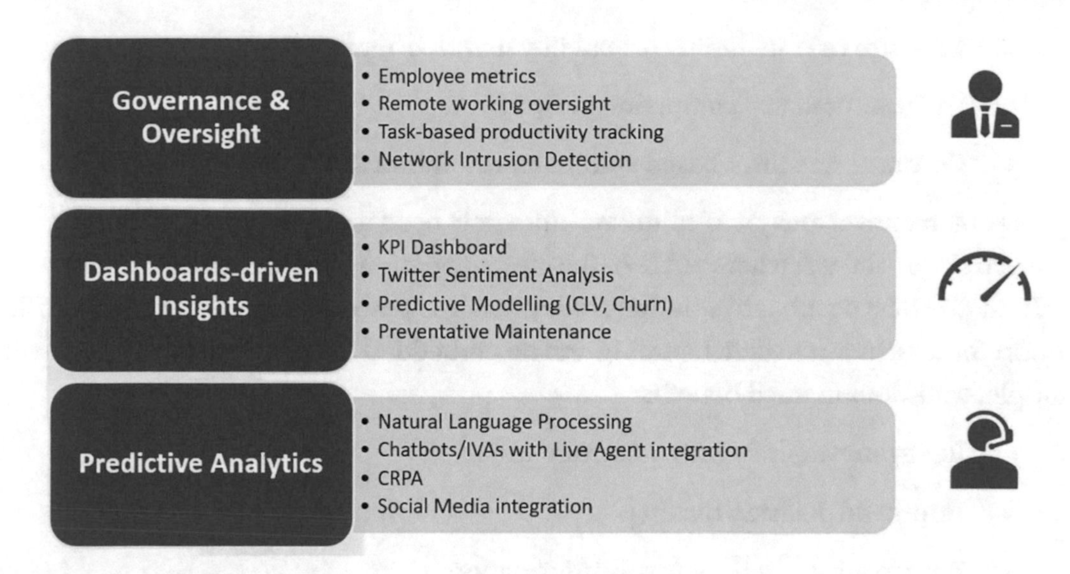

Figure 8-5. *Telco AI solution categories*

Real-time Dashboards

Churn is one of the main metrics by which a telco company is measured with industry analysis[1] showing customer churn drivers for Mobile services varying by:

- Price 45%
- Customer Service 45%
- Network Coverage 36%

Ideally, businesses want a single place to track progress across multiple client metrics and internal KPIs and prevent churn. Making data-driven decisions requires a 360° perspective on key customer metrics including customer complaints tracking and key SLA metrics.

Real-time, compelling dashboards can support the supervisory process with key metrics, monitoring, and automated reporting on customer complaints:

- Examine customer profiles
- Evaluate customer assessments
- Monitor indicators

[1] See e.g. `www.analysysmason.com/`

- Measure risks including probability that risks materialize
- Unstructured text complaint analysis
- Resource and time-based complaint resolution KPIs

As shown above, today's AI solutions ultimately need to extract insights from increasingly pivotal **unstructured** text. Besides integration with multiple data sources on cloud, the best dashboard solutions today (such as PowerBI and Looker) come with support for unstructured data handling and an embedded "Single Customer View," with multiple, well-documented benefits:

- Real-time analysis with quick navigation
- Improved decision making
- Easy to access and shareable information
- Know your business (or customer) with key metrics reporting
- Cross-sell and up-sell opportunity identification
- Fast answers to business queries

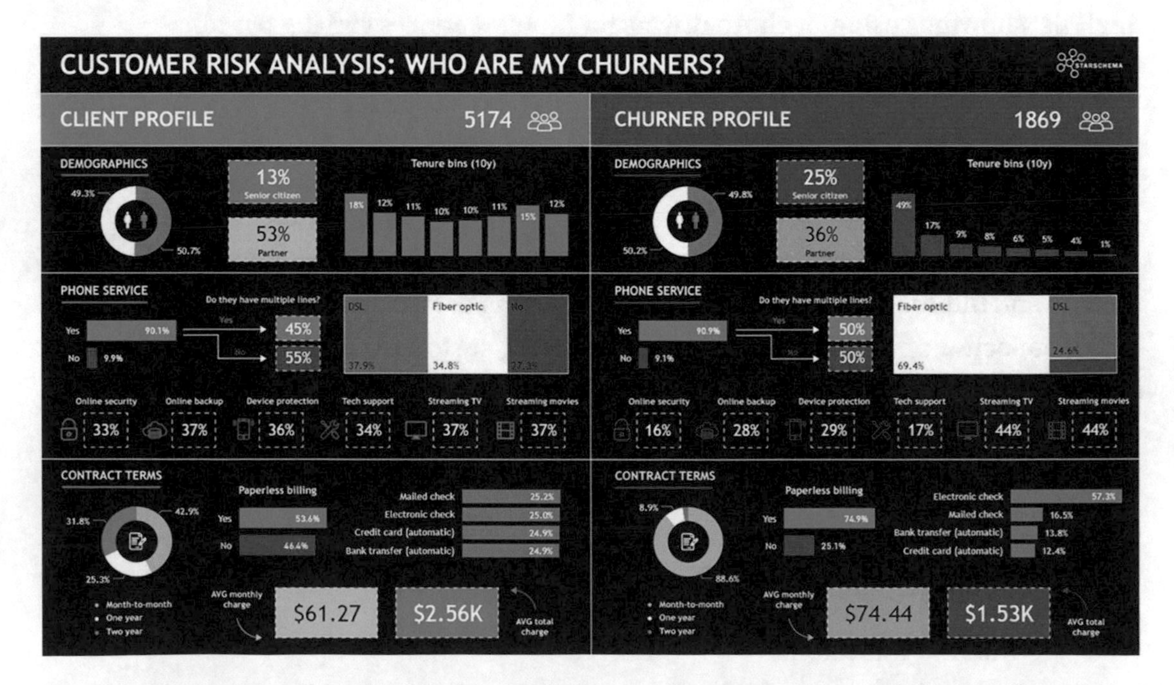

Figure 8-6. *Example Telco dashboard view – churners vs. non-churners (Source: Microsoft PowerBI / Starschema)*

Sentiment Analysis

Sentiment Analysis using Natural Language Processing has become a valuable tool for companies in the Telco industry to better understand the voice and opinions of the customer. But typically customer "sentiment" is just one of the key tracking criteria (KTCs) that a company is interested in gleaning from its customer engagement channels. Other KTCs include customer themes or concepts, emotions, entity, and keyword analysis

We will look below at how to implement a Sentiment Analysis solution for Telco using the Python Twitter API below, but in terms of a project approach these KTCs are a good starting point in which to create a solution design for displaying social media metrics relevant to the business and upon which marketing teams can segment customers and act.

Typically the best-of-breed sentiment analysis solutions are full-stack, combining analysis of feature-rich Telco datasets and under-the-hood KTC analysis of the latest Twitter feeds (or Facebook Insights) with a front-end dashboard displaying KPIs in real time.

Figure 8-7. *Telco datasets are famously "feature-rich"*

Predictive Analytics

Besides dashboards and sentiment analysis, more generic "Predictive Analytics" solutions leveraging Machine Learning are sought after by Telcos to achieve productivity improvements.

The use cases are varied, from optimal planning and scheduling (e.g., of broadband repairs) through push-button forecasting of customer demand (by segment) and competitor pricing, support for what-if scenario analysis (e.g., modeling rollout of a new service), automation of customer service delivery including installations and service upgrades to document search capability/probabilistic document retrieval (e.g., based on contract terms).

A natural fit for these challenges is an "in-house" Machine Learning as-a-service capability that can tackle these myriad issues, transforming into solvable machine or deep learning problems, often unsupervised at first before moving to a scheduled (e.g., monthly) supervised learning approach (until a sufficient volume of data is captured and auto-updated through a pipeline). The key here for delivering production-grade solutions in Telco, as in all sectors, is performance benchmarking against legacy processes with either

- A demonstratable ROI from the underlying predictive analytics solution
- A cost reduction or
- Incremental revenue benefits

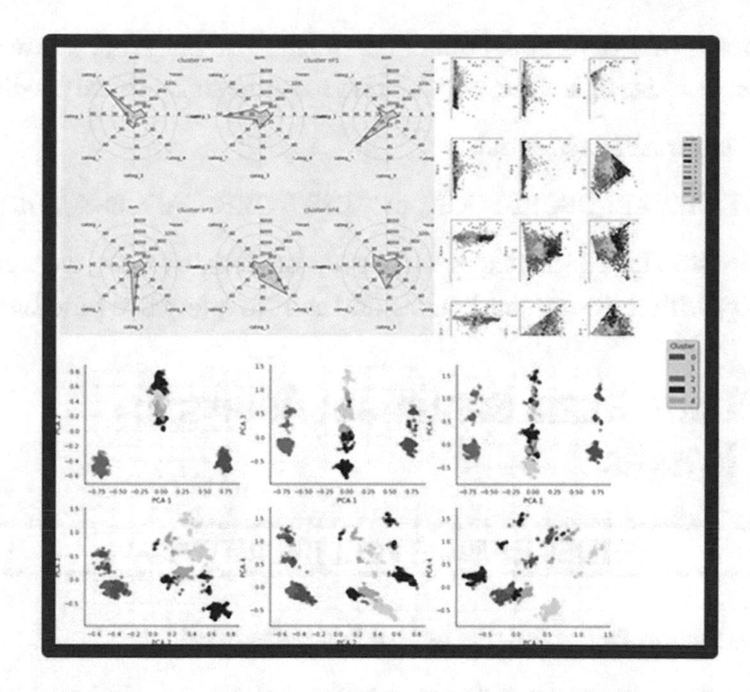

Figure 8-8. *Use of PCA – a valuable tool for data reduction on "feature-rich" Telco datasets*

Connecting to the Twitter API from Python

We wrap up this section with a look at how to connect to the Twitter API from Python – a prerequisite for our first hands-on lab in this Chapter.

Before we can import the twitter library in a Python notebook, there are three overriding setup requirements we need:

1) A twitter account
2) A twitter developer account
3) Approval from twitter for use cases

The process is as shown below:

- Set up a twitter account
  - `https://twitter.com/i/flow/signup`
- Set up a Twitter Developer Account
  - `https://developer.twitter.com/en/portal/petition/use-case`

  - There is a 5-10 minute application form to fill out - make sure to clarify use cases are for upskilling on NLP and to understand social media sentiment
- Get Twitter API Credentials
  - API_KEY, API_SECRET_KEY, ACCESS_TOKEN, ACCESS_TOKEN_SECRET

Opening a Python notebook (e.g., with Jupyter or Google Colab) we can now import the twitter library with the command below and start to query live (and historic) tweets.

Twitter API and Basic Sentiment Analysis: Hands-on Practice

GETTING FAMILIAR WITH THE TWITTER API

Accessing the Python Twitter API, the goal of this lab is two-fold:

a) Analyze the twitter feed in Python with Jupyter Notebook

b) Perform basic sentiment analysis on the latest tweets with tweepy and nltk

a) Analyze the twitter feed:

1. Make sure you have followed the set of the Twitter API as described under "Connecting to the Twitter API from Python" above
2. Download the notebook from GitHub repo below:

   `https://github.com/bw-cetech/apress-8.2.git`

3. Run through the "TwitterFeed.ipynb" notebook and using Figure 8-9 as a roadmap:

   a) **Get tweets** – Import Libraries and credentials, define twitter search criteria, and fetch tweets

   b) **Process the tweets** – Collect tweets in Python list and filter out retweets

   c) **Analyze tweets** – Who is tweeting? Isolate locational metrics and convert list to a pandas dataframe

4. Finally, complete the advanced twitter query Exercise.

b) Perform basic sentiment analysis

1. Run through the "TwitterSentimentAnalysis.ipynb" notebook following the steps/substeps shown in the image below
2. Make sure to complete the exercises in the notebook for creating analytics on the processed tweets

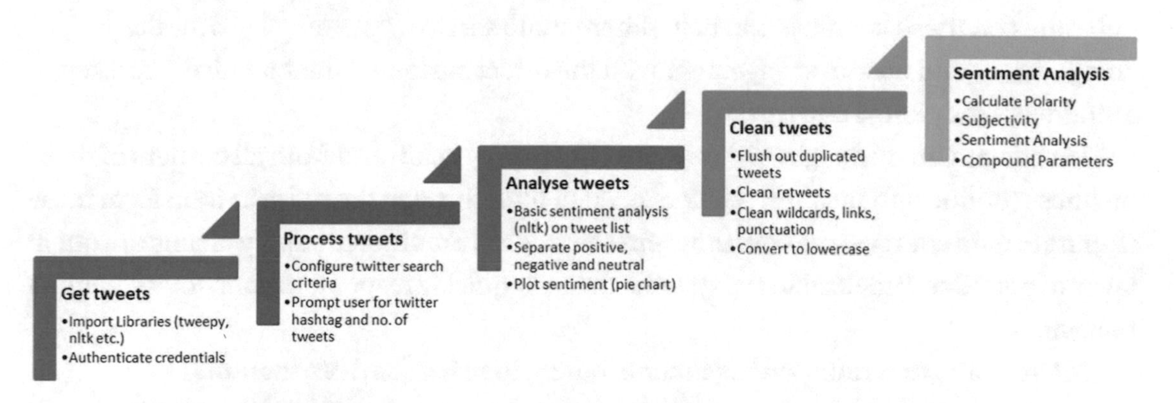

Figure 8-9. *Sentiment analysis with the Twitter API*

Retail Solutions

Many of the same challenges exist in the retail sector as in the Telco industry, but there are also certain unique challenges, related to products and customer experiences that distinguish digital disruption from more service-driven industries such as Telco. This section takes a look at how AI solutions are helping retailers overcome these challenges.

Challenges in the Retail Industry

Digital disruption to the retail sector and the high street has been in evidence since the start of the 2010s and has accelerated in the last five years, exacerbated further by the Pandemic.

As markets have rapidly evolved, so too have consumer expectations, such as in the provision of **unique in-store experiences**, with statistics[2] suggesting that more people are likely to shop in-store than online if a personalized experience is offered. Synchronized, seamless online and store shopping experiences not only help to create more "immersive shopping experiences" but can also address operational inefficiencies. With customers **increasingly reliant on social media,** retailers leveraging model outcomes on these channels can help differentiate services/give an edge over the competition, and ensure engagement with the direct customer, their relatives, and key influencers they know and trust.

Faced with an onslaught of data across the supply chain and from all corners of their business (online and instore), retailers need to filter through the noise to transform these disparate data sources into consumer-first strategies. Traditional supply chains are out in favor of adaptive and flexible ecosystems that can quickly respond to consumers' shifting behaviors.

Meanwhile, **dynamic pricing** is contributing to **price (and promotions) transparency** and helping some retailers define their brand value. At the same time, the constant undercutting (and commitment to delivery dates) by online retailers has presented numerous challenges to stores looking to compete.

All of the above trends in retail (and to a certain extent transport and aviation sectors) are driven by data and increasingly innovative predictive modeling techniques. Although one of the leading sectors in terms of AI uptake, the retail sector by virtue of its size remains a key battleground for AI solutions in 2022, particularly those with better data pipeline integration and supporting cloud (storage and compute) infrastructure.

Churn and Retention Modelling

The above unprecedented changes to consumer shopping behavior and their preferences have brought an increasing focus on the "customer journey" from first purchase to (potentially) disenchantment.

In today's fiercely competitive markets, churn rate metrics reflect a group of customer's (customer segment) response toward the product, service, price, and competition. For most retail businesses measuring the existing churn rate is critical

[2] For example, Wiser's survey of consumer shopping preferences suggests that 29% of respondents are much more likely and 33% of them are more likely to shop in-store than online if a unique experience is offered.

to prevent profit erosion or negative marketing of the brand/store. More than that, an active approach to predict and prevent customer churn is a necessity to retain existing customers and drive dependable revenue streams.

A Best Practice Approach to Modelling Churn

Where to start from a modeling perspective? The data generated during the sign-up (**demographic** data) and sale process (**transactional** data) is key, but on its own that is insufficient in today's more mature, data-savvy retailers.

The Data Science/AI value-add for retailers today is in mining customer data to isolate **behavioral and attitudinal** features, such as via product and product category purchases made via a loyalty card (and therefore searchable) or in visits/page opens/ page hits on specific products sold or services marketed on the retailer's online platform.

While there is no hard and fast rule, a good working churn model will typically involve at least 50 features per customer, involving several data types (numeric continuous such as total transaction value, numeric categoric such as type of product purchased, text/string such as customer reviews and date-like, e.g., time since last online visit).

In churn modeling, a "holdout" data sample is created from the customer base and set aside for inference at the end of the model training and validation phase.

During model training and cross-validation (with multialgorithm benchmarking), the most important predictors are identified (and nurtured/included in subsequent runs) while feature elimination can address initial high variance (overfit) models. The model is thus iteratively improved, sometimes with more transactional, promotional, or behavioral data to increase performance or using recursive feature elimination techniques such as multicollinearity and chi-square tests.

Models with intrinsic commercial value tend to prioritize precision over recall due to the need to reduce costs and the relative cost overhead of a high recall but poor precision-based model.

Finally, the holdout sample defined at the start is used as the basis for churn propensity (i.e., does the trained model generalize well to new data – does it identify well at-risk customers, etc.) and A/B testing is performed using two control customer samples where model outputs are either used or omitted and comparing, for example, the subsequent income streams for the month ahead. A periodic (e.g., quarterly) "ROI review" of model value should also compare the likely cost or commercial impact of targeting all customers vs. targeting only our predicted churners.

Model Design and Outcomes

The best churn models are platform (and cloud) agnostic and (where necessary) connected to real-time customer transactions, for example, via a Data Lake with raw OLTP/OLAP data connectors.

This kind of "extensible" model should also facilitate reporting across multiple cashflow at-risk scenarios (by region, business division, product or service line, and customer segment) with a front-end hook-up clearly identifying the multifaceted opportunities and threats:

- Tangible churn prevention/retention impact
- Ranked customers
- Revenue at risk
- Highest (segment) propensity to churn
- Greatest value at-risk consumers

Given the end customer of a churn model is often a marketing team looking to support campaign ideas with data-driven analytics and model outcomes, marketing (and media) "hook-ups" can help automate engagement with the most at-risk customers.

Online Retail Predictive Analytics with GCP BigQuery: Hands-on Practice

PREDICTIVE RETAIL ANALYTICS

Using Google Cloud Platform BigQuery ML, the goal of this lab is to train a machine learning model to forecast average duration for bike rental trips

1. Sign up for a GCP Free Tier account at the link below:

   `https://console.cloud.google.com`

2. Login to GCP sandbox below (free even after three-month expiry of GCP Free Tier) and create a project

   `https://console.cloud.google.com/bigquery?ref=https:%2F%2Faccounts.google.com%2F&project=gcp-bigquery-26apr21&ws=!1m0`

3. Add the Public Austin Bike Share dataset
4. Exercise: perform a couple of quick SQL "EDA" queries to get (a) the number of records and the (b) the busiest bike station
5. Compose another SQL query to build an ML model using the features below:
    a. Starting station name
    b. Hour the trip started
    c. Weekday of the trip
    d. Address of the start station
    e. 2018 data only
6. Repeat the above for a second model but add "bike share subscriber type" and remove the two features:
    a. Weekday of the trip
    b. Address of the start station
7. Evaluate the two models
8. Exercise: Predict average bike duration for the best model

Predicting Customer Churn: Hands-on Practise

CHURN MODELLING WITH PYTHON

Using Python in Jupyter Notebook, the goal of this exercise is to predict the likelihood of an active customer leaving an organization and identify key indicators of churn

1. Clone the GitHub repo below containing a churn model Python notebook and "customer_churn_data" dataset

    `https://github.com/bw-cetech/apress-8.3.git`

2. Run through the notebook in Google Colab, importing the data then carrying out:
    a. Pre-processing
    b. Visual EDA

c. Feature Engineering

d. Modelling

e. Exercise – identify the main indicators of churn by looking at feature_importances

f. Exercise – carry out future predictions using the trained model

g. Exercise (Stretch)

i. rather than using a dataset with a predefined target (churn) variable, swap it out with a retail dataset containing customer order information (such as the dataset "AW-SOH.xlsx" also uploaded to the GitHub repo above) and define "churn" as a custom variable based on a target number of months (N) elapsed since a customer made a purchase.

ii. Aggregate order data by customer, engineering any new transactional features which could be indicative of churn (e.g. average purchase amount, length of time as customer etc.).

iii. Finally, split the data into a training and test set, train a predictive model in the same way, then predict whether new customers are likely to churn or not in the next N months

Social Network Analysis: Hands-on Practise

SUPERMARKET TWEET ANALYSIS WITH NETWORKX

Our final lab in this section takes a look at how to implement social network analysis on tweets. Here we take a look at tweets on four UK supermarkets and perform network analysis on the location of those tweets focussed around major cities

1. Clone the GitHub repo below to a local folder

   `https://github.com/bw-cetech/apress-8.3b.git`

2. Run through the notebook in with Jupyter Notebook or Google Colab, following the steps below:

   a. Install network as described

   b. Import the (static) dataset of tweets

c. Perform a quick EDA on the tweet information

d. Filter out specific locations (i.e. not "UK")

e. Create an exclusion list for further analysis

f. Create a major cities list

3. Finally create the network analysis showing tweets by supermarket and (major city) location

Banking and Financial Services/FinTech Solutions

Banking and Financial Services industries are the subject of our next section in this chapter.[3] We take a look first at industry challenges before presenting one of the main AI solutions – Fraud Detection.

Industry Challenges

Across multiple industries including Banking and Financial Services, AI is rapidly reshaping risk management practices and transforming client and internal services. Key areas of disruption are shown below:

- Governance, Risk Management and Compliance (GRC)
- Fraud detection and prevention
- Personalized banking
- Trading
- Process automation
- Customer complaints

[3] The closely coupled Insurance sector is covered separately at the end of this chapter

But it's not just in risk control and cost prevention where AI-as-a-service can add value in Banking and FinTech. AI innovation is also helping to:

- Create an engaging and differentiated customer experience (retail banking)
- Change business relationships with customers from money movers to money managers (digital banking)
- Increase the lifetime value of customers (CLV) from global access enablement to banking, investment, and stock market trading services (such as via Robo Advisory or algorithmic trading services and "day trading"/CFD apps)
- Decrease customer support costs through the use of chatbot and IVA technologies and automated systems including document search, retrieval, and archiving
- Build trust and loyalty through robust fraud detection measures and periodic personalized offers (loans, cashback incentives, etc.)

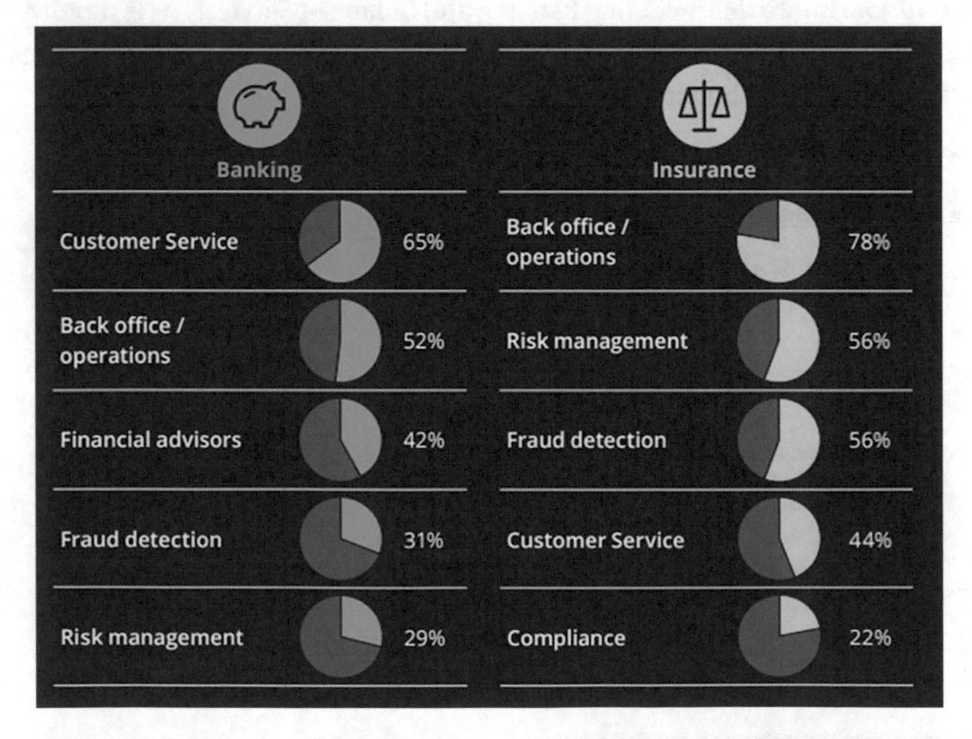

Figure 8-10. *Deloitte – key focus B&FS use cases from C-Level Executives*

While there are clearly many areas where Predictive Analytics is adding value, our main focus in this section is on the "flagship" AI use case for Banking and FinTech – Fraud Detection.

Fraud Detection

Business and organizational fraud risk goes beyond a simple identification of large transactions – a robust solution must address key operational risks too. With fraud an ongoing problem costing businesses billions of dollars annually, applying learning capability, as opposed to simpler, rule-based methods, is perceived as an existential need in the highly disruptive B&FS sector.

The main Fraud Detection applications of Machine and Deep Learning today are able to achieve better sophistication in model outcomes and by virtue of cloud-connectivity have embedded auditability to prevent fraud out-of-the-box. These applications can detect a wider range of fraud incidents by combining machine learning with an advanced rule engine, identifying along the way:

- Duplicate payments
- Duplicate invoices
- Unusual payment amounts
- Unusual payment description
- One-time payments
- Entities of interest

Besides the above, Fraud Detection bound to a company's internal Data and AI Strategy and fully integrated with internal and external data sources – claims systems, watch lists, third-party systems is a matter of course, as are model features including dynamic approaches to authentication flows in an age of multifactor authentication (2FA, 3FA), addressing specific fraud challenges associated with mobile channels, plus encoding of unstructured text in the feature engineering process.

Case Study: AWS Fraud Detection

Amazon Web Service's industry leading Fraud Detection Solution uses Automated Transaction Processing with SageMaker and is the subject of our hands-on lab below. SageMaker itself sits on a fully managed EC2 (compute) instance that runs Jupyter Notebook

Essentially the Fraud Detection solution loads a "live" credit card transaction dataset, training two models to recognize fraud patterns before deploying as two separate endpoints. The initial model performs Anomaly Detection using unsupervised learning (Random Cut Forest algorithm) and is then followed up by supervised training (xgboost algorithm) for Fraud Detection using labeled data.

The deployed solution automates detection of fraudulent activity, flagging fraud activity for review.

As shown in the architecture diagram below, integration of SageMaker is with

a. Two Amazon Simple Storage Service **(S3) buckets** for data storage, one for storing input data to the model, and one for storing downstream (postmodeling) analytics data[4]

b. **Lambda** for processing a continuous stream of prediction requests (triggered by **Amazon (REST) API Gateway and**

c. **Kinesis Firehose** which loads processed transactions into the second S3 bucket for downstream BI in **QuickSight**

In the lab that follows we will make use of an **AWS CloudFormation** template to provision the multiple resource requirements. Note these resources do incur costs:

- **$1.50** – one-time cost for Amazon SageMaker ml.c4.large instance
- **$0.65/hour** – cost to process transactions using AWS example dataset

[4] In the hands-on lab that follows the first S3 bucket contains example (PCA'd) credit card transactions, while the second bucket is the post-modelling data for downstream analytics in QuickSight (AWS's BI platform)

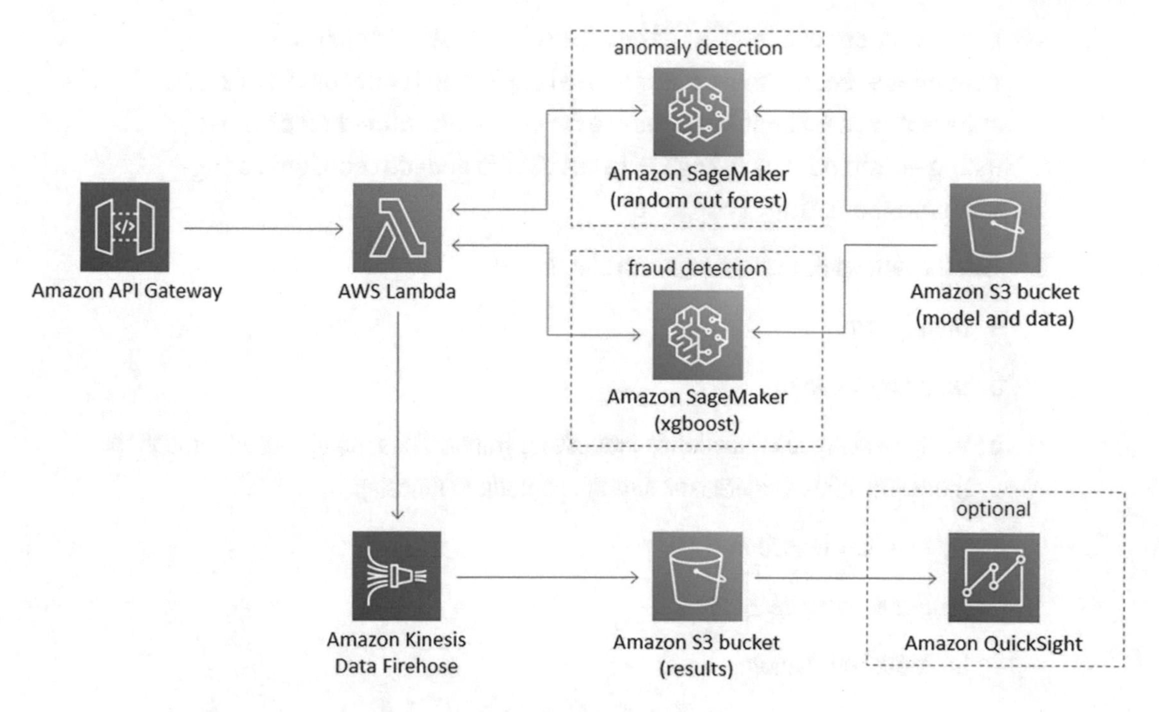

***Figure 8-11.** AWS Fraud Detection - architecture*

AWS Fraud Detection with AWS SageMaker: Hands-on Practice

FROM TRANSACTION MONITORING TO DASHBOARD ANALYTICS: FRAUD DETECTION ON CLOUD

Our goal in this lab is to build an end-to-end implementation of the AWS Fraud Detection solution, from resource setup via a Cloud Formation template to creation of dashboard analytics via QuickSight.

The operational process for this lab, from Data Import to QuickSight Analytics is extensively described in the section above and summarized in the image process map below:

1. Create AWS Fraud Detection Resources by going to the link below and launching an AWS Cloud Formation template:

`https://console.aws.amazon.com/cloudformation/home?region=us-east-1#/stacks/new?&templateURL=https:%2F%2Fs3.amazonaws.com%2Fsolutions-reference%2Ffraud-detection-using-machine-learning%2Flatest%2Ffraud-detection-using-machine-learning.template`

2. Run the notebook, performing each step in turn:
    a. Data Import
    b. Data Exploration
    c. Verify the Lambda Function Is Processing Transactions then proceed through the modeling steps (SageMaker through Lambda to Kinesis):
        i. Anomaly detection
        ii. Fraud event detection
        iii. Model evaluation
3. Check the S3 bucket results and the Kinesis Stream
    a. Data Pipeline and Model deployment (API)
    b. CloudWatch logs
4. Finally view results in QuickSight and show Fraud Detection analytics as a Dashboard
5. Stretch Exercise – improve the model performance by using SMOTE (Synthetic Minority Oversampling Technique) for balancing the minority class (here fraudulent transactions)

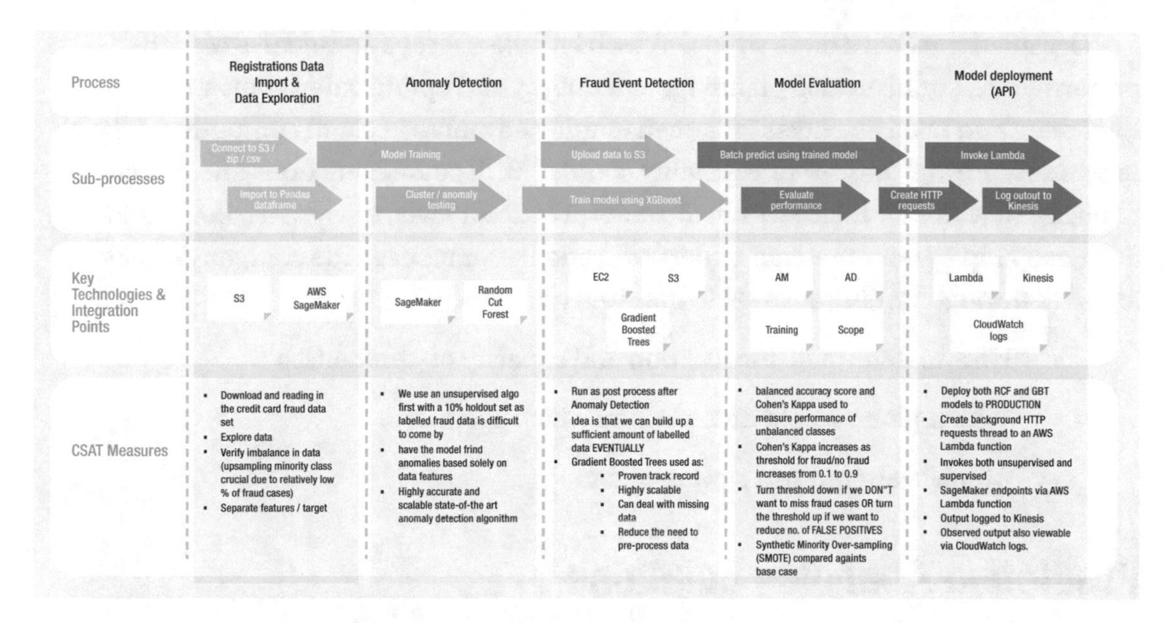

Figure 8-12. *AWS Fraud Detection – seamless automation from transactional data import to (QuickSight) Dashboard Analytics*

Supply Chain Solutions

Fraud of course doesn't just impact financial institutions. Supply chains are impacted as well, and there are further disparate challenges faced by vertically integrated companies, not least in dealing with inflationary pricing, managing or shipping scarce resources and caring for the environment amid a destabilizing global climate.

This next section takes a look at how AI solutions are being used in SCM (Supply Chain Management).

Challenges Across Supply Chains

"Supply Chain" is of course not a specific vertical industry or sector, rather a means by which many businesses and organizations source and deliver their products and services. Planning, logistics, and procurement activity within Supply Chain Management (SCM) stretches across multiple industries including manufacturing wholesale, distribution and retail, transport (haulier, fleet delivery) and oil and gas industries and healthcare and public services.

Organizations have often struggled with knowledge gaps around use of high-performance, sophisticated planning, scheduling, and optimization software but AI Engineers and Data Scientists today can employ advances in cloud computing, "visual" machine learning[5] and solver capability coupled with compelling front-end design to engage a wider audience within the enterprise. Moreover, today's AI solutions for SCM can help operational staff understand and better manage the company's delivery logistics through finding answers to questions such as

- How to determine the location and capacity of warehouses
- Which plant should manufacture which product
- Allocate aircraft and crew to flights

Predictive Analytics Solutions

Before taking a look at a niche end-to-end solution for Supply Chain Optimization, there are plenty of cases of Predictive Analytics implemented in-house to manage key areas across the supply chain.

These include risk managing vendor/supplier/customer relationships with transactional anomaly detection, procurement engines, and fraud detection models, customer support solutions like chatbots and IVAs or document search and retrieval, logistics using inventory and contract management tools with "Edge AI" solutions to enable more granular predictive analytics (e.g., supply/demand matching at local rather than regional level) as well as asset optimization and forecasting for specific assets and business divisions.

[5] Drag and drop style Data Science or Machine Learning, or the use of No/Low code UIs

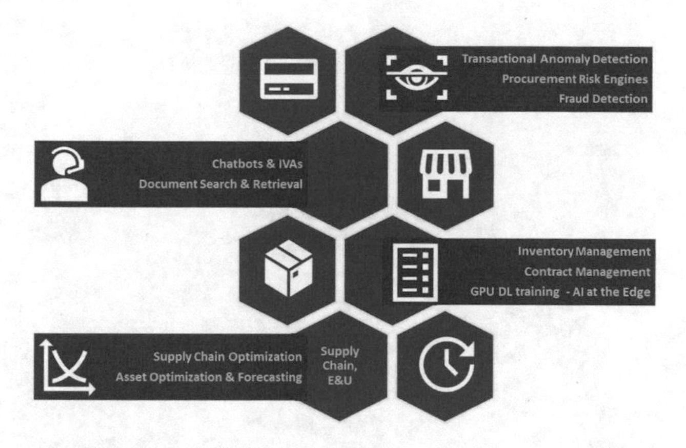

Figure 8-13. *SCM Predictive Analytics solutions*

Supply Chain Optimization and Prescriptive Analytics

Predictive Analytics on its own is insufficient for many Supply Chain Optimization problems as its focus is on what may happen (as in expected demand in a demand forecast). Prescriptive Analytics extends the solution space to "the way to make it happen," as in matching supply to expected demand as shown in the picture below.

Many supply chain problems (e.g., routing, scheduling, warehouse optimization, and dispatch) involve **matching supply and demand** and require the use of high-performance solvers to ensure optimal scheduling, routing, and dispatch solutions applied to targeted vertical industries. CPLEX, Gurobi, and FICO are industry-leading solvers for these types of planning and scheduling problems, and we will make use of CPLEX in the hands-on lab for this section below.

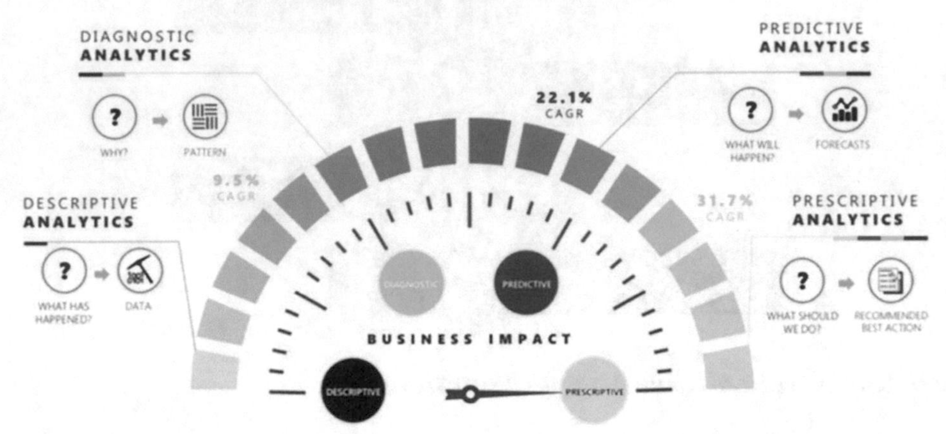

Figure 8-14. *-way Analytics (Source: iotforall)*

Implementing robust prescriptive analytics algorithms or "solvers" while following a best practice "DataOps" delivery model can ultimately bring about some key value differentiators for an organization:

- Optimal planning and scheduling
- Support for what-if scenario analysis
- Cost reduction and incremental revenue benefits
- Performance benchmarking against legacy processes

Supply Chain Optimization with IBM CloudPak/Watson Studio: Hands-on Practice

ORCHESTRATING DECISION OPTIMIZATION AND CPLEX WITH WATSON STUDIO AND IBM CLOUD PAK

In this lab, we will deploy an end-to-end boilerplate Supply Chain Optimization project in IBM CloudPak for Data:

1. Sign up/log in to IBM Cloud Pak

   a. `https://dataplatform.cloud.ibm.com/exchange/public/entry/view/427846c7e99026edd5fa0022830bc002?context=cpdaass`

2. Create a project
3. Create a deployment space
4. Run each of the Jupyter notebooks below, one by one:
    a. **Data preprocessing:** preprocess supply-demand data
        i. Suppliers
        ii. Manufacturers
        iii. Warehouses
    b. **Shiny Data Prep** notebook
    c. **LSTM Demand Forecasts** (not included)
    d. **Decision Optimizer**: Run a Prescriptive Optimization
        i. Load in preprepared LSTM Demand Forecasts
        ii. Optimize cost to ship parts (from four suppliers to three manufacturers) and deliver products (from three manufacturers to four warehouses)
5. On completion of the Prescriptive optimization process with Decision Optimizer, run the (Shiny) Dashboard App and interrogate the Inventory and Demand and Supply and Logistics views
6. Exercise – change the product categories on the dashboard to names rather than numbers
7. Stretch Exercise – schedule a once per month retrain and model run with new data

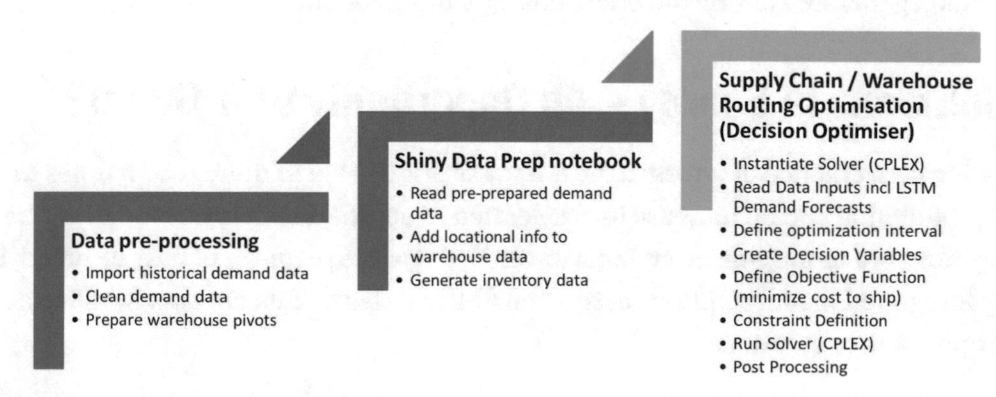

Figure 8-15. *IBM CloudPak for Data – Supply Chain Optimization*

Oil and Gas/Energy and Utilities Solutions

The Energy sector encompasses many different businesses across the value chain, from oil and gas majors to distribution companies, electricity generators and renewable operators (RECs), utilities and energy retailers. We look at the opportunities for AI below.

Challenges in Energy, Oil, and Gas Sectors

In the current climate of reduced supply, tight supply chains and inflated prices, the energy sector could be forgiven for paying little attention to AI as an enabler for revenue creation and cost optimization. But there are a wealth of opportunities for AIaaS to unlock value, particularly in the distributed energy chain and the renewables sector. AIaaS is also adding value in the oil and gas sectors too, as finding safe, economic, and sustainable methods of exploration and production is a key concern, as is reducing the environmental impact of oil production.

Long before the current energy crisis, corporates have been burdened by legacy (ETRM) tools and poor innovation practices. Part of the slow pace of change in energy stems from a statistically aging workforce coupled with a bloated tier of middle managers fearful of the impact of transformation on their jobs and livelihoods. Others have ethical concerns about the use of emerging technologies, even though many of these tools have long ceased to be "emerging."

Changing workforce demographics are beginning to fuel demand for data-driven innovation and the use of AI to break free from backward, inefficient business processes. Indeed, there is no shortage of data across the energy value chain - the challenge is what to do with it all and how to process the vast amounts of exponentially increasing IoT sensor data generated across the entire energy supply chain.

AI Solutions in Energy – An Opportunity or a Threat?

A growing convergence of forces in the energy sector has led in more recent times to rapid prototyping and an increase in accelerated AI solution delivery. The current trend is away from old monopolistic systems toward a disparate platform of best-of-breed, but highly integrated digital solutions across the AI Ecosystem; often cheaper, smaller, and more open source solutions.

Smaller-scale, or "Tiny AI" solutions are faster to deliver while single-cloud "Big Tech" solutions have become in some sense "too big" and too costly. The productivity benefits of AI have become too difficult to ignore and might be expected to give way to a vicious culture of disruption especially if performance benchmarking against legacy processes suggests a compelling case.

Risk/asset management tools such as customer risk engines with embedded machine learning to "push-button" customer/load/price forecasting with fbprophet or recurrent neural networks are starting to become commonplace, if not as production-grade solutions, then as protype python/Dash scripts. Back office CRPA and document retrieval solutions are in evidence, for example, for gas/electricity policy search, settlements, and billing as well as contract management. And trade support solutions including algorithmic trading and automated market analysis and intelligence with support for "what-if" scenario analysis are being used to boost front-office profitability.

Many of the same Predictive AI tools mentioned above in the retail sector are also proving valuable in the energy sector such as "live" customer decision engines with social media sentiment analysis, chatbots and IVAs, automated customer service delivery and dashboards and reporting solutions reflecting wholesale market dynamics and energy news. Similarly, the same supply chain optimization solutions with robust solver capability mentioned in the previous section are often applicable to delivery planning, optimized scheduling, and routing from upstream energy sources through transmission and distribution channels to suppliers and down to customers.

Edge AI is another exciting application area which can in time capitalize on increasingly, decentralized, distributed asset generation and consumption (e.g., wind turbines, solar panels, heat pumps, and small-scale battery storage).

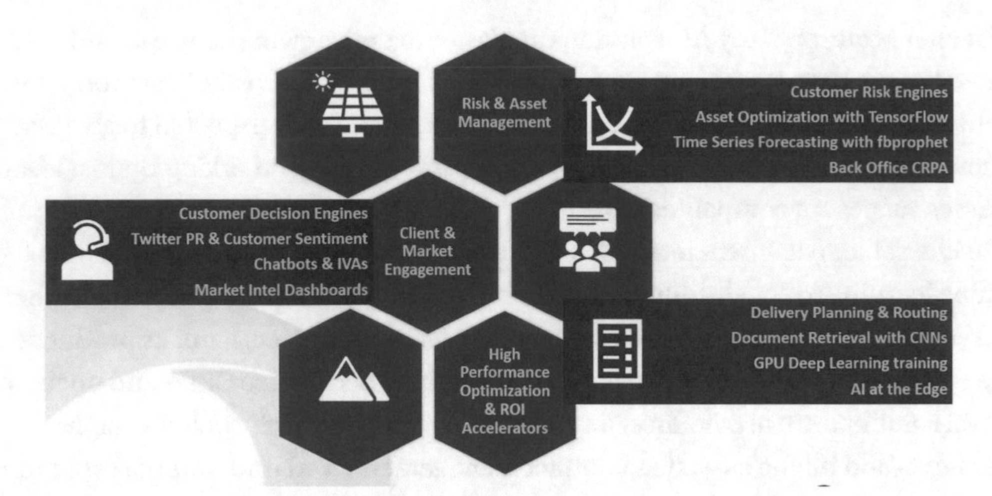

Figure 8-16. *AI solutions in the Energy sector*

Healthcare and Pharma Solutions

While AI deployments in energy may lag some other sectors, the same cannot be said healthcare and pharma industries. Consumer-driven engagement with, and enthusiasm for, Health Tech or Digital Health products and services as a means to monitor, compare and improve individual health and fitness has led to substantial AI innovation in this space.

Healthcare – The AI Gap

The Healthcare industry moves at a break-neck speed and it can seem at times there is little time to investigate better ways of doing things. By nature, care is prioritized over innovation, which can lead to a failure in ground-breaking technological solutions reaching the doctors, consultants, and patients that may need them.

The rapid pace of digitalization has opened opportunities to tap into low latency, easy-to-use, highly visual, interactive, and informative predictive analytics support for healthcare professionals, helping to close finance and efficiency gaps.

Particularly in healthcare support functions, an AI-led cultural shift away from ad hoc, poorly executed operational processes, overly complicated nonstandard processes, subjective/biased/arbitrary decisions, and human error can help unlock value, whether to help reduce or control ballooning healthcare procurement costs, or enable patient-targeted marketing, for example, of pharma/natural health products.

Healthcare and Pharma Solutions

From NHS support services to mental health trials, chatbots and IVAs are being more noticeable across the healthcare sector. Conversation Assistants or Chatbots have come a long way since the days of Cortana with IVAs coupled with NLP and document retrieval providing automated knowledge transfer to (and from[6]) health managers and consultants helping to achieve a more holistic decision support system and better patient outcomes/"pulse-checks."[7]

Natural Language Generation is also an interesting AI innovation which can simplify pre and post patient treatment paperwork as well as administrative activity and appointment scheduling. GPU or cluster-leveraged machine Learning-enabled procurement cost forecasting, conducted using, for example, Databricks, Apache Spark, and fbprophet over multiple healthcare facilities and across multiple drug product lines or equipment necessary for patient operations can also dramatically reduced local hospital or pharmacy budgeting costs.

Forecasting is just one of the applications of machine learning in the Healthcare and Pharma industry, data-driven drug discovery, clinical trials, patient diagnostics, and (deep-learning) image classification techniques applied to x-rays or MRI scans are providing valuable "second-opinions" to support doctors and nurses.[8]

HR Solutions

For our final focused AI use case, our attention now turns to a key business function served by AI in 2022 – Human resources.

[6] For example, Chatbot corpus creation via consultant annotation of medical publications and reports

[7] Frequent patient trials coupled with recurring model retraining can also lead to continual improvement in IVA-supported patient outcomes

[8] In certain cases delivered as services by AI startups disrupting the healthcare sector, such as Atomwise and Deepcell. See e.g. `https://venturebeat.com/ai/6-ai-companies-disrupting-healthcare-in-2022/`

HR in 2002

The HR Analytics market is forecasted to grow into a $6.3b market @ 14.2% CAGR as employers grapple with an uncertain postpandemic future and awakening employee sentiment and attitudes to work. But only 21% of HR leaders believe their organization is effective at using data to inform HR decisions.[9]

In an era of shifting demographics and uncertainty wrought by digital disruption, Covid and now spiraling living costs HR executives are looking for accessible data and HR analytics solutions for better talent management and support, reducing bias in recruitment and in the workplace and to help improve employee performance and attrition rates.

If Enterprise AI is to be successful, a company could do worse than starting with challenges in Human Resources and where Predictive Analytics can help. As shown in Figure 8-17, we see these current challenges (and associated HR analytics solutions) falling into one of three areas mirroring the employee journey: Recruitment, Talent management/Human Capital Management (HCM), and Employee Experience.

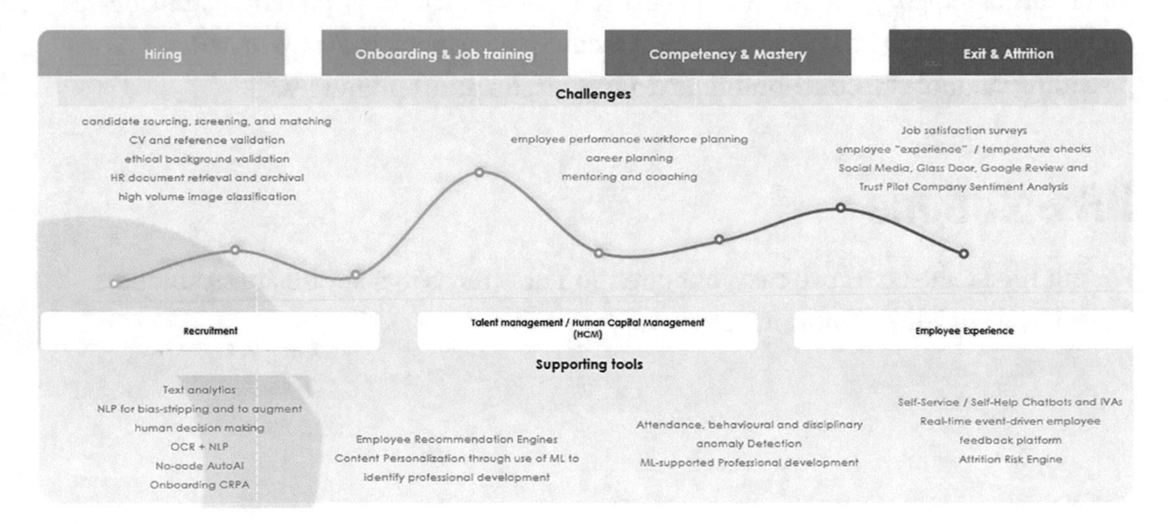

Figure 8-17. *AI in HR 2022*

[9] Gartner

As we shall see below, managing the data deluge and ensuring performance metrics are nonintrusive is the main challenge to creating a 360-degree employee analytics solution today. AI solutions for Human Resources can help both employees manage a balanced workload and by informing HR functions who is proactive, who wants to do more, who is close to burnout, and who needs more help or support. If used in the right way, these metrics can also lead to employee task-based efficiency and improved focus, through defining achievable goals and helping employees achieve them.

Sample HR Solutions

Dashboards are everyone's favorite analytics tool, and especially for HR Execs, vital to cover various aspects of the employee journey, including hiring KPIs, employee on-the-job metrics including workforce productivity and remote working,[10] attendance[11] and disciplinary oversight as well as employee churn and company feedback.[12]

One of the best HR Dashboards on the market is from Agile Analytics - with 20+ descriptive and predictive PowerBI built-in reporting views.

As shown in Figure 8-18 there are multiple machine learning applications including OCR, NLP, and text analytics applied to candidate information (LinkedIn, CVs). Deep Learning can help with job description (JD) scanning (or NLG-support JD creation) and drawing up employee contracts while NLP-supported bias stripping can also help ensure equitable recruitment policies.

[10] Integrated with, for example, Teams and able to display, for example, sign in/off times, email, calls and meetings KPIs, application usage periods, employee location/dual-location, multi/cross-platform productivity

[11] Including holiday/sickness calendar integration

[12] For example, Glass Door

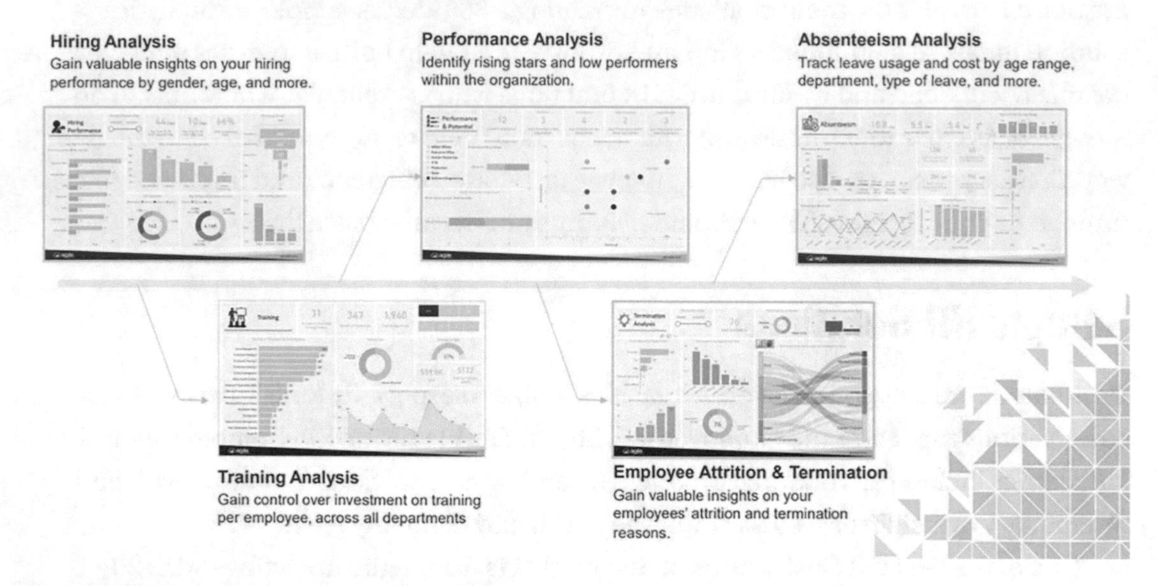

Figure 8-18. *HR Analytics (Agile Analytics)*

No/Lo code UIs are growing in popularity among HR managers, giving nontechnical, intuitive access to predictive analytics and employee modeling. Chatbots and IVAs can help with real-time "Employee Experience," feedback on organizational changes and company culture, take employee mood and sentiment "temperature-checks" and identify potential employee attrition. Todays' IVAs can also come with search automation[13] and are integrated with back-end systems for, for example, internal job vacancies or company policy retrieval.

For onboarding, internal content personalization for employee professional development can be supported with recommendation engines based on demographics, skillsets, and psychometric scores. Anomaly detection can help identify amber or red flags in employee attitudes and behavior.

[13] For example, Watson Search as a value-added service on top of Watson Assistant

HR Employee Attrition: Hands-on Practice

GETTING A HANDLE ON EMPLOYEE CHURN

Using Python in Jupyter Notebook, the goal of this exercise is to run AutoAI to perform feature engineering and algorithmic benchmarking on an employee attrition dataset:

1. Login to IBM Cloud Pak at the link below and create a project

   `https://eu-gb.dataplatform.cloud.ibm.com/home2?context=cpdaas`

2. Create an AutoAI experiment
3. Add the dataset emp_attrition.csv
4. Configure the AutoAI experiment to predict employee attrition
5. View performance of the model training process and test results
6. Implement a Flask UI to run predictions for employee types
7. Exercise – test whether the predictions make sense, for example, by entering data for an employee on a relatively low monthly income, with limited years of service

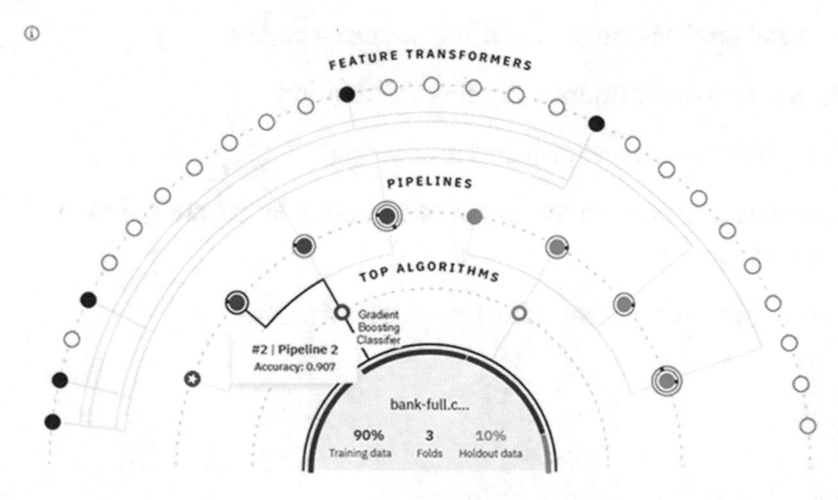

Figure 8-19. *Performing AutoAI on Employee Attrition data with IBM Cloud Pak*

Other Case Studies

Our final section in this chapter takes a look at the key areas ripe for AI innovation within some of the other important industries and subsectors we have thus far not mentioned.

Public Sector and Government

Estimates for the use of Artificial Intelligence in government varies market from $4.9b in government[14] to a $1b Global AI governance market.[15] While ethics and risk aversion are the primary concern, governmental departments are increasingly adopting data-led analytics and AI approaches to deliver public services. Many of the same approaches are also used in the media for advertising and gauging public perception.

Besides key public engagement applications such as dashboard-driven insights into sector performance and the public mood, chatbot 2.0, IVAs and sentiment analysis for "voice of the public" and to provide personalized public services, data project are often undertaken to help back up or steer policy and to boost trust and public perception.

AI adoption is also helping Public Sector agencies and organizations to drive efficiency improvements via, for example, document processing and automation, and tackle a surge in fraudulent activity, aided by the proliferation of IoT devices. And at the same time, more granular analytics for supporting policy planning and development and productivity improvements are in evidence, including:

- Sectoral analysis, policy creation, planning and delivery
- Admin and departmental process automation
- Policy document retrieval and processing
- Anomaly detection including public fraud, health issues, and the benefits system
- Econometrics Forecasts and Time Series Analysis

[14] researchandmarkets.com

[15] marketsandmarkets.com

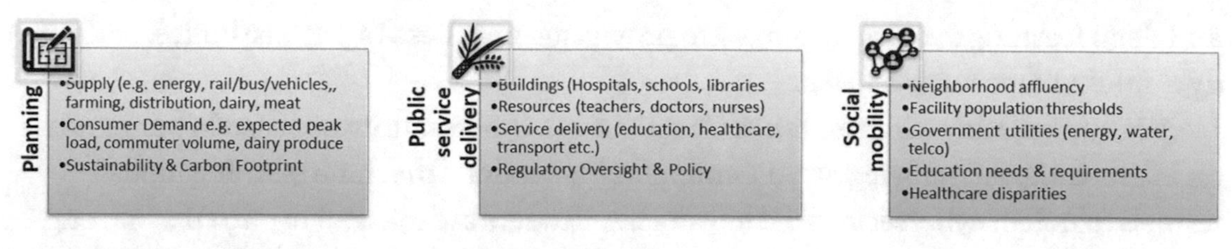

Figure 8-20. *Challenges for AI in Government*

Manufacturing

Much of the use of AI across Supply Chains described earlier also covers manufacturing use cases with Manufacturing 4.0 intended to bring together IoT, cloud, and analytics to revolutionize the production and distribution process.

Safety stock levels, delivery schedules, supply chain logistics, and underlying expenses are all optimized parameters under an aspiring, if not functioning Smart Manufacturing Solution. Warehouse use of blockchains is also increasing, with the intention of improving auditability and making supply chains more efficient and reliable.

AI simulation in theory should help reduce product recalls by reducing manufacturing processes rigidity, dynamically correcting product flaws and encouraging product improvisation. Similarly knock-on impacts of equipment/machinery failures on delivery schedules, budgets, and reputation are better managed today using AI applications. Predictive Maintenance in particular, coupled with big (often IoT) data statistical analysis, leads to better scheduled maintenance (without impacting delivery) and reducing future failures.

Given this digital reliance and big data dependency, no Manufacturing 4.0 solution today is complete without addressing cyber security concerns - the subject of our next subsection.

Cybersecurity

There is no greater challenge to business and organizations in 2022 than Cybersecurity.

The Internet has well and truly penetrated into every industrial system today, but at a local (production) level and across global supply chains. With it has come a huge increase in risk of cyberattack from hackers trying to gain access to information or worse, take control of sometimes highly strategic facilities tied to national security. Besides critical Defense in Depth layers of security and provision of robust firewalls, machine, and deep learning is being leveraged to protect against these highly disruptive, real-time, dynamic cybersecurity attacks.

We have already taken a look in Chapter 7 at a cybersecurity use case in our hands-on lab looking at deploying an API endpoint for a trained machine learning model trained to detect cybersecurity DDoS attacks. Readers are referred back to that lab and in particular the model training process (step 4) which shows how various network parameters are used as features to identify patterns in network activity that might constitute, for example, raindrop malware from a network intrusion event.[16]

Insurance/Telematics

Insurers' business models are changing with AI rapidly reshaping the sector and transforming multiple client and internal services:

- Risk assessment and risk management to risk-based pricing
- Customer segment targeting
- Personalized services
- Bespoke premiums and underwriting expenses
- Claims Management and Operational Process automation

Augmenting traditional actuarial-based methods, dynamic risk-based pricing drawing on the deluge of (customer behavior-rich) big data from in-vehicle telematics is a particularly exciting area in the automobile/fleet insurance sector. However, the use of AI across all Insurance sectors is evident from tapping into the voice of the customer via social media and creating a sensitive, engaging, or differentiated customer experiences based on captured customer attitudinal data.

Legal

The legal sector is experiencing something of a transformation from the use of AI technologies, particularly in relation to research activity and contracting.[17]

---

[16] Raindrop Malware was discovered in the Dec-20 SolarWinds supply chain attack

[17] See e.g. `https://legal.thomsonreuters.com/en/insights/articles/ai-and-its-impact-on-legal-technology`. LegalAI `www.legalai.io/` are also automating litigation through Artificial Intelligence – we all look forward to that

An exhaustive analysis is beyond the scope of this document, however we lay out in Figure 8-21 one key integrated application which is being used to help digitalize legal documents and support the search and retrieval and archive process - Document Classification and Text Extraction with OCR (optical character recognition).

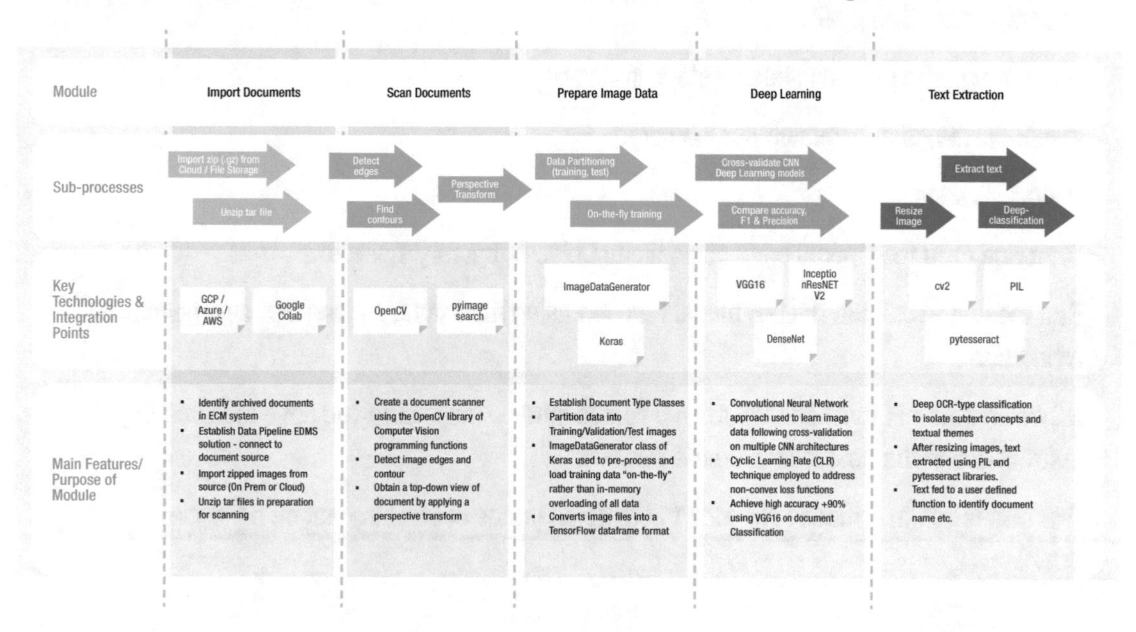

Figure 8-21. *Legal AI: Digitalizing Documents through Document Classification and Text Extraction with OCR*

DALL-E for the Creative Arts: Hands-on Practise

GENERATIVE AI – IMAGINING DIGITAL IMAGES FROM NATURAL LANGUAGE

We have one more lab in this chapter – walking through implementation of one of the most exciting developments in AI – OpenAI's DALL-E.

DALL-E is composed of an autoencoder (see Chapter 5) and transformer (Chapter 8) – essentially by typing in natural language we can conjure any image at will. Although we haven't discussed applications in the creative arts industry in this Chapter, there are obvious applications here: from e.g. design and fashion to architecture.

Register an account with OpenAI `https://openai.com/join/` and access the API Key provided under "Personal" in the top right corner of your dashboard.

Clone the below GitHub repo:

```
https://github.com/bw-cetech/apress-10.9.git
```

Copy and paste the OpenAI API Key inside the double quotes of the string defined in the openai_credentials.py file

Run through the python notebook steps in Colab:

Install openai, then comment out the code

Import libraries

Drag and drop the OpenAI credentials file to Colab temporary storage

Run the cell to call the DALL-E model function using the example given of a cow holding a frying pan

NB calling the API function requires a paid account[18] – make sure you have followed the instructions in the notebook to create this first

Exercise (stretch) – modify the code to create an image of an astronaut on a bicycle

Wrap-up

The legal perspective above brings this chapter to a close. From a look at the digital (and cybersecurity) challenges faced in Telco and Retail sectors, Financial Services, Insurance and Government, across manufacturing, supply chain and energy industries, in healthcare organizations and across HR functions our trawl through AI use cases here sets the scene for our final chapter[19] on productionizing AI.

Despite the very different nature of these vertical industries, AI solutions are often surprisingly similar, "horizontal" in their application, and recyclable. Adopting economies of scale makes sense. As such, this is the approach we will take in the next chapter – not just to describe the end-to-end processes and implementation actions to develop, build and deploy robust plug-and-play AI solutions, but additionally in our final two end-to-end deployment labs in this book.

[18] Costs are modest and lower for smaller scale images (256x256 pixels)

[19] Chapter 10 on NLP is our last chapter, but Chapter 9 will complete the main theme of this book on Productionizing AI

CHAPTER 9

Deploying an AI Solution (Productionizing and Containerization)

Productionizing an AI model is not easy. Many AI projects get stuck at Proof of Concept (PoC) phase, with Gartner suggesting 50% of IT leaders will struggle to move their AI projects past demo/prototype and into production. There is often organizational confusion between the two aims of creating a POC and a production-grade Enterprise AI solution, partly from the lack of expertise from the rest of the business. There is no point, after all, in productionizing if the rest of the workforce isn't, at least at a basic level, upskilled on AI, how to use it and its benefits and importantly, how to interface or engage with it or how to use it.

AI implementation requires a practical mindset and discipline. Contrary to the hype, many issues actually stem from attempts to apply AI to "everything." AI is Augmented Intelligence, meaning it's applied to solving a specific business problem with a "human-in-the-loop." Other project issues have arisen from the Data Scientists of the 2010s taking liberties to build "Kaggle-grade" models, building up "technical debt" and losing sight of the practicalities, limitations, and collective learning required in a business/organizational context.

The sheer volume of data involved also means trying to solve the biggest problem first doesn't work in AI – effort estimation should be factored into upfront planning. It's often better to start small and "stay niche" – you don't always need "Big Data" to train a model – and ask questions such as is the incremental performance improvement worth the storage and/or pipeline latency overhead?

B. Walsh, *Productionizing AI*, https://doi.org/10.1007/978-1-4842-8817-7_9

Our second last chapter attempts to ask these questions and provide a practical look at "joining the dots," addressing the barriers and simplifying the challenges to full-stack deployment and productionization of Enterprise AI on Cloud.

Journeying from beta application to production and ultimately hosting apps on cloud in our hands-on labs, we start by revisiting the project lifecycle and agile techniques in development, delivery and testing phases of an AI project. We look at mapping the user journey with best practice and define frameworks for success, as well as process optimization and integration of the leading AI tools.

After another look at distributed storage, parallelization, and optimizing compute (and storage) in the context of AI application scaling and elasticity, we end the chapter with a couple of hands-on labs around two of the best ways to productionize an AI solution – containerizing an AI app on Azure and hosting on Heroku.

Productionizing an AI Application

We start with the big picture and look at barriers to deployment, decisions around cloud/infrastructure, and the importance of walking before we run.

Typical Barriers to Production

There are in essence two distinct phases of an AI project – experimentation and industrialization.

The goals for both are different; for experimentation, it's about finding a model that answers a (business) question accurately and as fast as possible while for industrialization it's about running the application reliably and automatically.

Whether the client/company is small (start-up) or large (corporate) will impact our solution. As shown in Figure 9-1, for start-ups experimenting, the lack of data and budget is a constraint, while a (data) engineering resource gap is the main barrier to productionization. For mid-tier organizations often dedicated servers and cloud infrastructure needed to run POCs are often lacking while data silos[1] and the difficulty of recreating a multisystem production environment present significant challenges to corporates. When coupled with devolved stakeholder responsibility and immovable legacy systems, these issues can prevent any meaningful productionized (Enterprise AI) solution.

[1] Data buried away in one department

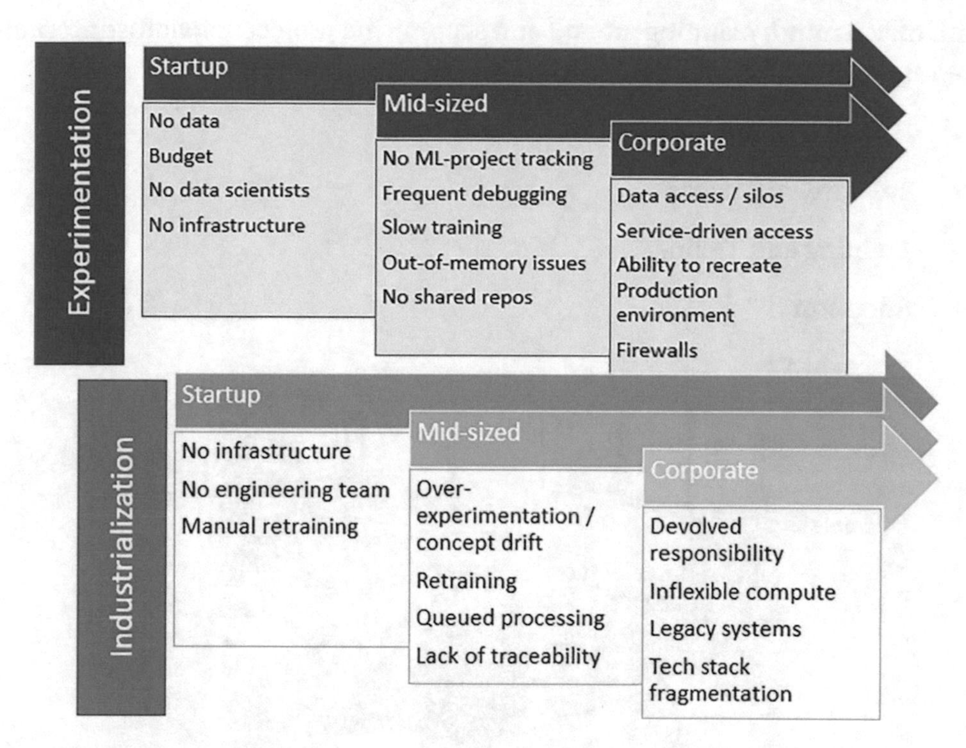

Figure 9-1. *AI Experimentation vs. Industrialization (Source: `towardsdatascience.com`)*

Cloud/CSP Roulette

Whether it's a start-up, SME, or corporate, implementable Enterprise AI solutions generally have to overcome five key constraints:

- Winning over stakeholders
- A clear data-handling strategy
- Cloud storage and compute
- Cost of solution persistence
- Innovation

Most AI solutions today have significant "solution concentration" around the three main CSPs of AWS, Azure, and GCP (Figure 9-2). Whether it's storage or compute, there is likely to be one or two services/resources provisioned on Amazon, Microsoft, or Google.

With this in mind, planning around and scoping the project in four distinct phases can help steer an AI project toward an agreed stakeholder solution:

- Design and Implementation
- 3rd Party Interfaces
- Training and Testing
- Adoption

Figure 9-2. *CSP market power*

Simplifying the AI Challenge – Start Small, Stay Niche

In scoping out our solution, and in particular toward the goal of the last phase above (achieving AI adoption), it's a good idea to prepare for development and implementation around three Data/MLOps maturity levels:

1. **Manual building and deploying of models**

   Make use of simple tools, for example, Jupyter notebook, Colab, AutoML, drag and drop (No Code) to demonstrate quick business/organizational value

2. **Deploy pipelines instead of models**

   Besides use of (pretrained) model registries, this means handling each subprocess below as a "pipeline" in order to achieve increases in stability and traceability within the AI project:

   a) Data preprocessing

   b) Algo tuning

   c) Training and evaluation

   d) Model selection

   e) Deployment

3. **CICD integration, automated retraining, concept drift detection**

   Essentially full automation including batch inference to streaming API inference and full integration into a CI/CD triggered re-training flow

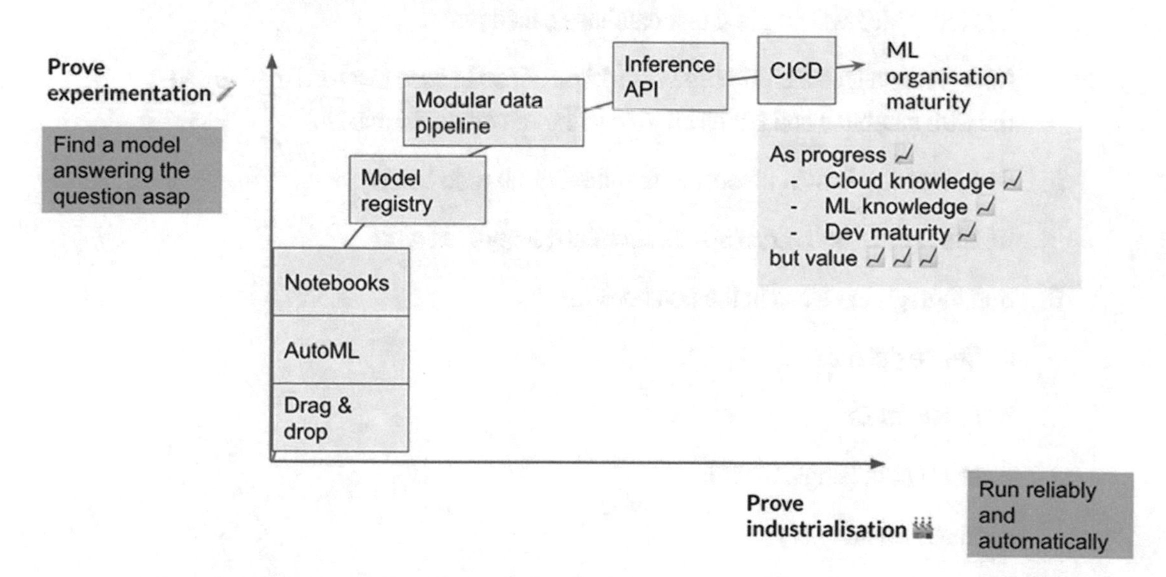

Figure 9-3. *Model maturity from POC to Production*

Database Management in Python: Hands-on Practice

SQL DATABASE READ/WRITE FROM PYTHON

Integrating two important components of an AI solution – Data Storage and Python, the goal of this lab is to use Python to write to and then read back from a SQL database:

1. Install SQLite by going to `www.sqlite.org/download.html` and downloading the precompiled binaries for Windows (both sqlite-dll-win64-x64-3360000.zip and sqlite-tools-win32-x86-3360000.zip)

2. Create a folder on your local drive C:\sqlite and unzip the two files above

3. Add C:\sqlite as a system path environment variable

4. Test the install in terminal by typing:

   ```
   sqlite3
   sqlite3 test.db
   ```

   The latter step will create a test database called test

   Now (optionally) install DB Browser `https://sqlitebrowser.org/` to see the SQL database and tables created in the remainder of this lab

5. Download the Python notebook from the GitHub repo below

   `https://github.com/bw-cetech/apress-9.1.git`

6. Run through the steps in the notebook to:

   a. Create database

   b. Create table

   c. Insert data (single record)

   d. Run a select query

   e. Import the "HRE-short.csv" file downloaded from the GitHub link above to Python and then bulk export to SQLite

   f. View the data in DB Browser and

   g. Finally, convert the data to a DataFrame in Python

App Building on GCP: Hands-on Practice

DEPLOYING A MACHINE LEARNING MODEL ON GCP

In this lab we look at deploying a simple machine learning model on Google Cloud Platform

1. Create a project on GCP `https://console.cloud.google.com`
2. Create an app using App Engine at the link below: `https://console.cloud.google.com/appengine`
3. Start GCP cloud shell and connect to the project
4. Clone (in Cloud Shell) the sample model at the GitHub link below:

   `https://github.com/opeyemibami/deployment-of-titanic-on-google-cloud`
5. Initialize gcloud in the project directory by running

   gcloud init.
6. Deploy the app following the steps in the link below:

   `https://heartbeat.comet.ml/deploying-machine-learning-models-on-google-cloud-platform-gcp-7b1ff8140144`
7. Download Postman Desktop version from the link below: `www.postman.com/downloads/`

And test the connection with the app

PowerBI – Python Handshake: Hands-on Practice

FRONT-ENDING PREDICTIVE ANALYTICS

This short lab takes the previously completed "Python Data Ingestion – Met Office weather data" lab from Chapter 3 and creates a PowerBI front-end from the Python-scraped output

1. If not already installed, download PowerBI desktop from the link below:

   `https://powerbi.microsoft.com/en-us/downloads/`

   and follow the steps here to set the Python path in PowerBI:

   `https://docs.microsoft.com/en-us/power-bi/connect-data/desktop-python-scripts`

2. Paste the completed script from the Met Office Data Ingestion lab in Chapter 3, under Data > More > Other > Python script in PowerBI

3. Create a line chart showing the minimum and maximum temperatures for the next 5 days. Note you will need to:

   a. Convert the imported columns (forecast values) to a whole number

   b. Format the chart into a "presentation-quality" visualization in PowerBI

AI Project Lifecycle

Design Thinking Through to Agile Development

Having taken a look at what works in moving from experimentation to productization, how to go about implementing an approach from scratch?

Intense competition has driven the need for organizations to change strategy quickly, forcing companies to look closely at their capabilities and processes and identify barriers and benefits to adaptability and innovation. Adoption of initial **Design Thinking** approaches helps to navigate this disruptive landscape, ensuring that business stakeholders, processes, tools, systems, and data touchpoints are brought in-scope

Design Thinking as creative problem-solving process is focused on people, rather than tools and, when done well, a key outcome is a set of projects to deliver the "right things." These projects themselves are not enough to guarantee a successful deployment - but often feed into one of the cornerstones of DataOps - **agile** development and delivery focused on "building the thing right." Figure 9-4 illustrates this concept:

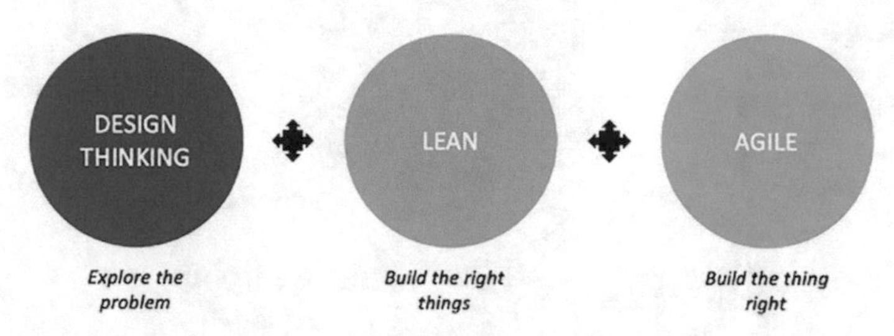

Figure 9-4. *Design Thinking, Lean and Agile (Source: Jonny Schneider)*

Driving Development Through Hypothesis

How does agile development look in an AI project? **Hypothesis-driven development** is a prototype methodology that allows Solution Architects (as well as Data Scientists and AI Engineers) to develop, test, and rebuild a product until it's acceptable by the users.

Highly iterative, the process takes assumptions defined during the project and attempts to validate them with users/client feedback. More evolutionary than "traditional" requirements capture which give rise to errors similar to those in "waterfall" approaches, these hypotheses recognize that world is complex, changing, and often confusing.

As shown in Figure 9-5, the general idea is to experiment early and often, solicit feedback from customers and discard features that provide little benefit.

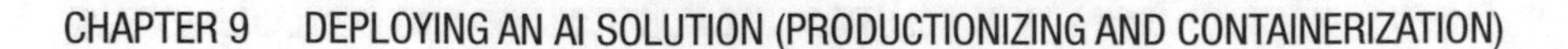

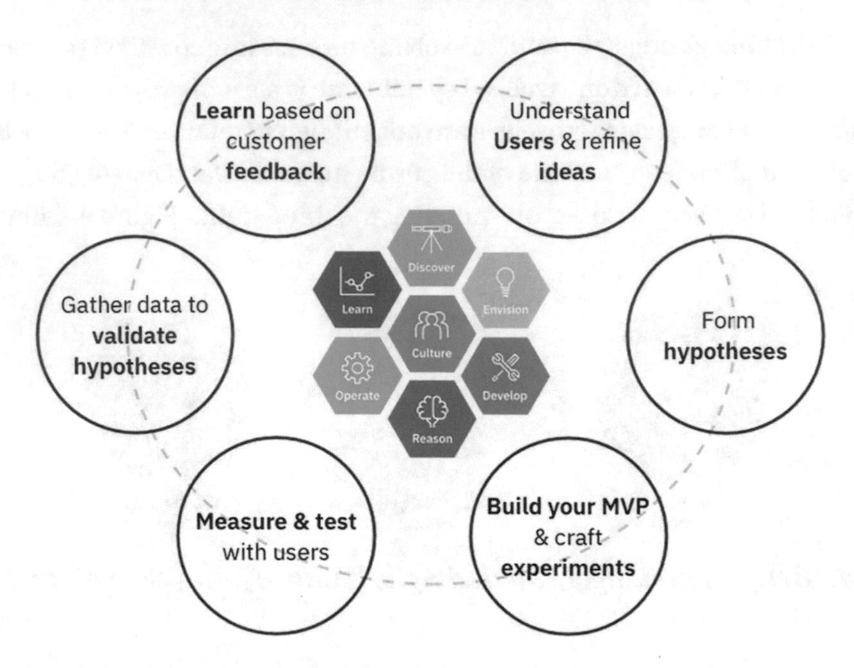

Figure 9-5. *IBM Garage hypothesis-driven development*

Collaborate, Test, Measure, Repeat

As mentioned in the introduction to this chapter, a lack of foresight from companies employing Data specialists in the 2010s has led to an enormous accumulation of **technical debt.** Rather than fully integrated AI (stakeholders!) and solutions, the result has been systems deployed without the processes and tools to maintain, monitor, and update them.[2]

Continuous testing in DataOps is one means to help address these issues. Ostensibly it's about building trust in data upfront and reducing the time it takes to identify and fix problems before trust is lost:

Version control – using git, GitHub to improve collaboration

Automate tests – automating tests: to help accelerate innovation lifecycle and change processes

Measure errors – tracking and lowering monthly production error rates

Track productivity – benchmarking process improvements and time to deliver

[2] Source: DataKitchen

Continual Process Improvement

How exactly do we track productivity and benchmark process improvements in an AI project?

Business Intelligence dashboards are perfect for reporting DataOps KPIs, particularly when automated orchestration is embedded to collect and display metrics.

A **CDO[3] dashboard,** for example, can be built to track and monitor metric progress such as

- **Team collaboration** - creation of "kitchens" for each team project
- **Error rates** - over time these should decrease as testing matures and the number of tests increases
- **Productivity** - measured by number of tests and data pipeline steps (keys)
- **Deployments** - tracking, for example, mean deployment cycle times, trend reduction in SLA breaches
- **Tests** - volume of tests should increase over time, as should prevalence of more robust quality control

[3] Chief Data Officer

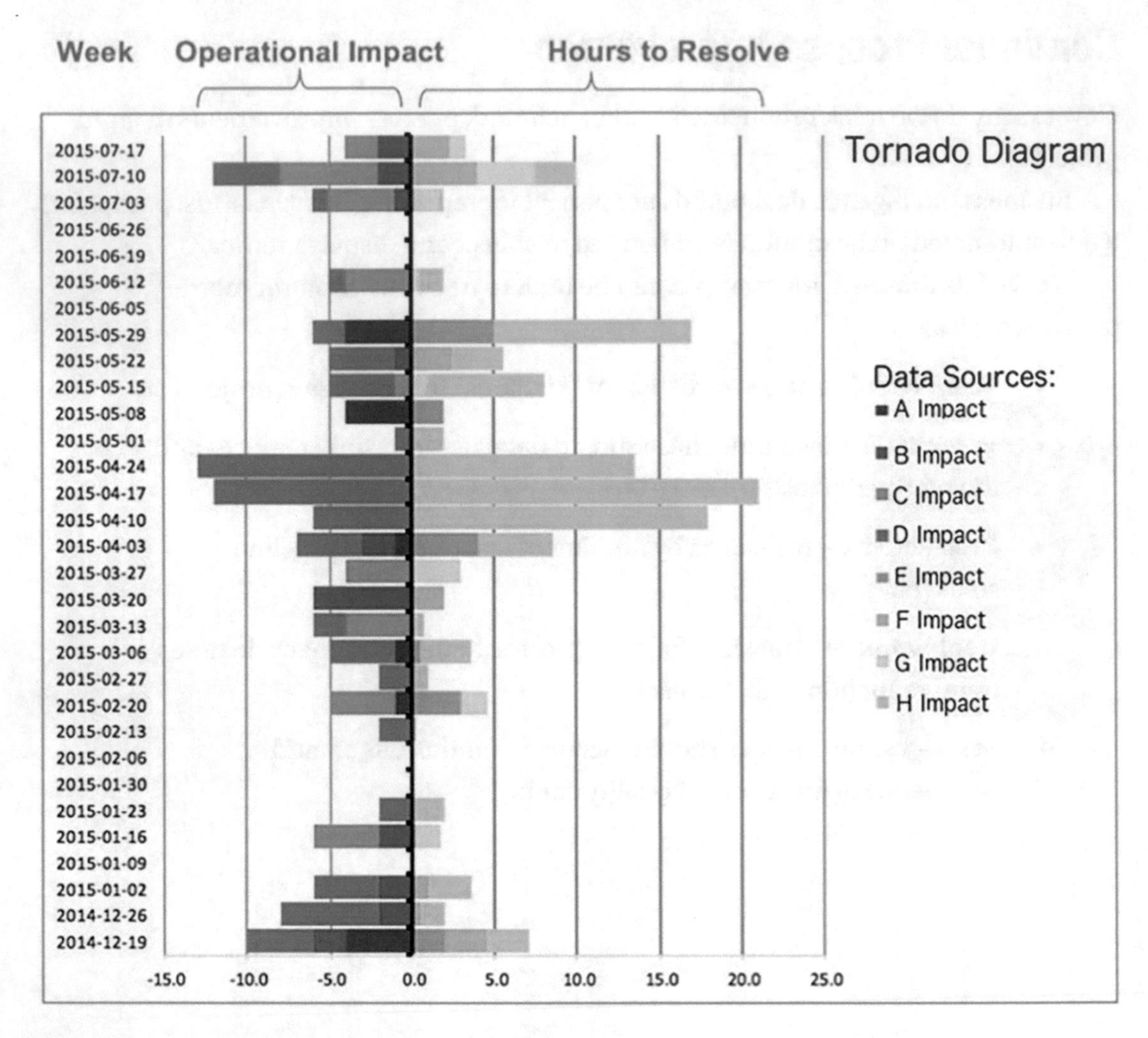

***Figure 9-6.** DataKitchen Tornado chart showing operational impact of production issues and the time required to resolve them*

Data drift

Any continuous improvement cycle for delivering and deploying an AI solution also needs resiliency in the face of post deployment interface/data changes, including robust data drift mitigation and model re-training.

Degradation in model performance over time is usually attributed to one of (a) data (or covariate) drift or (b) concept drift, that is, when data distributions have deviated significantly from those of the original training set:

- **Data drift** refers to feature drift and the possibility that relationships between features change over time
- **Concept drift** refers to changes in the target variable

The recommended way to manage model drift is to implement DataOps/MLOps best practice and track changes to underlying data including correlations between features, either, for example, in a separate Python tracking script or an operational (as opposed to strategic/CDO) dashboard.

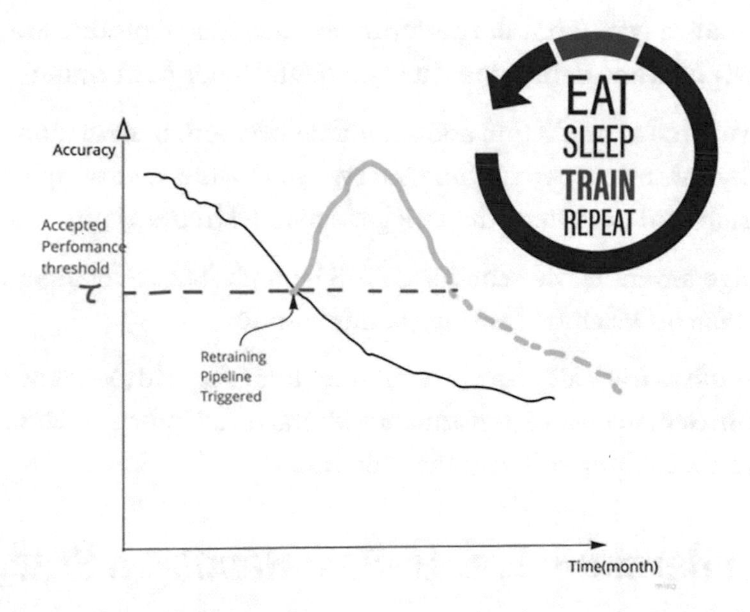

***Figure 9-7.** Model degradation over time as a result of data drift (Source: `www.kdnuggets.com`)*

Automated Retraining

Once we have started tracking data drift, we need to implement a process to ensure our model remains valid/performant. Ideally, this should be automated – to implement an automated retraining process based on data/model drift, there are two main options:

Scheduled retraining

1. In an end-of-day (EOD) process, archive the oldest day from the training set, and add in the new day of data
2. Schedule a retrain process each month (or each day if especially dynamic data)

3. Evaluate new performance and check whether performs better than previous run(s)

4. Use a cron job/job scheduler or, for example, AWS CloudWatch Event to trigger a Step Function orchestration of data prep, training, evaluation, and deployment

Performance-based/Dynamic Retraining

1. Collect new daily training data

2. Automatically monitor the performance of model in production on new data and determine if it is suddenly under-performing

3. Measure mean prediction and standard deviation. If prediction falls by 10% over a certain time interval, or outside, for example, 2 or 3 standard deviations then trigger a parallel (retrain) run

4. Manage retraining via scheduled AWS Lambda/Step Functions, or IBM Watson Machine Learning continuous learning

5. Continue to track all historic retraining runs (e.g., with MLFlow in Databricks[4]) to ensure performance (fbeta, recall/precision, loss/accuracy, etc.) doesn't degrade over time

Hosting on Heroku – End-to-End: Hands-on Practice

FROM DEV TO PRODUCTION

Our next lab takes a boiler plate app and pushes it to cloud as a Heroku-hosted application:

1. Sign up on Heroku at `www.heroku.com/`

2. Create a new app in Heroku in the Europe region, for example, my-heroku-app

3. Install the Heroku CLI

[4] MLFlow (`https://mlflow.org`) is an open source platform developed by Databricks to help productionize models by managing the complete machine learning lifecycle with enterprise reliability, security and scale.

4. Clone the boilerplate application at the link below:

   `https://github.com/bw-cetech/apress-9.2.git`

5. Set up a Virtual Environment as described in Chapter 7 and install the dependencies

6. Exercise – try to run the app locally

7. After cd'ing into the cloned app on your local drive, login to Heroku from your terminal with "heroku login" then push to Heroku by entering the following commands in sequence:

```
git status # local repo should already be initialised
git add .
git commit -am "updated python runtime to heroku supported stack
version 3.9.7"
git push heroku master
```

8. Finally, open your Heroku app by going to the url `https://my-heroku-app.herokuapp.com/` or by typing "heroku open" in terminal

9. Exercise: still in terminal, try to rename your app

Enabling Engineering and Infrastructure

People, processes, and tools are the three cornerstones of any best practice framework including productionizing an AI application. Starting with the AI Cloud Stack, we take a look in this short section at bringing together as a coherent AI solution some of the engineering and infrastructure tooling discussed in previous chapters.

The AI Ecosystem – The AI Cloud Stack

As seen in Chapter 3 successful AI requires an end-to-end Cloud Infrastructure and especially agile (scalable and elastic) services for handling Big Data, from data ingestion, collection, processing, storage, querying through data visualization. Besides data handling, "scale-up" scalability and "scale-out" elasticity are crucial supporting resource requirements in the handling of user traffic, and specific patterns of usage on the end AI application.

Whether our solution is served by a thick client application[5] or a thin client,[6] key storage (or data), such as a Data Lake or NoSQL database, and compute, such as a Virtual Machine or Apache Spark will serve the underlying AI services and tools we use in developing and ultimately deploying our project. Figure 9-8 describes this ecosystem of supporting infrastructure for AI.

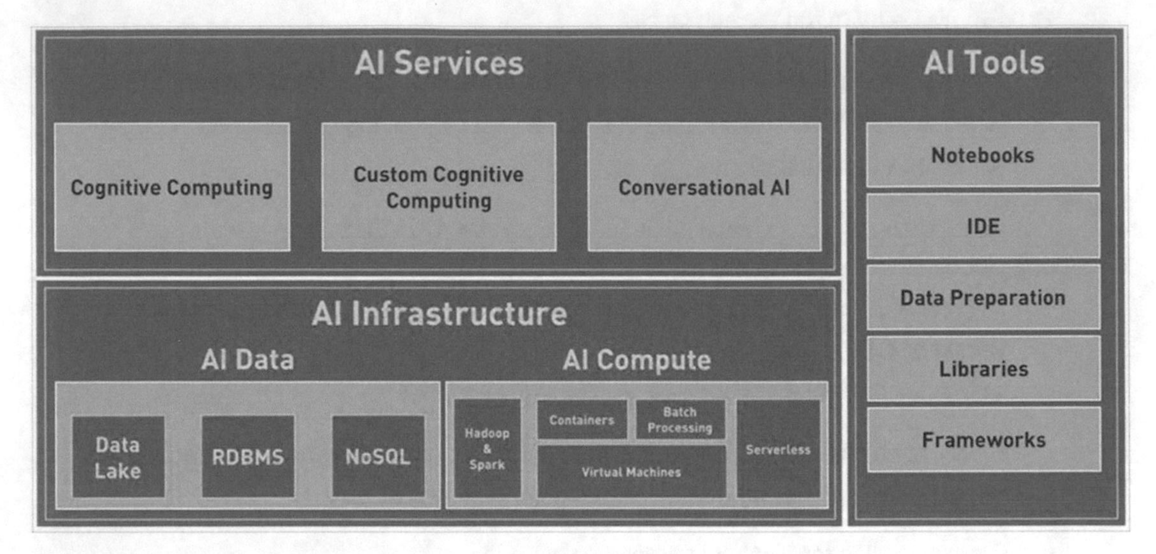

Figure 9-8. *AI Ecosystem: Infrastructure, Tools, and Services*

Data Lake Deployment – Best Practice

Many companies looking to implement Enterprise AI today strive to architect a Data Lake that constantly refreshes its multiple data streams yet optimizes performance and topology.

Not every company can afford the cost of building a Data Lake, but even if one single tool is out-of-reach, replicating as far as possible a data lake architecture is the best way to achieve production-grade agility while at the same time avoiding multiple (and often redundant) data flows to and from resources such as data warehouses, relational and nonrelational databases, big data engines, machine learning tools, and log files. To address these issues, dremio, for example, recommend the following best practice design:

[5] For example, a local install, Jupyter notebook, GitHub Desktop, etc.

[6] Web browser, Colab-based notebooks, GitHub, etc.

Employ a Data Lake-Centric Design – view the data lake as the single source of truth

Separate Compute from Data – achieve cost savings

Minimize Data Copies – reducing governance overhead

Determine a High-Level Data Lake Design Pattern – support user hierarchies and compliance

Stay Open, Flexible, and Portable – keep architecture change/future-proof, support multicloud

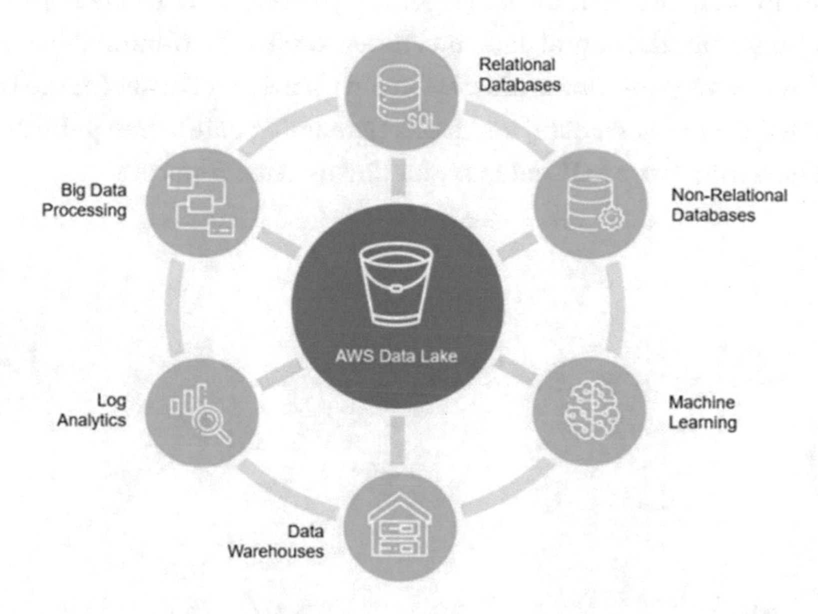

Figure 9-9. *Data lake as a centralized single source of truth (dremio)*

Data Pipeline Operationalization and Orchestration

To see how this might work in action, it's worth taking a look at a Case Study. The below architecture, with Azure Data Lake Storage (ADLS Gen2) at its heart, shows how a gaming company might run sophisticated analytics.[7]

[7] In production, whereas development and staging environments may include connectors to legacy data sources

The underlying business scenario is a gaming company collecting petabytes of (user) data from game logs and wanting to analyze these logs to gain insights into customer preferences, demographics, and usage behavior. Typically an on-prem data store will hold reference/master data such as (sensitive) customer details, game IDs, and marketing campaign data.

Secondary goals for this company might include identifying up-sell and cross-sell opportunities, developing compelling new features, driving business growth, and providing a better customer experience (CX).

The target architecture[8] in this case (Figure 9-10) uses Azure Data Factory (ideally suited for its cloud-based ETL and data integration services) to (automatically) orchestrate data-driven workflows/pipelines, spin up a Spark cluster (Azure HDInsight) feeding off ADLS Gen2 (batch data) and Kafka (streaming data), then publish transformed data into Azure SQL database for downstream reporting.

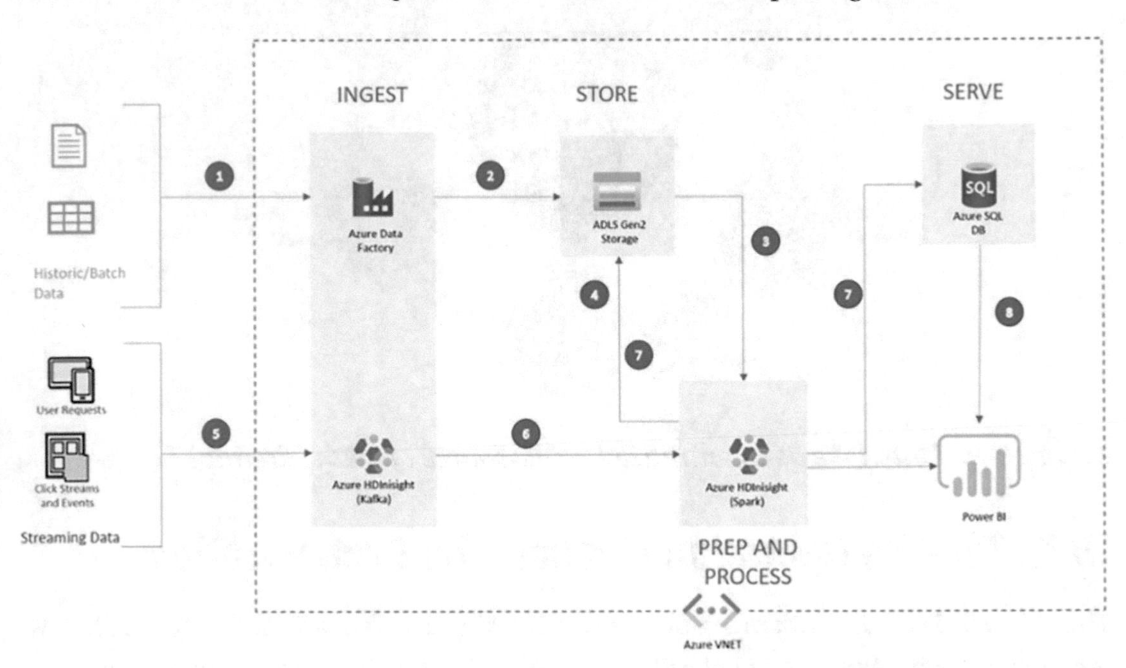

Figure 9-10. *Modern architecture for peta-scale data and analytics handling (Azure)*

[8] "To-Be" production architecture, whereas development, staging, and initial production environments may include stopgap connectors to legacy data sources

Big Data Engines and Parallelization

We have already looked at distributed processing and the use of clusters in earlier chapters as well as[9] Apache Spark as a Big Data processing engine in Chapter 5.

As the de facto standard for distributed, scale-out data processing, Apache Spark is naturally suited for AI applications that require (petascale) Big Data processing, but its Scala-based syntax and usability, even via PySpark (Python library/wrapper for implementing Apache Spark), is somewhat more complicated than standard python libraries like pandas.

There are however alternatives to using Apache Spark directly, such as Koalas – which combines user-friendly pandas-style dataframe manipulation with performant Apache Spark DataFrame distribution[9].

Dask - a platform for scaling Python which, like Spark, can scale from single node to thousand-node clusters is another alternative, which we will cover now.

Dask

Dask has fewer features than Spark and is smaller and therefore more lightweight than Spark. It also lacks Spark's high-level optimization of uniformly applied computations but Dask does have its advantages:[10]

- Written in Python rather than Scala
- Strong integration with other Python libraries such as Pandas and Scikit-learn
- Python integration focus rather than Apache project-integration
- Moves computation to the data, rather than the other way around

Dask works with two key concepts: delayed/background execution and lazy execution for stacking transformations/compute for parallel processing. We will look at both of these in a hands-on lab below.

[9] See `https://www.databricks.com/blog/2019/08/22/guest-blog-how-virgin-hyperloop-one-reduced-processing-time-from-hours-to-minutes-with-koalas.html` for a neat comparison of Pandas, Koalas and PySpark runtime

[10] There are some potential implementation challenges, for example, Dask DataFrames are used to partition data and split across multiple nodes in a cluster but calculation with compute() may run slowly and out of memory errors can occur when the data size exceeds the memory of a single machine

Leveraging S3 File Storage: Hands-on Practise

APPLICATION FILE STORES WITH S3

Using Python, the goal of this exercise is to interface to one of the most important storage resource on cloud: Amazon Simple Storage Service (S3):

1. Clone files from the following GitHub link, and unzip into a local directory:

   `https://github.com/bw-cetech/apress-9.3.git`

2. Create a publicly accessible S3 bucket called e.g. “my-s3-fs' in an AWS region closest to you

3. Create AWS Access Keys and download:

   `https://console.aws.amazon.com/iam/home?#/security_credentials`

4. Download the AWS CLI installer `https://awscli.amazonaws.com/AWSCLIV2.msi and install AWS Command Line Interface`

5. Push (upload) your local files to your S3 bucket on AWS by following the steps below:

   a. open terminal / command prompt)

   b. type “aws configure”

   c. enter access and secret key downloaded above

   d. enter your Default AWS Region e.g. eu-west-2

   e. specify default output format as “json”

   f. cd into folder where the local (unzipped) files have been extracted

   g. sync the data to your S3 bucket with:

   ```
   aws s3 sync . s3://pv-s3-fs
   ```

   The download GitHub data has now been pushed to AWS and is stored in your S3 bucket

6. Exercise - try to connect to S3 from Python to download the images. Check your answer with the notebook ”AWS-S3-Download.ipynb provided in the GitHub link above

Apache Spark Quick Start on Databricks: Hands-on Practice

BIG IOT DATA PROCESSING WITH APACHE SPARK

In this lab we revisit Databricks and Apache Spark to compare runtimes on an IoT dataset

1. In Databricks Community Edition `https://community.cloud.databricks.com/login.html`

   create a blank notebook and spin up a Cluster

2. Following the steps in the notebook below: `https://community.cloud.databricks.com/?o=765164012049213#notebook/1443608314106734/command/3293421293983457` complete the steps as shown:

   a. Import IoT data from the link below: `https://raw.githubusercontent.com/dmatrix/examples/master/spark/databricks/notebooks/py/data/iot_devices.json`

   NB due to the 61 MB file size, you may need to copy and paste the json file contents into a .txt file and save locally as a Json first

   b. Using Scala perform EDA and Wrangling on the dataset

   c. Query the data using SQL

   d. Compare runtime Spark Cluster vs. Jupyter

Dask Parallelization: Hands-on Practice

DASK PARALLELIZATION

Following the steps shown in the process diagram below, the goal of this exercise is to familiarize ourselves with (Python-based) big data processing using Dask:

1. Clone the GitHub repo git clone `http://github.com/dask/dask-tutorial`
2. Create a conda environment
3. Launch either Jupyter Lab or Jupyter Notebook
4. Walk through the code sample "dask.delayed.ipynb" to get a feel for how Delayed Execution works in Dask

   NB to see how complex extract, transform and load (ETL) over a cluster works see the gif in the follow-on notebook:

   `https://github.com/dask/dask-tutorial/blob/main/01x_lazy.ipynb`
5. Exercise – do the same with Lazy Execution, that is, run the 01x_lazy.ipynb notebook
6. Compare runtime reading the README.md file word by word vs. line by line

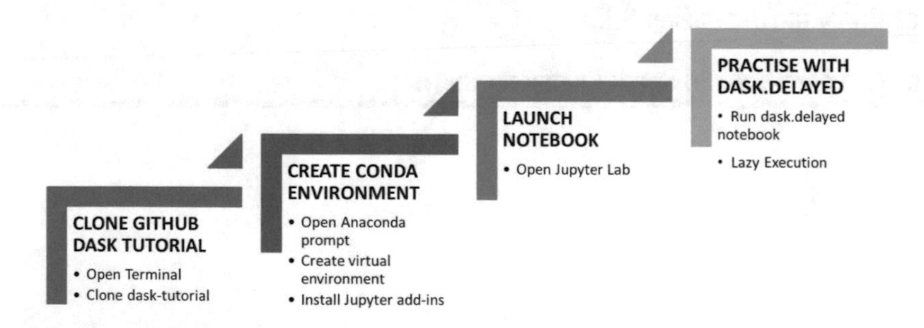

Figure 9-11. *Introduction to Dask – lab process*

Full Stack and Containerization…the final frontier

And so we arrive at the last section in this penultimate chapter where we address packaging a "full-stack" deployed solution. We revisit our full stack AI application lab from Chapter 7 before wrapping up on the use of containers, specifically Docker[11] - often the final piece of the jigsaw to simplify and successfully deploy an AI app.

Full Stack AI – React and Flask Case Study

At the end of Chapter 7, we deployed our first AI application using react.js, Plotly Dash, Flask, and TensorFlow. The process used is a great step-by-step guide to implementing any AI application from training a model and exporting the model as a hierarchical data format (.h5) file, setting up a virtual environment, creating a back-end for model integration, then a front-end UI before running the app locally first, then as a hosted endpoint solution.

This deployment process is shown graphically in Figure 9-12 for reference.

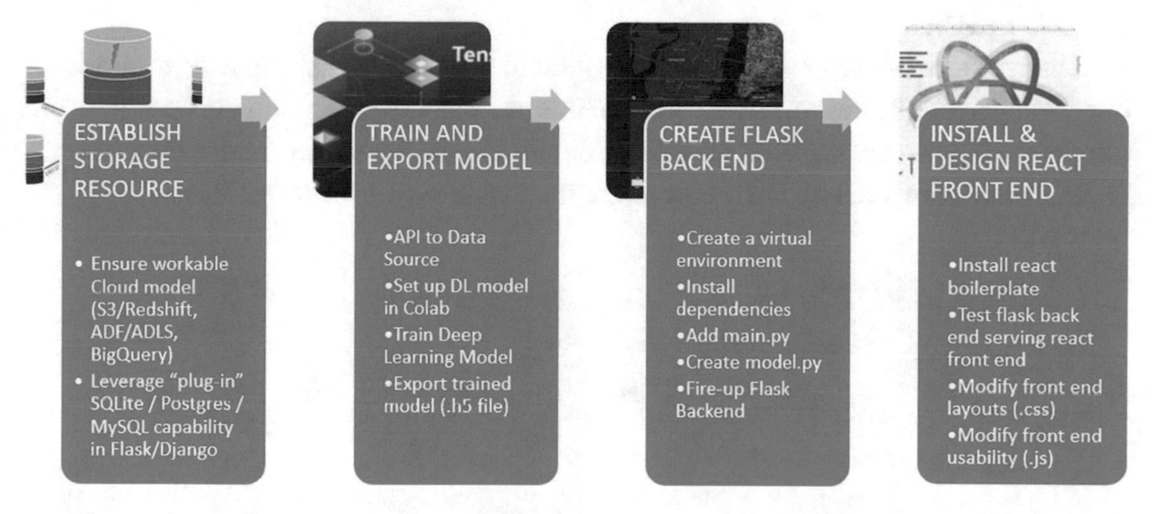

Figure 9-12. *Full-Stack Deployment revisited – react.js and Flask*

[11] See also Chapter 2 on DataOps

Unfortunately building an AI app in this way can be rather cumbersome due to the high dependency on local file and library configuration. This is where containers can help - providing a means to ringfence dependencies and ensure there are no conflicts when deploying across different end-user systems.

Deploying on Cloud with a Docker Container

The subject of our final hands-on lab in this chapter, deploying on cloud with Docker, is the recommended route for productionizing AI. As shown in the graphic below, this starts by training a model before exporting (this time as a pickle file).

We then download a boilerplate solution (source code From GitHub), and can optionally test the steps for Full-Stack AI described above, to carry out an (a) local (standalone) install (complete with local file dependencies).

Following a successful local install, we proceed to (b) a local containerized docker instance, testing that before finally (c) authenticating the cloud (in this case ACR - Azure Container Registry), rebuilding our container image and pushing our Docker instance to ACR.

Lastly, we create an (Azure) Web App, point to the Docker image in ACR, and view our deployed application. The real value of this productionized solution is that all our dependencies are installed via a requirements.txt file within the internal Docker environment, ringfenced from potential conflicts with outside (external) files.

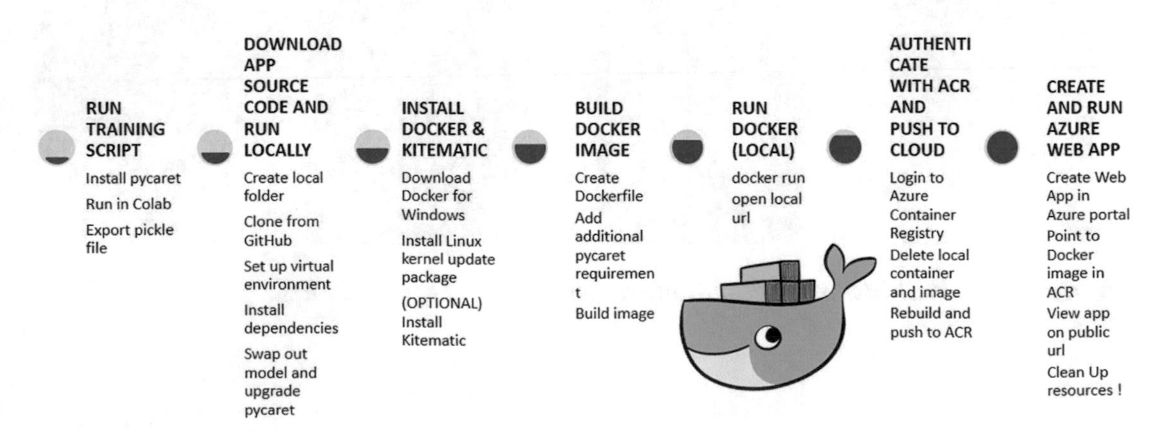

***Figure 9-13.** Deploying a containerized Docker AI app to Azure*

Implementing a Continuous Delivery Pipeline

We finish off by returning to our distinct AI project phases from "Typical barriers to Production" above. Experimentation around solution development and design is a means to an end, that is, toward our goal of productionizing or industrializing our AI application.

As DataKitchen put it, we have an innovation pipeline and a value pipeline:

- **Innovation Pipeline** – AI models are developed, designed, tested, and deployed into the Value Pipeline
- **Value Pipeline** – data is fed into AI models producing analytics that deliver value to the process for new model creation

Each pipeline in itself is a set of iterative stages typically defined by a set of files and implemented using a range of tools, from scripts, source code, algorithms, html configuration files, parameter files, and containers.

And code of course controls the entire data-analytics pipeline from end to end, effectively from ideation, design, and requirements capture through training, testing, deployment to operations, and postproduction maintenance.

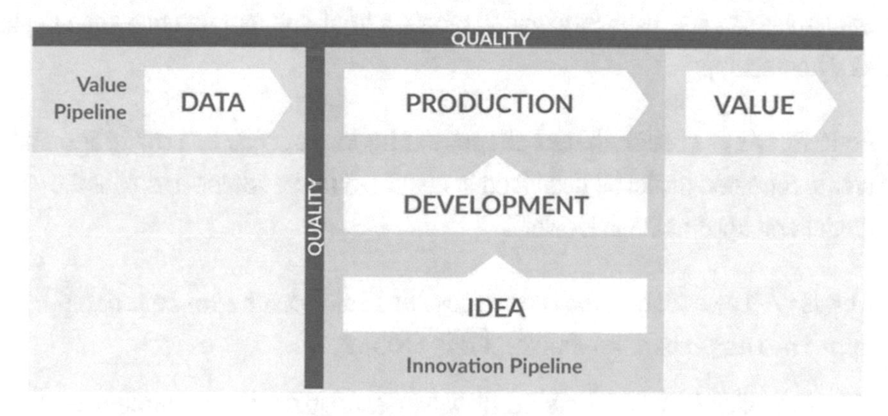

Figure 9-14. *The continuous delivery pipeline*

Wrap-up

With that look at continuous AI delivery, we have reached the end of our journey through how to deploy an AI solution. The two labs that follow are intended to give a more "immersive" experience of productionizing AI, with the end goal being "full-stack" AI application.

Although this chapter brings us to the end of the main theme of this book, we do have one further chapter to go. Largely due to step-changes in enabling technology (particularly transformers) and heightened focus on unstructured data, Natural Language Processing has carved off as its own field of interest and is the subject of our final chapter.

DL App deployment with Streamlit and Heroku: Hands-on Practice

DL HOSTING WITH STREAMLIT, GOOGLE, AND HEROKU

Our last two labs are end-to-end application deployments, this one trains a model first with Google teachable Machine, uses Streamlit to create a front-end for inference, then deploys to Heroku as a hosted app:

1. Go to `https://teachablemachine.withgoogle.com/train/image` and create a trained model as described in steps 1-5 under "Steps to create the model and app" at the link below:

   `https://towardsdatascience.com/build-a-machine-learning-app-in-less-than-an-hour-300d97f0b620`

2. Download files from "C:\Users\Barry Walsh\Testing\xray-automl\Streamlit-Heroku-setup.zip and unzip to a test folder on your local drive

3. Install all requirements in a virtual environment including pillow, tensorflow and scikit-learn

4. Run the command below and check that the app is working correctly locally

   ```
   streamlit run xray.py
   ```

5. Exercise: finally repeat the steps in the lab above "Hosting on Heroku – End-to-End: Hands-on Practice" to deploy as an endpoint solution on cloud

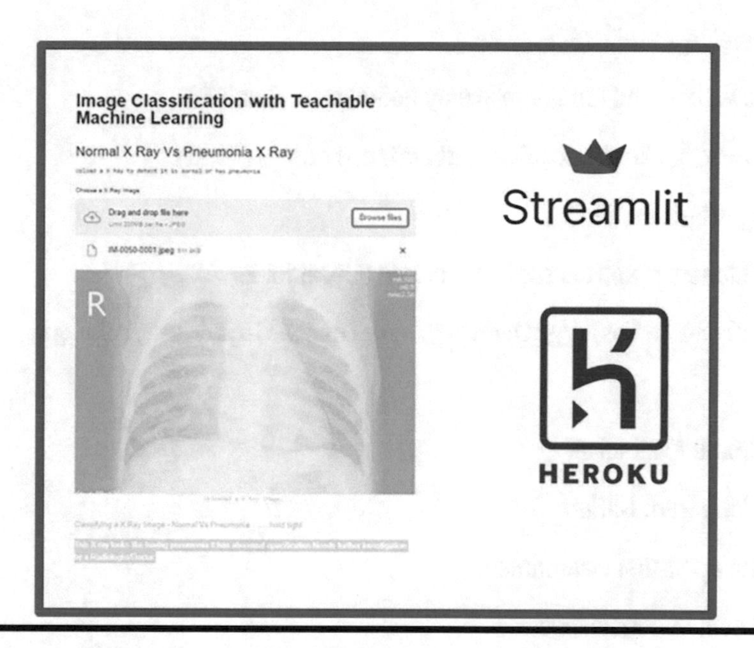

Deploying on Azure with a Docker Container: Hands-on Practice

CONTAINERIZING AN ML MODEL FOR PREDICTING INSURANCE PREMIUMS

Using PyCaret in Colab, our final "marathon" lab trains an AutoML model to predict insurance premiums, then exports the resultant pickle file. We then attach the trained model to a boilerplate Flask app, run locally before Dockerizing as a container app and pushing to Azure Container Registry for running as an (Azure) web app.

The process is described further in the section above "Deploying on Cloud with a Docker Container" and is summarized below:

1. First, we train a machine learning pipeline (AutoML) to predict insurance charges:

   a. Install pycaret

   b. Download and run the following notebook in Colab:[12]

   `https://github.com/bw-cetech/apress-9.4.git`

   c. Export pickle file

2. Next clone the app source code, create a local folder

   ```
   git clone https://github.com/pycaret/deployment-heroku.git
   ```

   and

   a. Create local folder

   b. Clone from GitHub

   c. Set up virtual environment

   d. Install dependencies

   e. Swap out model and upgrade pycaret

3. Install Docker

   a. Download Docker for Windows

   b. Install Linux kernel update package

   c. (optional) Install Kitematic

4. Build the Docker Image

   a. Create Dockerfile

   b. Add additional pycaret requirement

   c. Build image

[12] Note the first part of this lab is our PyCaret AutoML Introduction to AI run through from the last lab in Chapter 1

5. Run Docker locally
    a. Docker run
    b. Open local url
6. Authenticate with Azure Container Registry (ACR) and push/deploy our container solution to cloud
    a. Login to Azure Container Registry
    b. Delete local container and image
    c. Rebuild and push to ACR
7. Create and run Azure Web App
    a. Finally, we create a Web App in the Azure portal
    b. Point to Docker image in ACR
    c. View app on public url

NB don't forget to clean up (stop and delete) resources (Azure Web App, ACR instance, and any VMs), to preserve cloud credits/prevent running up cloud costs!

CHAPTER 10

Natural Language Processing

Any book on operationalizing Artificial Intelligence today cannot ignore growth in Natural Language Processing (NLP), with market size projected to be $43b in 2025.[1] Ostensibly the branch of AI concerned with programming computers to understand both written and spoken text, NLP applications in 2022 are being pushed further to do much more. Qualitative analysis, contextual/domain-specific reasoning, and thought leadership creation[2] are all in scope and performance improvements are dramatic, even potentially seismic if you believe Google's Engineers.[3]

Perceived by many as a critical skill in the race to digitalize, much of the attention on NLP is taking techniques and best practice learned from structured data wrangling and predictive modeling processes and expanding the scope to unstructured data. The implicit goal is to transform unstructured data into machine readable formats where we can then carry out similar, if not identical processes carried out in standard ML/DL.

While the overwhelming focus in this book is on applications of machine and deep learning in their own right, we cover in this last chapter the main themes for Natural Language Processing, basic NLP theory, and implementation before addressing all the important tools and libraries for deploying NLP solutions.

[1] www.statista.com/statistics/607891/worldwide-natural-language-processing-market-revenues/

[2] See e.g. https://hbr.org/2022/04/the-power-of-natural-language-processing

[3] Advances in Google's chatbot LaMDA have led to a former senior engineer claiming the technology is "sentient." Sign up to test it here: https://aitestkitchen.withgoogle.com/

B. Walsh, *Productionizing AI*, https://doi.org/10.1007/978-1-4842-8817-7_10

Many of these tools rely on machine and deep learning techniques described elsewhere in this book already but we will go through in this final chapter the use of the main Python NLP library, NLTK, alongside other libraries such as PyTorch and spaCy used for solving NLP problems. We will also cover some of well-known APIs which leverage NLP (such as the Twitter API) which are being applied to highly relevant customer journey and public perception use cases in 2022.

We will also take our usual hands-on, lab-based approach, in this last chapter at automating the process of understanding complex language by identifying and splitting (parsing) words and extracting topics, entities, and "intents" – core subprocesses for many NL applications today, including sentiment analysis and chatbots/conversation agents.

Introduction to NLP

We start with a brief historical context and a look at a basic definition of Natural Language Processing, its place in the wider "AI Ecosystem" and interaction with machine and deep learning processes.

Our first section then proceeds to address how and why NLP is being applied at scale in businesses and organizations worldwide before wrapping up on the NLP lifecycle and in particular, a best practice roadmap for the sequence of tasks central to delivering a successful NLP implementation.

NLP Fundamentals

Natural Language Processing is the branch of AI that deals with the interaction between computers and humans using natural language. The objective is to read, decipher, understand, and make sense of language in a manner that is valuable in some way to end users and organizations as a whole. To do this, two key linguistic techniques are adopted: syntactic analysis and semantic analysis.

Importantly, and linking back to the main theme of this book, NLP then applies Machine and Deep Learning algorithms to unstructured data, converting into a form that computers can understand. This "overlap" is shown in Figure 10-1.

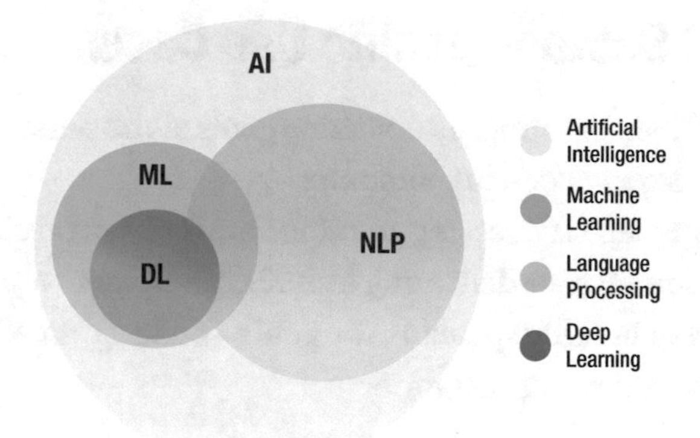

Figure 10-1. *NLP interface to Machine and Deep Learning*

Historical Context and Development of NLP

We mentioned above, linguistics study underpins use of Natural Language Processing. Early applications of NLP involved rule-based approaches, limited by the scope of rules that existed and computational scale and speed as the size of the dataset increased. Probabilistic modeling approaches used in statistical NLP have since allowed NLP solutions to scale, leveraging machine and deep learning techniques to more dynamically interpret natural language. One of the main examples of this is an N-gram, where we can train an NLP model to **probabilistically predict word sequences**, such as the bigram "Machine Learning."

The power of Deep Learning in particular to generate sophisticated predicted outcomes from big datasets fits the overall problem formulation we have with linguistic models, where the underlying structure and meaning of words and characters is complex - in effect we are dealing with another "Big Data" dataset but the difference here is that the underlying grammar (in any language) has an inherent sequence or order. For this reason, Recurrent Neural Networks/specialized LSTMs have been adopted to propagate word order as state through a neural network architecture.

As we will see, **word embeddings** or **vectorization** is the essential mechanism by which Recurrent Neural Networks read, transform, and iterate over syntax and semantic meaning of the underlying language - with words being expressed in an N-dimensional vector space.

NLP Goals and Sector-specific Use Cases

As with machine and deep learning, establishing goals at the outset is critical to ensuring a project is likely to deliver successful outcomes.

Not all Natural Language Processing solutions are focused on classifying text, generating sentiment metrics, or delivering a chatbot, and there are many nuances across sectors. However, broadly speaking the goal of building an NLP application falls into one or more of the following categories:

- Information Retrieval and Extraction (IR/IE)
- Text Classification
- Topic Identification/Detection
- Named Entity Recognition (NER)[4]
- Translation
- Text Summarization
- Sentiment Analysis

The above are often confused with actual NLP applications – however in reality, businesses or organizational applications tend to incorporate one or more of the above goals while the marketing or product names given to these NLP solutions often blurs the line between goal and technique.

As an example, the business value from using Text Classification is typically driven by the downstream customer – there is both "macro" value in, for example, sales emails falling into a classification system (e.g., based on customer segment, product line, or proposal stage) or "micro" value from the email being re-routed to specific departments based on keywords identified in the email. Equally the same text classification may trigger automated sending of a customer KBA or a chatbot response.

Key Industrial Applications

We will cover in the rest of this chapter several hands-on labs on the main NLP applications in use today. From rather primitive beginnings (spam filters, web search applications, text summarization, etc.) the scope of NLP solutions has grown to the extent that these applications have become key differentiators for many companies.

[4] Also a data transformation step – see section below

Natural Language Processing is the driving force behind many well-known applications including language translation applications such as Google Translate, word processors such as Microsoft Word and Grammarly that employ NLP to check grammatical accuracy of texts or perform autocompletion tasks and applications which leverage social media sentiment analysis.

Intelligent Virtual Agents (IVAs) or Chatbot 2.0/3.0 as well as Interactive Voice Response (IVR) applications used in call centers now respond to targeted users' requests, while personal assistant applications such as OK Google, Siri, Cortana, and Alexa are mainstream household or mobile devices.

And natural language generation promises much more, bringing full automation of short report or potentially large-scale publication writing closer to adoption. Before we go further, we will take a quick look at the technical perspective and how NLP uses a common development framework to deliver this diversity of industry solutions available.

The NLP Lifecycle

From scraping and data collection through character splitting, the removal of stop words, tokenization, lemmatization, and embedding the development of an NLP app requires every bit as strong a framework as with any approach in dealing with structured AI data.

From Parsing to Linguistic Analysis

Data Wrangling unstructured data can be problematic, but as with machine and deep learning, best practice is to follow a certain process, where underlying "parsing" subprocesses get us to a final format which we can feed into the downstream vectorization (encoding) and modeling process. In Python, the main libraries for wrangling data are NLTK and the regular expressions (re) library.

Depending on the task, some of these subprocesses may not be required, and in some cases additional domain-specific wrangling tasks may be needed, such as extrapolating FX information from a banking report. For now we list below the main NLP wrangling subprocesses for reference, broken down into three core processes: preprocessing, linguistics and transformation, and premodel encoding/word embedding.

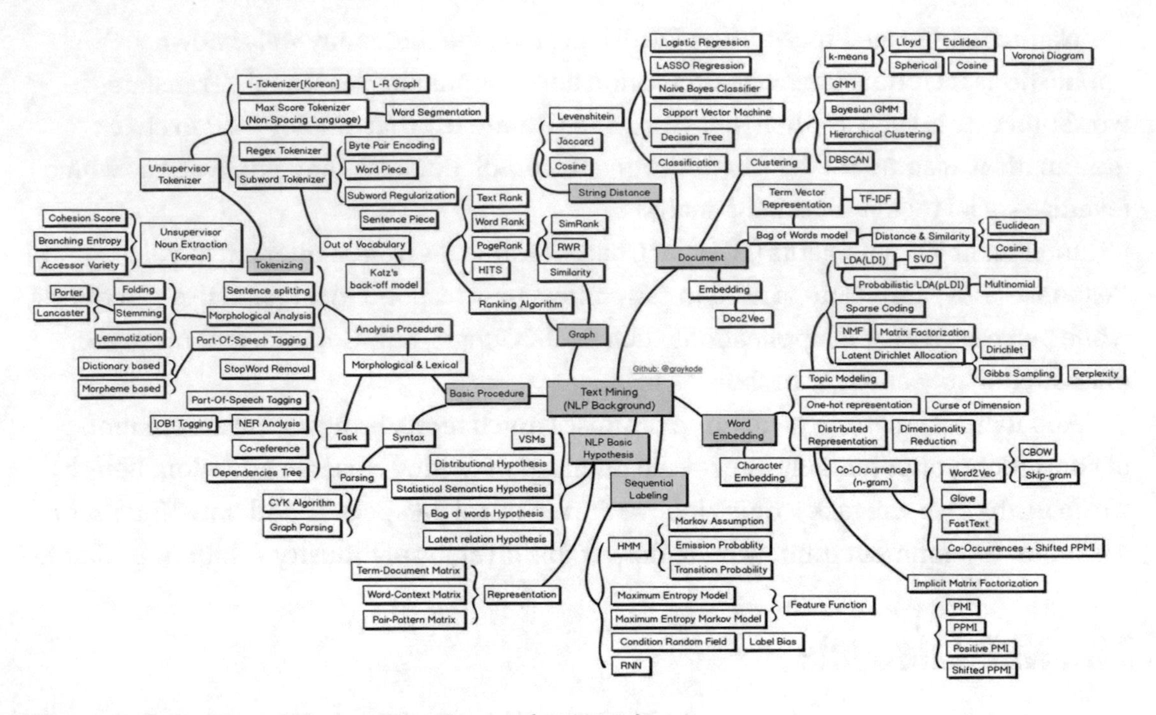

Figure 10-2. *Text Mining Roadmap (GitHub)*[5]

Preprocessing/initial cleaning

- Remove special characters
- Regular expressions: remove symbols (e.g., "#" and "RT" from tweets)
- Regular expressions: remove punctuation
- Strip HTML tags

[5] See `https://github.com/graykode/nlp-roadmap`

Linguistics and Data transformation (Figure 10-3)

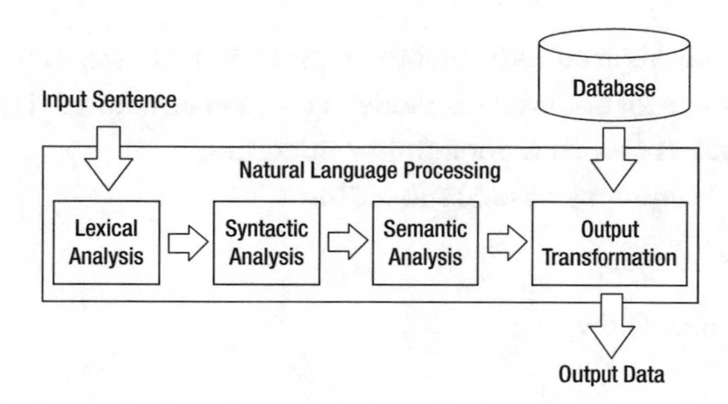

Figure 10-3. *NLP premodeling processes*

Lexical analysis:

- Filter out "stop" words
- Tokenization

Syntactic analysis:

- Switch to lower case
- Part of Speech (POS) tagging
- Named Entity Recognition (NER)
- Handling contractions[6]
- Stemming

Semantic analysis:

- Lemmatization
- Disambiguation (determining the most probable meaning of a specific phrase)
- N-grams

[6] Such as ensuring slang expressions are converted to their full equivalent: I'll = I will, You'd = You would, etc.

Word Embeddings to Deep Learning

In fact, semantic analysis extends further to vectorization and word embeddings – the final preparation steps for undertaking modeling with unstructured data

Text Vectorization / Word Embeddings / Encoding

Rule-based: / Frequency based Embedding

- One hot encoding
- Bag of words (BoW)
- TF-IDF

Prediction-based Word Embedding:

- Word2Vec (Google)
- GloVe (Stanford)
- FastText (Facebook)

In the next section, in addition to Word Embeddings, we will cover modeling essentials for text summarization, topic modeling, sequence modeling, and transformers/attention models. It is important to note that there is no single "catch-all" modeling process for the entire myriad of NLP solutions – the approach is highly dependent on the problem formulation, industry context, overriding goal-driven outcomes, and performance threshold.

Much like in Deep Learning, we are outlining at the outset a best practice "roadmap" which can hopefully be useful to the reader in parallel with an acceptance that there are likely to be many iterative steps along the way, in both the premodeling parsing and vectorization processes and in the actual model training and inference and performance benchmarking follow-up.

Creating a Word Cloud: Hands-on Practice

ZODIAC-STYLE CHINESE CHARACTER WORD CLOUD

One of the simplest applications of NLP, widely used by marketing teams across the globe, is a Word Cloud. Although basic, the techniques used here are replicated in many of the high-profile industrial applications currently in existence – this lab provides an introduction to those techniques:

1. Clone the GitHub repo below to your local drive:

   `https://github.com/bw-cetech/apress-10.1.git`

2. Go to your local folder where you have cloned the GitHub files and set up a virtual environment using the commands below (one after the other):[7]

   ```
   python -m venv env
   env\Scripts\Activate
   ```

3. Run the Python script in your virtual environment with

   ```
   python WordCloud_中文.py
   ```

   NB the Chinese characters in the name will not be rendered in terminal but the above command will still run

4. If prompted, install the dependencies one by one with

   ```
   pip install wordcloud
           pip install jieba
   ```

5. A word cloud will generate with the Chinese characters from the China Daily news extract provided in the GitHub folder

6. Exercise: run the code using instead the English data example (pasta recipe)

7. Exercise: replace the pig image template with a different Chinese zodiac image (use images from, for example, `www.astrosage.com/chinese-zodiac/`)

8. Exercise (stretch): update the code to generate the word cloud based on the current Chinese Year (tiger, rabbit, dragon, etc.)

---

[7] See e.g. Chapter 7 lab "Running Python from Terminal: Hands-on Practice" for further help with this

Preprocessing and Linguistics

Because we are dealing with the most part with unstructured text data, natural language processing is inextricably linked to linguistic structure.

Whether it's taking source data and performing initial preprocessing and cleaning tasks through the use of regular expressions, applying syntactic or semantic analysis or implementing word embeddings to convert data into word vectors, having an understanding of linguistics helps in both framing the journey from a raw to model-ready data format and ultimately extracting insights.

This section covers the key NLP concepts, referenced to linguists and broken down as described in the NLP lifecycle.

Preprocessing/Initial Cleaning

Regular Expressions

Regular expressions or "regex" are the bread and butter of unstructured data "string" searches. Often employed as a first step after scraping/data import, the idea is to quickly clean the data by searching for specific patterns in the data that we want to remove. As tokenization and vectorization naturally follow this step, the aim is to remove punctuation, symbols (such as smiley faces or emoticons from text messages or hashtags from a tweet), and special characters (such as currency symbols or brackets) that are unhelpful for converting encoding text as word vectors.

The below example shows an example implemented in Python – "re" is the library used[8]

```
import re
pattern = '^a...s$'
test_string = 'abyss'
result = re.match(pattern, test_string)
if result:
  print("Search successful.")
else:
  print("Search unsuccessful.")
```

[8] See `https://docs.python.org/3/library/re.html` for a list of regular expressions

The "^" matches the start of the string, while the "$" matches the end of the string, so in this case "Search successful" is returned

Text Stripping (e.g., HTML tags)

Python comes with an abundance of built-in functions which can simplify the process of cleaning unstructured data/searching for regular expressions. Particularly useful for text "stripping" is the Python .strip method. More sophisticated html stripping, to extract, for example, a news headline from an html tag is performed using .text.strip() method chaining with the BeautifulSoup library.

The .split function in Python is often applied after stripping/cleaning underlying text data – .split effectively tokenizes text into separate words.

Linguistics and Data Transformation

After preprocessing our data, while the unstructured data may be "clean" in the sense that we have removed redundant characters, it is not yet in a state that can be "vectorized" for modeling.

As with the data wrangling process in machine and deep learning, a number of transformation steps are required to prepare the unstructured data, each of which is underpinned by the linguistic structure of the data. Broadly these are arranged into lexical, syntactic, and semantic analysis steps.

Lexical Analysis

The first of these steps, lexical analysis refers to the process of conversion of the text data into its constituent building blocks (words, characters, or symbols according to the underlying language).

Removing Stop Words

Stop words are often quite frequent words in a sentence that add no value to the underlying meaning of a sentence. In most cases they should be removed, but for specific applications such as machine translation and text summarization they should be retained.

Words such as ("a," "the," "is," "at," etc. are removed in Python using the stop words module in the nltk library.

Tokenization

We already mentioned one way to parse text when stripping text during the preprocessing step. More commonly tokenization is performed using one of the tokenize methods from the NLTK library, typically .word_tokenize to split sentences into words, but depending on the goal, phrase tokenization may be required such as in the German example below:

```
import nltk
german_tokenizer = nltk.data.load('tokenizers/punkt/german.pickle')
german_tokens=german_tokenizer.tokenize('Wie geht es Ihnen?  Gut, danke.')
print(german_tokens)
```

The above returns the question and answer:

```
['Wie geht es Ihnen?', 'Gut, danke.']
```

Python-supported Hugging Face,[9] the open source Data Science platform is increasingly being used to scale tokenization processes on larger datasets, reducing training time (as well as environmental impact) by fine-tuning pretrained models rather than building from scratch.[10]

Syntactic Analysis

After obtaining our language building blocks from lexical analysis, the next step is to transform the data using syntactic analysis; extracting logical meaning from the text whilst considering rules of grammar.

Switch to Lowercase

The first step in syntactic analysis is to switch our data to lowercase using the built-in Python function lower(). Because the same word in proper case ("Hello") and lowercase ("hello") would be represented as two different words in vector space model, applying the lower() function addresses sparsity, reduces the dimensional problem we are solving and speeds up runtime.

---

[9] https://huggingface.co/docs/tokenizers/index

[10] While building from scratch is clearly computationally expensive, multiple GPUs can achieve savings via distributed training. See e.g. www.determined.ai/blog/faster-nlp-with-deep-learning-distributed-training

Part of Speech (POS) Tagging

POS tagging is central to information extraction, one of the key aims of an NLP solution mentioned in our first section. Using the NLTK .pos_tag method, the process involves categorizing words according to their grammatical form. The purpose of carrying out this task is to better equip the subsequent deep learning modeling process with sequencing information as to the probability of sequencing-specific words.

These probabilities are determined by the frequency of certain grammatical forms (e.g., a verb or a noun) following other grammatical forms (such as a verb)

```
text = word_tokenize("And now for something completely different")
nltk.pos_tag(text)
```

The output is a list of tuples:

```
[('And', 'CC'), ('now', 'RB'), ('for', 'IN'), ('something', 'NN'),
('completely', 'RB'), ('different', 'JJ')]
```

where: [11]
CC is a coordinating conjunction
RB is an adverb, like “occasionally” and “swiftly”
IN is a preposition/subordinating conjunction
NN is a singular noun
JJ is a large adjective
Essentially Markov Chains are being employed here, indicating (as shown in Figure 10-4) the probability of a specific grammatical term following another word.

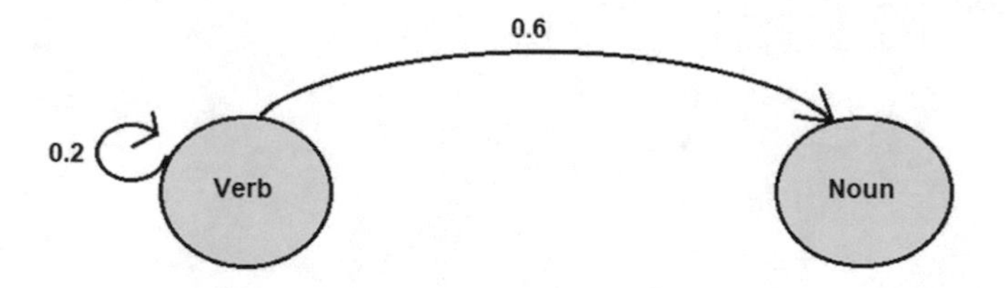

Figure 10-4. *PoS Tagging (Source: Towards Data Science)*

[11] A more extensive list of POS Tags is here: `www.guru99.com/pos-tagging-chunking-nltk.html`

Named Entity Recognition (NER)

Named Entity Recognition is by some measures an application of Part of Speech tagging. Often carried out after POS tagging, the idea is to identify topics from text based on locating named entities in text. The process is particularly efficient in the Python spaCy library,[12] where noun phrases can be extracted quickly by recognizing a determiner, followed by adjective(s) and then a noun, displaying a quick summary of the main entities as shown below:

```
[('European', 'NORP'),
 ('Google', 'ORG'),
 ('$5.1 billion', 'MONEY'),
 ('Wednesday', 'DATE')]
```

NLP applications like all AI apps are highly domain-specific, and while POS tagging and NER tasks can do the heavy lifting of annotating texts, they are always gaps with certain lexical terms. As such, both POS and NER tagging are often supported by a (manual) sector-specific entity curation and manual annotation process.

The below example shows how this process works with IBM Watson Knowledge Catalogue, where an approved group of subject matter experts are able to highlight words/terms and define entity types.

[12] The NLTK application of NER involves a more complicated use of regular expressions to be recognized, for example, noun phrase patterns.

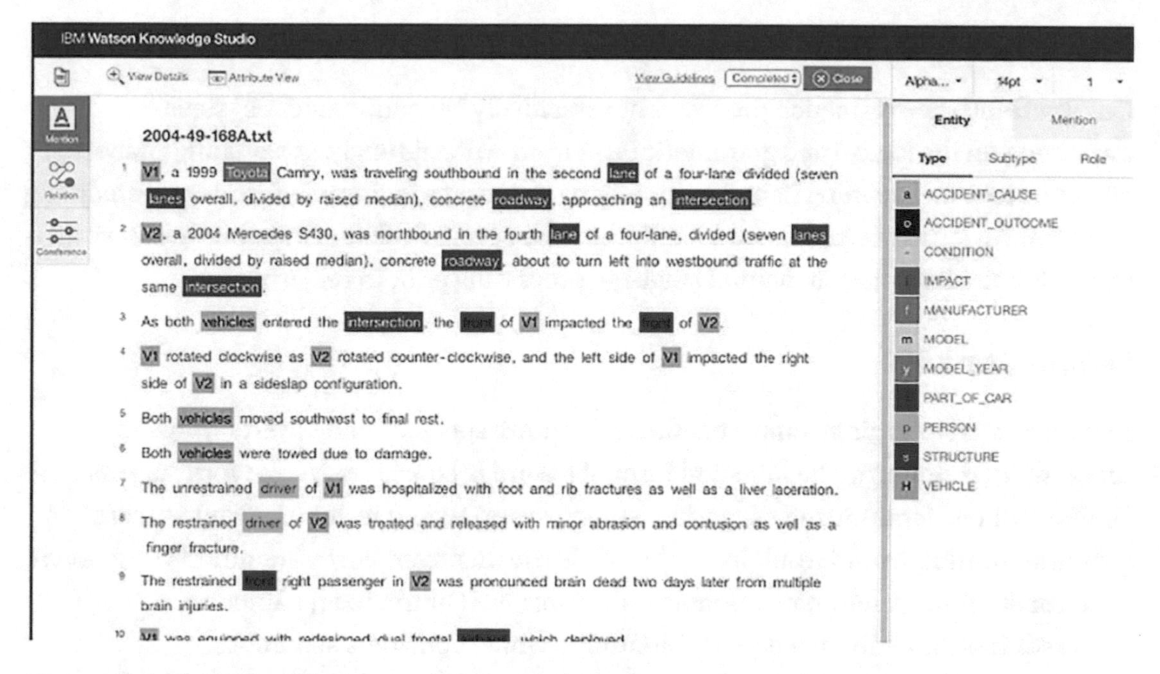

***Figure 10-5.** Manual annotation in IBM Watson Knowledge Catalogue*

Handling Contractions

Contractions in NLP refer to ensuring slang expressions are expanded to their full equivalent: I'll = I will, You'd = You would, etc. The purpose in this case is similar to lower casing in that removing contractions before vectorizing helps with dimensionality reduction.

Contractions can be implemented using the contractions library in Python.

Stemming

Stemming and lemmatization are closely coupled **text normalization** subtasks in natural language processing. Many languages contain words with the same underlying root or "stem" and stemming refers to the process of shortening these words to their root form, regardless of meaning - essentially here we are just stripping down the end characters to a common prefix, even if the prefix itself does not standalone itself as a grammatical term.

Stemming in useful for sentiment analysis as the stem word can convey negative or positive sentiment.

Stemming is typically implemented using nltk.stem.

Semantic Analysis

Our final NLP transformation processes are essentially "semantic analysis" steps – improving on the logical and grammatic tasks from syntactic analysis, semantic analysis allows us to draw meaning from the underlying text – interpreting whole texts and analyzing grammatical structure in order to identify (context-specific) relationships between lexical terms. It is the final stage in natural language processing prior to vectorization.

Lemmatization

In contrast to stemming, lemmatization is "context-specific" and coverts words to a meaningful root form. The inflected form of a word is considered, so a word such as "better" will be lemmatized as "good," while "caring" would be lemmatized as "care" (while stemming would result in "car"). While the nltk.stem PorterStemmer() function is used for stemming, lemmatization uses the WordNetLemmatizer() function.

Essentially an enhancement on stemming which considers semantics, lemmatization is more commonly used in more sophisticated applications of sentiment analysis, such as chatbots.

Disambiguation

Disambiguation is the process of determining the most probable meaning of a specific phrase when there is underlying ambiguity in its definition. On its own a word such as "bank" has several meanings,[13] the NLTK WordNet module allows us to identify probabilistically the actual meaning from its usage in a broader text, although Python also has a Word Sense Disambiguation wrapper (pywsd) which works with NLTK.

N-grams

The NLTK library also comes with an ngrams module. N-grams are a string of connected lexical terms – essentially a continuous sequences of words or phrases. "N" here refers to the number of connected terms or words we are referring to, a bigram (2-gram) would be "United States," a trigram "gross domestic product". The order is important as in matching "red apple" in a corpus, as opposed to "apple red."

[13] See this link for a number of different definitions (synset's) for the word bank: `https://notebook.community/dcavar/python-tutorial-for-ipython/notebooks/Python%20Word%20Sense%20Disambiguation`

N-grams are heavily used in natural language processing as typically we are creating features from series of words rather than individual words themselves

```
('this', 'is', 'a')
('is', 'a', 'very')
('a', 'very', 'good')
('very', 'good', 'book')
('good', 'book', 'to')
('book', 'to', 'study')
```

Text Parsing with NLTK: Hands-on Practice

WRANGLING WIKIPEDIA: SPACEX

Bringing together some of the techniques described in this section, the goal of this lab is to scrape typical unstructured web data (here the SpaceX Wikipedia web page) and walk-through preprocessing and lexical analysis steps in order to produce insights on word count:

1. Download the Jupyter notebook from this GitHub repo:

   `https://github.com/bw-cetech/apress-10.2.git`

2. Run the steps:

   a. Libraries and Connect to web page

   b. Parse the web page using Beautiful Soup

   c. Tokenize text

   d. Count word frequency

   e. Plot word frequency

3. Exercise – plot word frequency as a bar plot rather than a line plot

4. Exercise – continue the notebook analysis by performing further syntactic and semantic data cleansing/grouping steps[14]

[14] For reference see e.g. `www.kaggle.com/code/manishkc06/text-pre-processing-data-wrangling`

Text Vectorization, Word Embeddings, and Modelling in NLP

In the last section, we looked at specific linguistic/transformation processes undertaken in order to extract lexical, syntactic, and semantic aspects of the underlying unstructured data. Having established a process for preparing our data and implementing it, the focus shifts to the "encoding" or **text vectorization** process and deep learning modeling.

The first part of this next stage in the NLP lifecycle is to create *numeric* features from the text data using text vectorization and word embedding.[15] It is followed by a look at the main NLP modeling practices utilized in production-grade applications in 2022, rather than an exhaustive analysis of the entire ecosystem of techniques being applied.

On occasion, we reference specific tools and libraries which are described subsequently in our last section in this chapter.

[15] This "phase" of NLP modelling can be seen as the equivalent process to encoding a structured dataset for Machine Learning.

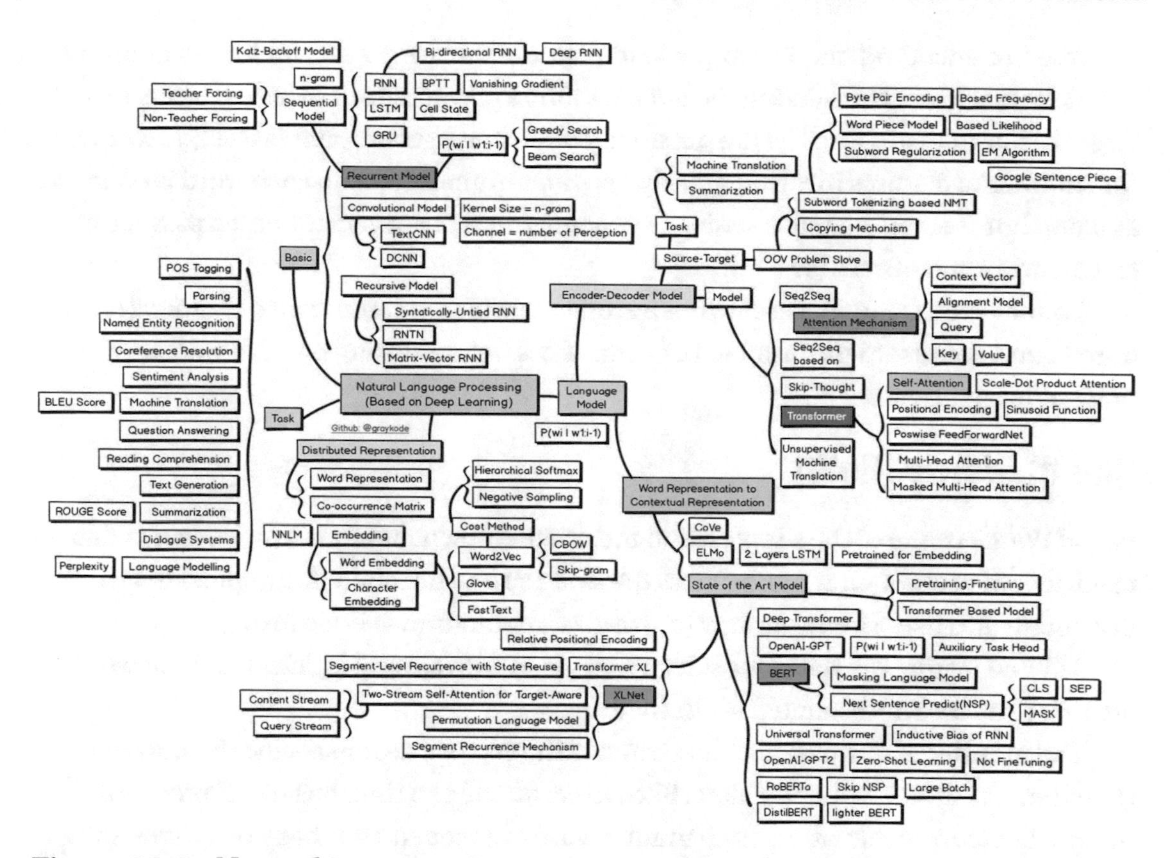

Figure 10-6. *Natural Language Processing Roadmap (GitHub)*[16]

Rule-Based/Frequency-Based Embedding

We start with simpler "frequency"- or "rule"-based approaches to vectorizing/encoding our data.

One Hot Encoding and Count Vectorization

One hot encoding methods should be familiar to readers from machine learning where nominal categorical text fields (i.e., those where the values have no inherent order[17]) are transformed by creating a new column for each separate value in the column.

[16] See `https://github.com/graykode/nlp-roadmap`

[17] Such as "gender"

One hot encoding may be simple to implement and works well for binary or discrete variables but when considering the almost infinite permutations of lexical terms in a large corpus of text, as well as the possibility of storing N-grams, not just single words, the number of features involved quickly becomes unmanageable. Each word is encoded as a one hot "vector," so that a sentence becomes an array of vectors, and a passage of text an array of matrices (so a tensor).

Count vectorization takes one hot encoded terms as columns in a **document term matrix** and counts up the number of occurrences before storing nonnull values for onward text summarization.

Bag of Words (BoW)

Bag of Words is one of the simplest and most well-known methods to encode text data as fixed-length vectors – it is more or less the same as one hot encoding applied to an entire document and uses simple numpy (or pandas) methods to encode text.

After collecting the data (consider a news feed with several articles), unique words are first extracted and counted (or hashed[18]).

The number of unique words determines the vector size for specific **documents** (here news articles) within the data. Each news article can then be represented as a unique boolean vector where each unique word is encoded as either 0 or 1 depending on whether the word appears in the news article.

Ultimately, there are many drawbacks with the BoW method in that there is no syntactic or semantic consideration for the order of the words or the context of the words and ultimately we end up with a huge sparsity problem from many documents containing no occurrences of less frequent words.

[18] Counted in sklearn with CountVectorizer or "hashed" with HashingVectorizer where the text tokens are mapped to fixed-size values. See e.g. `https://scikit-learn.org/stable/modules/generated/sklearn.feature_extraction.text.HashingVectorizer.html`

Latent Semantic Analysis (LSA)

LSA is an extension of the bag of words approach above which uses singular value decomposition (SVD)[19] to reduce data dimensions and thus sparsity. LSA takes a dataset of m documents and n words (**document-term matrix**) and remodels the text data in terms of r latent[20] features, where r is less than the number of documents we have.

TF-IDF

Otherwise known as Term frequency–Inverse document frequency, TF-IDF attempts to address the inherent weaknesses of the bag of words method by employing a numerical statistic to reflect how important a word is to a document in a collection/corpus.

TF here refers to how frequently a term appears in the document, but it is the IDF element, which measures how "important" that term is which distinguishes the method from the Bag-of-words model. For any specific word, its IDF is the log of the number of documents divided by the number of documents in which that word appears, so a word like "the" will have a low score as it will appear in all documents, and log(1) is zero.[21]

TF-IDF is the product of the Term Frequency and IDF score and tends to weight higher both ((a) words which are frequent in a single document and b) less frequent/rare words over the entire collection.

For all its advantages over the BoW model, these deterministic, frequency-based approaches are unable to scale to interpreting "context" - for these we need word embeddings.

A Word on Cosine Similarity

Whether it's BoW, LSA, or TF-IDF, cosine similarity is typically used for measuring similarity of documents in Word Embeddings.

---

[19] Instead of SVD, a variation of LSA, **Probabilistic Latent Semantic Analysis** (**pLSA**) builds a probabilistic model to generate data observed in the document-term matrix. Another alternative. **Latent Dirichlet Allocation** (**LDA**) is a Bayesian version of pLSA which approximates, and therefore, better generalizes document-topic and word-topic distributions

[20] Or hidden, "latent" here can be thought of as the underlying or implicit "theme" of a document

[21] The **Luhn Summarization** algorithm is based on TF-IDF, filtering out further very low frequency words as well as high frequency stop words which are not statistically significant

The measure is an enhancement on the simple Euclidean distance measure (straight line distance between two points) as word embeddings incorporate frequency of words in documents.

Two similar terms (e.g., "foul" and "infringement" in a football referee report) may be far apart in n-dimensional space by virtue of one word ("infringement") having a much lower occurrence in the document. Because cosine similarity measures the angle between the vector representation of these words, disregarding the magnitude of the vector, we manage to retain the concept of "sameness" as the angle between the words is small.

Word Embeddings/Prediction-Based Embedding

For all intents and purposes, Word Embeddings are a "probabilistic matching" technique which applies unsupervised machine or deep learning approaches to vectorize text data – effectively every word is converted into an n-dimensional "word vector," with words of similar meaning located in a dense "cluster."

Figure 10-6 shows how this works in reference to two documents – the outcome is groups of proper nouns, nouns, and verbs with similar meaning close together in dimensional space (i.e., Seattle and Boston, lecture and talk, had and gave, etc.).

Word vectors/Word Embeddings can be trained from scratch, but pretrained models like Word2Vec provide the main means to accelerate NLP application development.

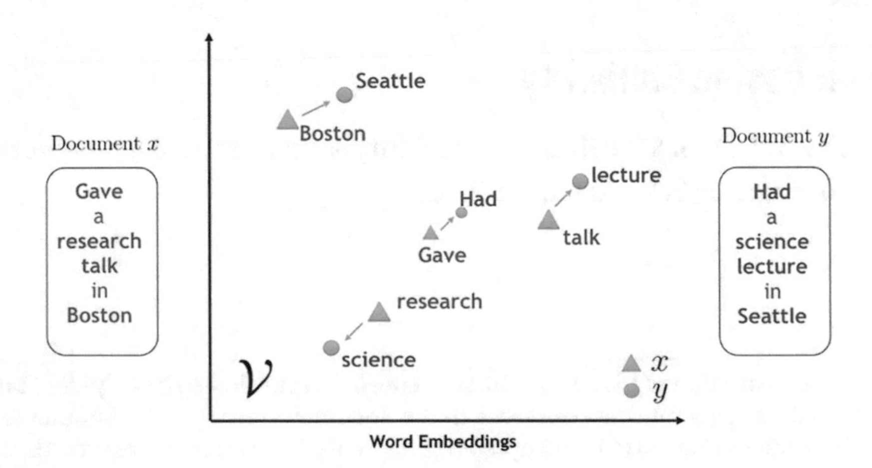

Figure 10-7. *Word embeddings example for two documents (Source towardsdatascience.com)*

Word2Vec (Google)

Word2vec is probably the most famous word embedding model – it is the process used by Google search engine to search similar text, phrases, sentences, or queries.[22] A two-layer neural net that processes text by "vectorizing" words, Word2vec is based on two architectures: **continuous bag of words or CBOW** and the **skip-gram** model.[23]

The CBOW architecture allows the underlying model to predict words based on similar-context words, but ignores the order of words, much like in the rule-based BoW model. The skip-gram model takes into account the order of the words and applies a greater weight to words closer in context in vector space. The value in CBOW is primarily the speed of execution, while skip-gram is more performant on infrequent words.

Document "similarity" for both Word2Vec and Latent Semantic Analysis is calculated using cosine similarity. Cosine similarity is....

Other Models

Word2Vec is by far the most common model for Word Embeddings, but there are many variants.

Additionally, open source projects **GloVe**, developed by Stanford and **fastText**, developed by Facebook also have their advantages.

Short for Global Vectors, GloVe model training uses word co-occurrence stats and combines global matrix factorization and local context window[24] methods. Its USP is in finding relationships between words such as company-product couples, but its reliance on co-occurrence matrix slows down runtime.

fastText is actually an extension of the Word2Vec model where words are effectively modeled as an n-gram of characters.[25] This approach means fastText performs better on rare words as the underlying character n-grams are shared with other words but it is also slower than Word2Vec.

[22] As opposed to the original keyword searches performed in Google's earlier days

[23] The two architectures/models can be viewed as variants to Word2Vec. **Lda2vec** is another variant which, as the name suggests, uses Latent Dirichlet Allocation where a document vector is combined with the key "pivot word" vector used to predict context in Word2Vec

[24] Similar to CBOW

[25] So for the word "apple," tri-gram vectors would be app, ppl, and ple word vectors and the word embedding for apple would be the sum of these

Other notable models include **LexRank** - an unsupervised graph-based approach for automatic text summarization.[26]

NLP Modeling

Although many of the processes described above implicitly leverage machine or deep learning in order to achieve vector representation of text, word embeddings are typically evolved to then perform natural language "predictions." In this last subsection, we take a look at the main predictive modeling techniques in NLP prior to moving onto our last chapter on Python implementation and the main NLP use cases today.

Text Summarization

Many Natural Language Processing applications involve Text Summarization, if not as a direct goal, then as an intermediate process. Text Summarization automates the process using either **extraction-based summarization** where keyphrases are pulled from the source document, or **abstraction-based summarization** involving paraphrasing and shortening the source document. Abstraction-based summarization is more performant, but more complex.

The algorithmic implementation for the extraction-based approach first extracts keywords using linguistics analysis (e.g., PoS), then collects documents with keyphrases[27] before employing a supervised machine learning technique to take the document samples with key phrases and build a model, with features such as the length, number of characters, most recurring word, and frequency[28] of keyphrases determined.

LexRank, Luhn, and LSA are all text summarization techniques previously mentioned and accessible from the Python sumy library, as is KL-Sum which uses word distribution similarity to match sentences with original texts.

[26] Where sentences are scored based on eigenvector centrality in a graph representation of sentences

[27] In practice documents with keyphrase (positive) samples and documents without keyphrases (negative) are included to fit a binary classification model

[28] Using, for example, TF-IDF for frequency-based summarization. Features can also be derived using Word2Vec for distance-based (vectorization) summarization

Topic Modeling

Closely coupled to text summarization is the concept of topic modeling. Many of the same algorithms mentioned above, specifically LSA, pLSA, LDA, and lda2Vec are used for topic modeling - the underlying goal is to recognize words from the *topics (or themes) present* in a document or corpus of data,[29] rather than more laborious method of extracting words from an entire document.

Sequence Models

Sequence models are machine learning models used to interpret word sequences in texts. Applications include sequence modeling of text streams, audio clips, and video clips, where, like time-series data, Recurrent Neural Networks (specifically LSTMs or GRUs) are used.

Sequence to sequence or **seq2seq** is probably the most well-known technique with applications in machine translation, text summarization, and chatbots.

seq2seq is in fact a special class of RNN which uses an **Encoder-Decoder architecture**. The encoder feeds the input data, sequence by sequence into an LSTM/GRU network, producing context vectors (hidden state vectors) as well as outputs.[30] The decoder (also an LSTM/GRU) initializes with the final states (context vector) of the encoder network and generates an output on one forward pass. The decoder trains by repeatedly feeding back into the decoder previous outputs to generate future outputs.

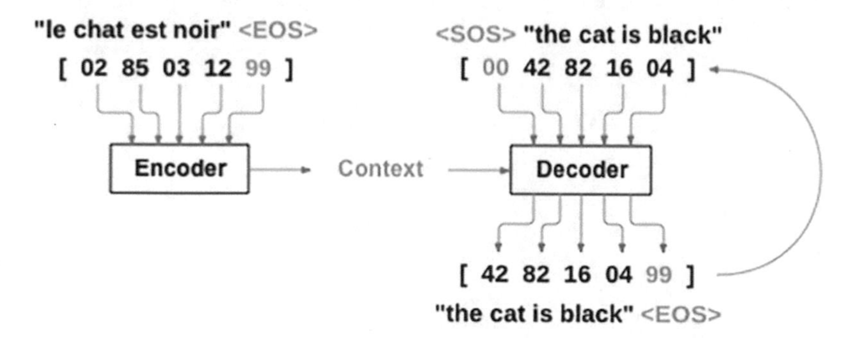

Figure 10-8. *PyTorch Sequence to Sequence Encoder-Decoder architecture showing Machine Translation*

[29] A corpus is the collection of unstructured data in the underlying data/document set

[30] Outputs are discarded

Transformers and Attention Models

The sequence-to-sequence model works well enough with smaller sentences but longer sentences tend to stress internal memory of the underlying encoder-decoder network. In recent years **Transformers** have been developed, where an augmenting **attention mechanism** or **attention model** is incorporated to force model attention on specific part(s) of the input sequence.

Essentially "massive" deep learning models, transformers can be implemented in Python using the Hugging Face library. Google's BERT (Bidirectional Encoder Representations from Transformers), OpenAI's GPT-3 (Generative Pretrained Transformer),[31] and the Allen Institute for AI's ELMo (Embeddings from Language Models) are three of the biggest.

There is a lot of hype around GPT-3 in particular, stemming from the ease at which the technology is able to leverage pretrained models on billions of parameters to write articles and essays[32] or even recreate a dialogue between ancient philosophers.[33] Google's decoder-only transformer language model LaMDA (Language Model for Dialogue Applications), referenced at the start of this chapter,[34] is arguably the next generation, pretrained on 137 billion parameters and trillions of public dialogue data.

The Python wrapper bert-extractive-summarizer utilizes HuggingFace PyTorch transformers library to produce shortened summaries of large text.[35] GPT-3 can be implemented in Python by installing the openai library and obtaining an API Key after registering a personal-use account with OpenAI.

---

[31] Transformers are also the other component of OpenAI's DALL-E Generative AI model shown in Chapter 8

[32] See `www.theguardian.com/commentisfree/2020/sep/08/robot-wrote-this-article-gpt-3`

[33] `https://betterprogramming.pub/creating-philosopher-conversations-using-a-generative-lstm-network-fd22a14e3318`

[34] And thought to be "sentient" by at least one Google Engineer

[35] See "Large Example" here: `https://pypi.org/project/bert-extractive-summarizer`

Word Embeddings: Hands-on Practice

WORD2VEC VISUALIZATION

We now take a look at some of the techniques discussed in this section, starting with a look at implementing Word2Vec to vectorize unstructured data and plotting the resultant Word Embeddings:

1. Clone the below GitHub repo:

   `https://github.com/bw-cetech/apress-10.3.git`

2. Run through the Python notebook steps in Jupyter or Colab to:

   a. Import the data from the intents.json file

   b. Clean the data using a combination of lexical, syntactic, and semantic techniques

   c. Tokenize the data and build a model with Word2Vec

   d. Display the model vocabulary

   e. View a sample word embedding vector

   f. Plot all the word embeddings

3. Exercise – improve the visibility of word labels in the word embeddings plot

4. Exercise (stretch) – swap out the json intents data with a larger 10+ page pdf and perform topic modeling before plotting word embeddings

Seq2Seq: Hands-on Practice

AN ENCODER-DECODER MODEL FOR LANGUAGE TRANSLATION

Using Keras and LSTM deep learning the goal of this lab is to implement a character-level sequence-to-sequence model to translate English to French[36]

1. Clone the GitHub repo below:

   https://github.com/prasoons075/Deep-Learning-Codes/tree/master/Encoder%20Decoder%20Model

2. Open the Encoder_decoder_model with Jupyter notebook
3. Run through the data prep cells:

   a. Display the number of sample translations for training

   NB you will need to change the path link to the training data from "fra.txt" to "fra-eng/fra.txt"

   b. Vectorize the data and display the number of unique input and output tokens and maximum sequence lengths

   c. Define the encoder-decoder architecture

4. Train the sequence model/LSTM on the training language samples – note this may take up to 5-10 minutes to run through 100 epochs

   NB if an error "NotImplementedError: Cannot convert a symbolic Tensor" is encountered, run the notebook instead in Colab (Open Notebook > GitHub > enter the GitHub source url above, opening the "Encoder_decoder_model" direct from GitHub. Note you will also need to drag and drop the "fra.txt" data file from the GitHub source "Encoder Decoder Model/fra-eng" folder to the default Colab temporary storage (content) folder

5. The model is saved in the previous step. Reimport this and set up the decoder model for sampling/testing the model

[36] See also www.analyticsvidhya.com/blog/2020/08/a-simple-introduction-to-sequence-to-sequence-models/#:~:text=Sequence%20to%20Sequence%20(often%20abbreviated,Chatbots%2C%20Text%20Summarization%2C%20etc for further context

6. Finally, test the model with a few sequences from the training set (up to 10000) to see how the model translates English terms
7. Exercise – adapt the notebook to support a basic (in notebook) user interface where the user's phrase is translated to French

PyTorch NLP: Hands-on Practice

TEXT CLASSIFICATION WITH PYTORCH

The goal of this lab is to predict which language a surname is from by reading in text examples as a series of characters, then building and training a recurrent neural network (RNN) with PyTorch

1. Upload the two files at the link below to Jupyter Notebook:

   ```
   https://github.com/bw-cetech/apress-10.3b.git
   ```

2. Rename the "PyTorch_Functions.py.txt" file without the .txt extension to convert to a Python script[37]
3. Open the .ipynb file
4. Install PyTorch with

   ```
   import torch
   ```

5. Run through the notebook, importing the PyTorch RNN functions (in PyTorch_Functions.py) to the notebook
6. Perform the exercises shown:
   a. Display a three-dimensional (random) tensor to test the PyTorch library import
   b. Download the training data dictionaries for this problem (surnames for 18 different languages), unzip and rename in your Jupyter working directory

[37] Email servers block python scripts when sending, hence the file has been uploaded to GitHub with a text file extension.

c. Display the last five names in the Portuguese dictionary to test the PyTorch custom functions

d. Observe the model performance on the test set (as shown in the graphic below) and export both the predictions and loss likelihood output (probability of the test sample names belonging to a certain language) to csv files

7. Exercise (stretch) – adapt the set of data dictionaries imported to solve a different NLP classifier such as predicting (a) names to gender or (b) character to writer or (c) city to country, etc.

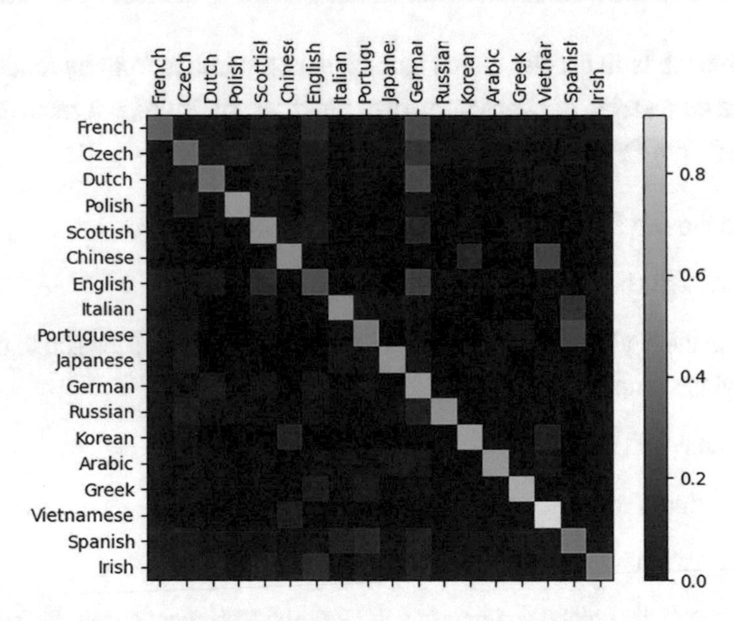

Figure 10-9. *Surname language classification with PyTorch*

Tools and Applications of NLP

Our final section in this final chapter takes a look at some of the better-known open source and industrial-scale Python libraries followed by an analysis of the main natural language applications and tools used today. We wrap up with our final hands-on labs addressing some of the key NLP use cases and tools used including an end-to-end chatbot deployment where we develop a user dialogue tree with Watson Assistant and push our app to IBM Cloud.

Python Libraries

What are the main libraries for implementing Natural Language Processing? We outline in Table 10-1 the main ones for general purpose NLP as well as a few of the specialist libraries used for targeted industry applications.

Table 10-1. *Python Libraries for NLP*

LIBRARY/ TOOL	DESCRIPTION/MAIN USE	USP	DRAWBACK
NLTK	Leading platform for NLP, sentence detection, tokenization, lemmatization, stemming, parsing, chunking, and POS tagging. UI to 50 corpora and lexical resources	Versality as the most commonly used Python library for NLP	Slow runtime, no neural networks
TEXTBLOB	Access to common text-processing ops through TextBlob objects treated as Python strings	Data prep for NLP/ DL, easy UI	Slow runtime, not for large scale production
CORENLP	Stanford-developed assortment of human language technology tools for linguistic analysis	Speed, written in Java	Requires Java install as underlying language[38]
SPACY	Designed explicitly for production usage for developers to create NLP apps that can process/understand large texts	Big data handling and multi-foreign language support	Lack of flexibility in comparison to NLTK

(*continued*)

[38] sumy can be used for most text summarization techniques including LSA, Luhn, LexRank, and KL-Sum

Table 10-1. (*continued*)

LIBRARY/ TOOL	DESCRIPTION/MAIN USE	USP	DRAWBACK
PYTORCH[39]	Facebook's open-sourced PyTorch is an API-driven framework for extending the Torch deep learning library[40]	Speed of execution, computational graph support	Core NLP algorithms complexity
GENSIM	Specialist library for topic modeling, document indexing, and similarity retrieval with large corpora	Memory independence supports large data sets > RAM[41]	Unsupervised text modeling limitation, requires integration with other Python libs

Others with specific strengths, such as **Pattern** and **PyNLPl** (pronounced Pineapple) are also used for their web data mining and file format handling capabilities respectively while **sumy**, **pysummarization,** and **BERT summarizer** are great for text summarization.[42]

Although not exclusively tools for natural language processing, two other important Python libraries are the Twitter API[43] and the facebookinsights wrapper for the Facebook Insights API. The use cases for those – social media sentiment (and metric) analysis will be discussed below while a hands-on lab is provided for the Twitter API in Chapter 8.

NLP Applications

Email/spam filters, word clouds,[44] auto-correct in word processors, and (code) auto-complete in programming are some of the earliest, and now established, applications of natural language processing. But it is only relatively recently that the above Python libraries, coupled with cloud, have unlocked higher-value natural language processing of VAST volumes of unstructured data.

[39] But provides an interface to python

[40] TensorFlow also does NLP although perhaps not as widely for this purpose as PyTorch

[41] See also Chapter 5

[42] Gensim also supports training your own word embeddings, see e.g. `www.analyticsvidhya.com/blog/2017/06/word-embeddings-count-word2veec`

[43] Import tweepy or import snscrape. The latter doesn't require a Twitter Developer Account.

[44] And as we have seen in the first lab in this chapter, relatively easy to implement

Whether its extracting business value from unstructured data, deep document information search and retrieval, acceleration of internal research or due diligence processes, reporting productivity increases and content creation or seeking synergies with overriding cognitive robotic process automation (CRPA) goals, companies are scrambling to establish in-house NLP capabilities to deliver value.

While many of the trending, "peta-scale" NLP accelerators may only just be starting to be employed, we take a look now current levels of sophistication around mainstream NLP business and organizational applications.

Text Analytics

Text analytics or text mining involves the extraction of high-quality information from data. Essentially an enabler for more complex unstructured data analysis including text-to-speech and sentiment analysis, a key value-add is the ability to augment domain-specific data/corpora in the training process[45] for enhanced classification tasks.

Microsoft is probably one of the leaders here, specifically with Azure Cognitive Services which encompasses a wealth of text analytics including Content Moderator and Language Understanding (LUIS). IBM Watson Knowledge Catalog is also a leading product.

Text-to-Speech-to-Text

Text to Speech (and Speech to Text) is now a well-established and somewhat crowded market. Market leaders include Amazon Polly and the API Cloud Service IBM Watson Text to Speech.[46]

Social Media Sentiment Analysis/Opinion Mining

Sentiment Analysis/Opinion Mining is contextual text mining which is used to identify and extract subjective information from source data. Overwhelmingly associated with social media channels, businesses (and governments/politicians) use the technique directly or indirectly to understand the social sentiment of their brand, the voice of the customer (VoC) or gauge public perception.

[45] See POS Tagging and NER in above section

[46] see e.g. demo @ `https://speech-to-text-demo.ng.bluemix.net/`

Sentiment Analysis today goes beyond just simple positive and negative sentiment metrics - listening in or monitoring (real-time) online conversation can trigger related key insight analysis on discussion categories, concepts and themes, and emotion detection.

All global retailers, FMCG industries, and the TELCO sector are reliant on sophisticated capture of sentiment analysis outcomes - with most apps reliant on accessing the Twitter API (via the Python tweepy or snscrape library) and/or Facebook Insights (via the facebookinsights wrapper). A hands-on lab for social media sentiment analysis is provided for the Twitter API in Chapter 8.

Figure 10-10. *Social media sentiment analysis*

Chatbots, Conversational Assistants, and IVAs

Probably the most famous application of natural language processing, chatbots today conduct interactive dialog, and come as standard with sophisticated speech recognition and text-to-speech capability.

Chabot technologies vastly improved since the earliest days of Cortana. Intelligent Virtual Assistants (IVAs) have AI training built-in, cognitive, self-learning capability, and adapt to context, leveraging the latest Transformer technology.

The business value in deploying chatbots today is focused around step-change improvements in customer journey and customer experience - with the obvious potential to resolve issues quickly, and at low cost.

Whether its Amazon Alexa, Apple's Siri, Google's IVA ecosystem (Meena/ DialogFlow/LaMDA), IBM Watson, Azure LUIS and QnA Bot or Rasa, Natural Language Processing and Deep Learning are combined to best-fit/learn user "intents" and "entities" to a dialogue "corpus."

NLP 2.0

The limitless potential of natural language transformers has already been discussed in the last section but we take a look below at other recent advances in state-of-the-art (SOTA) NLP technologies expected to become core organizational apps in the near future.

Natural Language Generation

Natural language generation is the production, rather than interpretation, of natural language. Although somewhat primitive, autocomplete is an example of NLG.

Extensive sectoral use cases include digital marketing content creation, financial/medical report creation and journalism, product labeling in ecommerce/retail, travel updates, and customer service optimization.

Although NLG has also been around for some years already, the increasingly massive parametrization capability of transformers has resulted in seismic improvements in outputs, particularly in relation to credibility of AI-generated report writing.

Google's Smart Compose, Arria, and WordSmith are three leading tools for supporting natural language generation[47] but as we will see in a hands-on lab at the end of this chapter, we can also use a GPT-3 transformer to generate text, in this case a cooking recipe based on ingredients.

Debating

Sophistication in computational discussion, argumentation, and debating technologies has reached a point where machines are able to credibly debate humans.[48]

IBM's Project Debater is pitched as the "first AI system that can debate humans on complex topics" and is composed of four core modules: argument mining, an argument knowledge base (AKB), argument rebuttal and debate construction, where the first two modules provide the content for debates.[49] The tool uses similar sequence to sequence

[47] Another tool, Automizy `https://automizy.com/` is free and uses NLG for email marketing content

[48] See e.g. `www.technologyreview.com/2020/01/21/276156/ibms-debating-ai-just-got-a-lot-closer-to-being-a-useful-tool/`

[49] Debater datasets can be found at the link: `https://research.ibm.com/haifa/dept/vst/debating_data.shtml`

and attention mechanisms intrinsic to transformers and has favorable ratings for, for example, an opening speech when compared with human (nonexpert) speeches and other NLP transformers such as GPT-2.

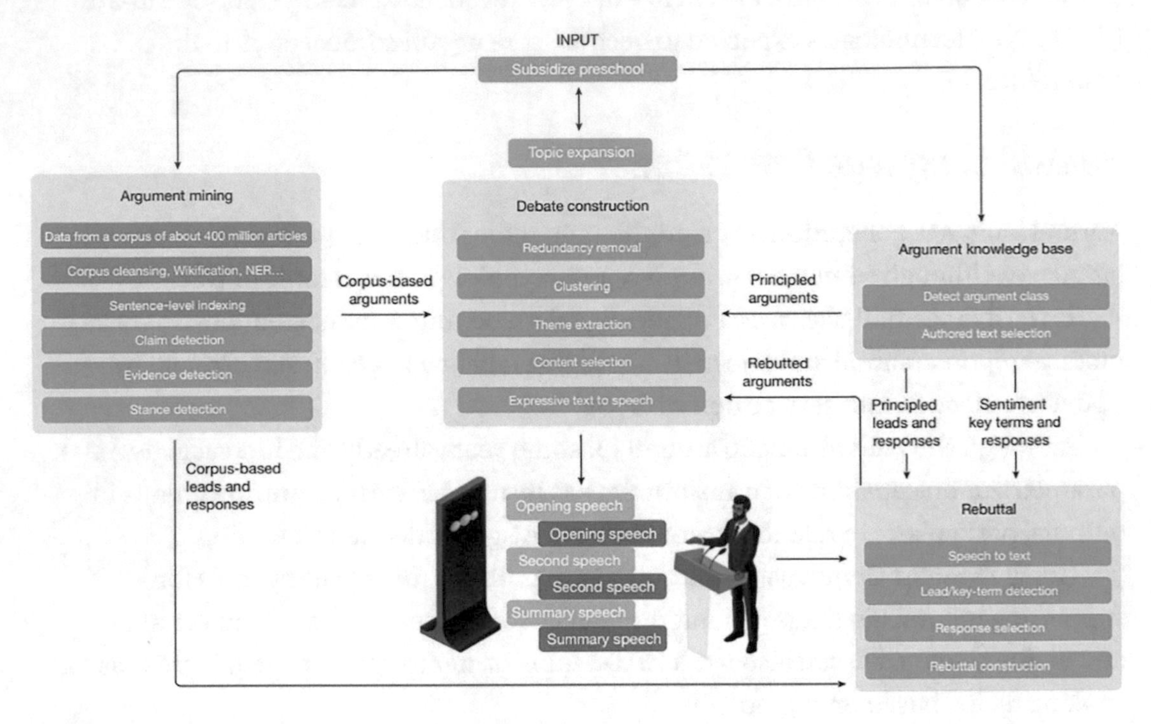

Figure 10-11. *IBM Project Debater System Architecture*

Auto-NLP

Naturally, given the trends toward full automation in machine and deep learning, there is much focus on automating the myriad of steps involved in the NLP lifecycle.

Hugging Face are a leader in this area, their AutoNLP tool integrates with the Hugging Face Hub's substantial collection of datasets and pretrained[50] SOTA transformer models. NeuralSpace are another, with multilingual support to train models with AutoNLP in 87 languages.[51]

[50] Including 215 sentiment analysis models. See `https://huggingface.co/blog/sentiment-analysis-python`

[51] `https://docs.neuralspace.ai/`. NeuralSpace specializes in local languages spoken across Asia, the Middle East, and Africa.

The auto-nlp Python library also provides an abstraction layer and low code functionality over existing Python NLP packages (such as spaCy) in much the same way as auto-sklearn provides low code automation for sklearn machine learning. AutoVIML (Automatic Variant Interpretable ML) is another Python library for Auto-NLP which automates the preprocessing, linguistic analysis (stemming and lemmatization, etc.), and vectorization steps.

Wrap-up

Moving on from these NLP trends, we complete this final chapter on natural language processing with a couple of hands-on labs focused on some of the dominant applications and tools in 2022. These labs bring us to the end of this journey and a practical conclusion to Productionizing AI solutions with Cloud and Python. We do though have some final words to close out on in our concluding few pages.

A recurring theme of this book has been the level of experimentation, and at times (cloud) cost-based workarounds, employed to implement AI. Not every company has budget to meet the costs of high-performance compute instances, or high throughput, secure storage. The current climate for AI solution implementation and the concentration of market power may mean these challenges remain for several years to come but we look in these last pages at the possibility of game-changing innovations that may allow the wider ecosystem to escape the gravitational pull of the CSPs.

WATSON Assistant Chatbot/IVA: Hands-on Practice

HR DIALOGUE CONFIG FOR CONVERSATIONAL AI

As a Gartner Magic Quadrant leader for Enterprise Conversational AI, and with IBM Watson users achieving a 337% ROI over three years,[52] Watson Assistant is one of the industry-leading tools for IVA automation.

The goal of this exercise is to create an HR IVA on Watson Assistant which uses growing user interaction to re-train and improve responses to user questions on job applications and internal company policy:

[52] IBM Watson users claims an 337% ROI over three years

1. Go to the link[53] `https://eu-de.assistant.watson.cloud.ibm.com/` and sign in with your IBM ID. If redirected to IBM Cloud, enter login details or sign up.
2. Select the Assistant option on the top LHS of the screen > Create Assistant
3. Add dialog skill > Upload Skill and upload the json dialogue from the link below

   `https://github.com/bw-cetech/apress-10.4.git`
4. Open the newly created Assistant and observe the user intents (these are the themes of the user's questions such as org structure, payroll, complaint, admin, humor, etc.) and **entities** (the subject of the user's questions such as team, people, services, pay, extras: stocks/shares/pension, etc.)
5. The dialogue is preconfigured – test it with the dialogue below[54]

   PART I (job enquiry)

   ```
   hello
   can you fill me in on my job application
   open position
   technical
   yes
   ```

   PART II (company policies):

   can you help me with share options

   is there a reduced price for employees?

   PART III (humor):

   I may be dead soon, how is my life insurance, salary and pension paid to my family?

   no

   goodbye

[53] Change "eu-de" in the url to uk or us-south region depending on nearest data center

[54] The "Preview" option for the Assistant can be used for this, but the equivalent Skill UI is more elegant - repeat the json file upload directly to a new Dialog Skill created from the Skill option on the top LHS of the screen, then after creation, select "Try it"

6. **Exercise** – swap the dialog for a typical customer support IVA for an online retailer
7. **Exercise (stretch)** – push to IBM Cloud by adding web chat integration to your Watson Assistant (API details are shown under "View API Details" on the Assistant UI)
8. **Exercise (stretch)** – add WhatsApp (with Twilio) integration to your Watson Assistant
9. **Exercise** – set up an app routine which automatically re-trains the NLP model each month based on the latest user conversations

Transformers for Chatbots: Hands-on Practice

NATURAL LANGUAGE GENERATION WITH GPT-3

In our final lab, we use OpenAI's pretrained GPT-3 (Generative Pretrained Transformer) models to (a) generate a cooking recipe based on ingredients entered by a user (using "zero-shot" training where no examples are provided to the model) and (b) implement a "sarcastic" chatbot (using "few-shot" training where a limited number of examples are provided to train the model):

1. Register an account with OpenAI `https://openai.com/join/` and access the API Key provided under "Personal" in the top right corner of your dashboard.
2. Clone the below GitHub repo:

   `https://github.com/bw-cetech/apress-10.4b.git`

3. Copy and paste the OpenAI API Key inside the double quotes of the string defined in the openai_credentials.py file
4. Run through the Python notebook steps in Colab:

   a. Install openai, then comment out the code

   b. Import libraries

   c. Drag and drop the OpenAI credentials file to Colab temporary storage

d. Define the GPT-3 transformer function – this will interface to a high performance GPT-3 "text-davinci-002" engine

e. Call the function to create a recipe from a basic recipe with apple, flour, chicken, and salt

5. Exercise – change the ingredients to e.g. fresh basil, garlic, pine nuts, extra-virgin olive oil, parmesan cheese, fusilli, lemon, salt, pepper, red pepper flakes, and toasted pine nuts[55] and call the function again to create a pasta recipe
6. Double-click the "receipe.txt" file that gets written to Colab temporary storage to validate the recipe
7. Exercise (stretch) – modify the code to ensure the recipe generated from the GPT-3 model isn't truncated
8. Using the same GPT-3 transformer model, proceed to run the last two cells which provide contextual examples[56] of "sarcasm" then calls the (same) GPT-3 transformer function. The function returns an (NLG-) generated "sarcastic" response to the (empty[57]) last question "What time is it?" in the contextual examples/chatbot text

---

[55] Just update the recipe variable to: recipe = 'fresh basil, garlic, pine nuts, extra-virgin olive oil, parmesan cheese, fusilli, lemon, salt, pepper, red pepper flakes, toasted pine nuts '

[56] That is, "few-shot" training - prompting a machine learning model to make predictions with a limited number of training examples

[57] Note that in the contextual examples the last response from "MARV" is deliberately blank – MARV's response is what we are trying to predict/generate

Postscript

Wrap-up

There are many trends waiting to happen in AI. The speed of change in the last ten years, since ImageNet,[1] has been phenomenal. Creating a summary of what is likely to commercialize in the next five years, never mind the next decades, is fraught with uncertainty. Rather than attempting to predict the future, and in line with the practical nature of this book, we will outline some of the more credible innovations having "disruptive" potential.

Next-generation AI or AI 2.0 is already here, addressing advancements in the technology that address portability, accuracy, and security challenges.[2] Hyperscalers, or Data Centre operators that offer scalable cloud computing services, are expected to leverage more transfer learning and reinforcement learning, and transformer networks are expected to make AI smarter and more mobile. The problem is hyperscalers tend to be the same group of Big Tech companies currently providing storage and compute services together perhaps with Alibaba AliCloud, IBM, and Oracle.

[1] See Chapter 1.

[2] See `https://www.forrester.com/report/AI-20-Upgrade-Your-Enterprise-With-Five-NextGeneration-AI-Advances/RES163520?objectid=RES163520`. The NLP 2.0 section from our last chapter showcases additional AI 2.0 advances specially in relation to Natural Language Processing.

B. Walsh, *Productionizing AI*, https://doi.org/10.1007/978-1-4842-8817-7

What may be more promising for companies/organizations looking to avoid vendor-lock or at least diversify cloud services is the use of generative models to create **synthetic data**[3] and managed services for reusing/sharing disparate model training processes (**federated learning**). The table in Figure 11-1 summarizes these innovations.

Technology	What is it?	What is it used for?	Who is using it?
Transformer networks	Giant pretrained, customizable, hyperaccurate, multitasking deep learning models	Any hard problem with a significant time or context dimension (e.g., understanding and generating text, software code, etc.)	Hyperscalers (Amazon Web Services, Google, IBM, and Microsoft), the advanced guard of speech and text analytics vendors, and many startups
Synthetic data	Generative models and simulated virtual environments used to create or augment existing training data	Accelerating the development of new AI solutions, improving the accuracy and robustness of existing AI models, and protecting sensitive data	Autonomous vehicles, financial services, insurance and pharmaceutical firms, and every computer vision vendor
Reinforcement learning	Machine learning approaches that test their way to optimal actions via simulated environments or a large number of micro-experiments	Constructing models that optimize many objectives/constraints or deciding on action based on positive and negative environmental feedback	Firms targeting particular B2C marketing tasks, optimizing repeatable manufacturing processes, and robotic learning
Federated learning	A managed process for combining models trained separately on separate data sets	Sharing intelligence between devices, systems, or firms to overcome privacy, bandwidth, or computational limits	Hyperscalers, AI-enabled application vendors, and consumer electronics companies
Causal inference	Approaches such as structured equation modeling and causal Bayesian networks that help determine cause-and-effect relationships in data	Business insights (e.g., attribution analysis) and bias prevention where insights and explainability are as important as prediction accuracy	Innovation teams at leading organizations (e.g., determining treatment efficacy for a given disease at healthcare providers)

Figure 11-1. *AI 2.0 (Source: Forrester)*

[3] Both training and testing data. Appen see this transformation as having a flywheel effect for businesses struggling to overcome data challenges for AI: `https://appen.com/solutions/training-data/`

While AI 2.0 is a work in progress, DataOps is meant to be the established framework for delivering current AI projects. But it is not as efficient as it can be, with dependencies on infrastructure, operating systems, middleware, or language runtime and data silos proving difficult to eliminate.

The trend toward unified **Data Fabric architectures** and enablement of frictionless access and sharing of data in a distributed multi-cloud environment is seen as one way to overcome these challenges. We have already taken a look at IBM Cloud Pak for Data, promising for its multi-cloud support and built-in data fabric-access to multi-source data.

Meanwhile, hyper-automation of cloud computing has created a **NoOps** environment where software and software-defined hardware are provisioned dynamically. These increasingly **serverless architectures** are decoupling IT functions from infrastructure management and freeing up organizations to further experiment (especially in relation to AI application development).

Epilogue

A great deal of frustration has gone into writing this book – frustration about the dominance of Big Tech but mainly frustration about the job market's sloping playing field. When you are out of a job and an industry and there's no money coming in, not good at self-promotion and you're not young either well.... let's put it this way...the world in 2022 doesn't care to notice, and I'm not sure it ever did. At times that frustration has turned to despair – like trying to clamber back into a boat in a turbulent ocean, at night-time.

Much of this job market pain has been channeled into writing this book. I hope as well that in some way the suggestions around avoiding obscured cloud costs and feeding the Big Tech machine further provide some practical help to readers. None of us should be charged for testing their services (it's my job – I don't have a choice!), while learning from poorly documented solutions (ok not all are produced by Big Tech), and having no successful use case at the end.

If there is any advice I can give to challenge cloud costs it's to always request itemized resource usage. Question that "EBS" instance you never knew you provisioned, ask what the hell is "D13 HDInsight," "D11 v2/DS11 v2 VM," "Basic Registry Unit," or "Premium All Purpose Compute." And why DO we have to empty S3 buckets before deleting them and I'm certain information about only the first 10 billing alarms on CloudWatch being free is not as visible as it could be[4]?

Certain tools and labs in this book are more problematic than others. Getting Kafka to consume events and Databricks to talk to AWS MSK comes with an abundance of errors and are both a complete pain in the butt. The arbitrary rent charged on cloud often feels no different from costs imposed by a cartel and support is generally designed to prevent raising a complaint. Raise a ticket if you can - but watch for the hard-coded default response to a question you didn't ask. And if you get through to the final form for submitting to a human, well expect to spend half an hour manually reducing the length of the complaint to fit the word limit and find and replace any characters that are not alphanumeric.[5]

At the end of the day, we all need them, but the new hyperscalers do get rich on this behavior, charging individuals costs for using services they need to survive in the Data market. When the purpose is for training, development, or testing,[6] shouldn't the service be free, especially when upskilling ends up requiring cloud native accreditation or drives further B2B cloud revenues, for which Big Tech already make a huge amount of money?

---

[4] So if you want billing alerts to ensure you are keeping an eye on cloud usage, you will be charged for those as well.

[5] If required, raise a billing dispute in AWS by going to `https://console.aws.amazon.com/support/home#/case/create` and selecting Account & Billing > Service: Billing > Category: Dispute a Charge. For Azure, choose new support request > billing > refund request at the following link: `https://portal.azure.com/#blade/Microsoft_Azure_Support/HelpAndSupportBlade/overview`

[6] And not for production - which should be obvious/auditable to the CSPs from resources being deleted shortly afterward.

Index

A

B. Walsh, *Productionizing AI*, https://doi.org/10.1007/978-1-4842-8817-7

B

E

F

G

H

I

J

K

L

M

N

O

P

Q

R

S

T